Clonal Forestry I

Genetics and Biotechnology

Edited by
M.R. Ahuja and W.J. Libby

With 32 Figures

Springer-Verlag

Berlin Heidelberg New York
London Paris Tokyo
Hong Kong Barcelona
Budapest

Dr. Mulkh-Raj Ahuja
Institute of Forest Genetics
Federal Research Center for Forestry and Forest Products
Sieker Landstrasse 2
W-2070 Grosshansdorf, Fed. Rep. of Germany

Dr. William J. Libby
Department of Forestry and Resource Management
Mulford Hall
University of California
Berkeley, CA 94720, USA

The chapters referring to "Conservation and Application" are published in Volume II of *Clonal Forestry*, 1993.

ISBN 3-540-52501-7 Springer-Verlag Berlin Heidelberg New York
ISBN 0-387-52501-7 Springer-Verlag New York Berlin Heidelberg

Library of Congress Cataloging-in-Publication Data. Clonal forestry / M.R. Ahuja, W.J. Libby (eds.). p. cm. Includes bibliographical references and indexes. Contents: 1. Genetics and biotechnology — 2. Conservation and application. ISBN 3-540-52501-7 (v. 1): — ISBN 3-540-55714-8 (v. 2) — ISBN 0-387-52501-7 (U.S. : v. 1). — ISBN 0-387-55714-8 (U.S. : v. 2) 1. Clonal forestry, I. Ahuja, M.R., 1933– . II. Libby, W.J. SD403.4.C56 1993 634.9′56—dc20 92-25603

Typesetting: Macmillan India Ltd., Bangalore 25

31/3145/SPS-5 4 3 2 1 0 – Printed on acid-free paper

Preface

In the past few decades, much progress has been made in creating the knowledge needed to perform clonal forestry, both in its theory, and in practice. This is reflected in the number and quality of recent scientific and professional meetings that have focused on cloning of forest trees and on clonal forestry throughout the world. In 1988, we decided that a book on clonal forestry would be timely, and we planned it to provide a broad coverage of clonal forestry. However, it later became apparent that such coverage was too much for a single volume, and we have therefore organized a two-volume treatment. The second volume surveys applications of clonal forestry for various purposes in various settings around the world, and this first volume includes much of the science behind those applications.

For both volumes, we set out to recruit authors from among the younger scientists in the field, from many countries and regions, to foster a diversity of thought and experience with respect to clonal forestry. When we could not resist recruiting some of the longer-established authors, we asked them to write on recent advances or on topics different than those appearing in their previous works.

In our view, cloning forest trees and clonal forestry are not identical. The concept of "full clonal forestry" is discussed in Chapter 1 of this Volume, and is developed in detail in Volume 2. In this respect, we consider that successfully cloning forest trees is a necessary prerequisite to clonal forestry, and that using cloned trees in forestry may serve to provide useful preliminary stages of clonal forestry. The information and understanding that allow the successful production of clones are the foci of several chapters of Volume 1.

Following the introductory chapter, Volume 1 is organized such that Chapters 2 to 9 address more or less classical genetics topics, and Chapters 10 to 15 address topics drawn from molecular genetics and biotechnology.

Chapters 2 to 9 address basic and theoretical topics fundamental to carrying out clonal forestry, or that fall into the category of topics related to this. Chapters are included on the theories of maturation, on the population biology of clonal deployment, on selection and breeding of extreme genotypes, on the useful interactions between clones and physiology, on genetic problems associated with tissue culture, and two methodology chapters on identifying clones, and on field-testing clones.

The chapters on molecular genetics and biotechnology contain many ideas that are both new and timely. If biotechnology indeed provides the tools for rapid and commercially feasible mass cloning of forest trees, and also creates new or modified plants best deployed as clones, the development of forest biotechnology will encourage forestry to change from seedling-based practice towards clonal forestry. Following Chapter 10, providing an overview of recent developments in forest biotechnology, the remaining five chapters of Volume 1 deal with vegetative propagation, rejuvenation from mature tissues, somatic embryogenesis, the molecular structure of forest trees, and DNA-level manipulations of forest-tree genomes.

Following an introductory chapter, Volume 2 includes 10 chapters on the historical, current and future applications of clones in forestry. Volume 2 concludes with a chapter on the organization and implementation of a clonal forestry program, and a chapter reviewing policies and laws affecting clonal forestry. We consider Chapter 3 in Volume 2, on conservation within clonal forestry, and on ways that cloning can and does contribute to the conservation of forest-tree germplasm, to be so ethically and politically important that conservation is given equal billing with application in the subtitle of Volume 2.

Each volume is indexed separately, and thus the reader pursuing a particular topic is well advised to check the Indexes of both volumes.

Both volumes have been prepared for a wide range of possible readers, including students, researchers, and forest managers interested in the development of clonal forestry.

Grosshansdorf, Fed. Rep. of Germany — M.R. Ahuja
Berkeley, CA, USA — W.J. Libby
February 1992

Contents

List of Contributors

Addresses are given at the beginning of the respective contribution.

1 Genetics, Biotechnology and Clonal Forestry

M.R. Ahuja[1] and W.J. Libby[2]

1 Interactions of Genetics, Biotechnology and Clonal Forestry

Early foresters were among the first to understand and apply principles of evolution to practical forestry (Matthew 1831, 1860). Matthew's first citation appeared 11 months before Darwin sailed on the *Beagle*, and earlier published observations on the effect of seed origin on the performance of planted trees go back to 1760 (reviewed in Larsen 1956).

Foresters were not as quick to embrace genetics, delaying nearly a half century after its important principles began to be understood before serious efforts were made to apply them (Larsen 1956). Once those efforts began, however, first in Scandinavia and soon elsewhere, the many branches of classical genetics were put into effective practical use (Zobel and Talbert 1984).

More recently, the intellectual excitement and great power of molecular genetics, and of its practical biotechnology arm, have attracted substantial interest and support in the forestry research community. Although these fields have already contributed some significant insights into fundamental tree biology, for example the patriclinal inheritance of organelle DNA in redwood (Harrison and Doyle 1990), substantial numbers of trees produced, modified, or engineered by workers in these fields have yet to appear in the forest.

Clonal forestry, whose applications and advantages are presented in *Clonal Forestry* Volume 2 (Ahuja and Libby 1993), has some significant problems that can be addressed using modern techniques. In addition, some of the great opportunities available to practical forestry through genetic engineering may be most effectively captured by deploying the genetically engineered trees clonally. This not only maintains and multiplies the genotypes that were favorably modified by genetic engineering, but it is likely to achieve substantial economies of scale by making many relatively inexpensive copies of plants that were initially very expensive. Thus, it appears that clonal forestry, molecular genetics, and biotechnology are coming of age at about the same time in the general field of forestry and, furthermore, that they may synergistically interact to produce mutual successes that would have been much more difficult or even impossible had each field developed in the absence of the others.

[1] Institute of Forest Genetics, Federal Research Centre for Forestry and Forest Products, Sieker Landstrasse 2, W-2070 Grosshansdorf, FRG

[2] Department of Forestry and Resource Management, Mulford Hall, University of California, Berkeley, CA 94720, USA

Clonal Forestry I, Genetics and Biotechnology
Ed. by M.R. Ahuja and W.J. Libby

Clonal forestry has also drawn theory and techniques from classical genetic fields. We begin this two-volume series on clonal forestry by addressing some contributions of classical genetics first, then considering biotechnology second, and then, in Volume 2, presenting case studies of historical, current, and possible future applications of clones in forestry, and of clonal forestry.

2 Classical Genetics

In addition to providing the theoretical and factual foundations for tree breeding, genetics is germane to clonal forestry at several levels. Clones can also contribute to various kinds of research. For some species, radiata pine for example, the use of clones in research preceded, and led the way to, the development of clonal forestry. In others, poplars for example, their widespread use in clonal forestry preceded their important use in research. Chapter 6, Vol. 1, presents many of the advantages of clones for research in general, and for physiological research in particular.

At the population level, genetic applications involve understanding and predicting how clones will perform when deployed in various configurations (for example, in monoclonal blocks, in clonal mixtures within monospecies plantations, or in multi-species mixtures) (see Chaps. 4, 5 and 9, Vol. 1). Such performance includes not only harvestable productivity, but also how the broader ecosystem develops within and around such plantations. Within this latter, it is important to consider how specific components of these ecosystems behave, for example whether pathosystems develop endemically or epidemically.

At the organism level, developmental genetics includes such crucial processes as maturation. An important idea developed in Chapter 3 is that readiness-to-flower (reproductive competence) may be more than an index of a qualitative difference between juvenile and mature trees. Rather, the progression from late juvenile through early adolescent, late adolescent, and early mature to fully mature may be tracked using male/female ratio as one type of sensitive observation. This question of maturation state is a major issue in other chapters as well. In Volume 1, Chapter 11 reviews the role of maturation state in vegetative propagation, particularly with respect to rooting of cuttings, and Chapter 7 relates maturation state to the success of various types of tissue culture. Chapter 12 reviews maturation from some different perspectives than those of Chapter 3, sometimes arriving at rather different pictures of both phenomenology and causation. In keeping with the theme of Volume 2, Volume 1's Chapter 12 also addresses the possibilities of manipulating maturation state. The obvious manipulation is from mature (when the tree or clone can more confidently be evaluated) to juvenile (so that clonal copies will proceed through the same developmental ontogeny as did the evaluated trees). Bonga and von Ardekas properly focus on this giant step from mature back to juvenile; but there are other options. In Chapter 9, Frampton and Foster are alert for changes in maturation state from embryonic to something more advanced among tissue-culture-origin plantlings of loblolly pine. Such rapid maturation leads to the possibility of early testing of members of the next breeding generation for characteristics (such as resistance to some diseases) that may be more important in later stages of maturation than in the juvenile and early adolescent phases. Finally, in Chapter 8 (Vol. 2), Gleed reports on how a maturation state that is between juvenile and mature is the propagule of choice in radiata pine for some purposes in New Zealand. In the sense of maturation as a phenomenon of developmental genetics, the manipulation of maturation state is a new and perhaps highly useful tool for applying an additional form of genetic leverage in clonal forestry. If maturation proves to be, as Greenwood and Hutchison suggest, a change in the reading frame of a given DNA library, then picking the maturation state of a particular clone for mass-propagation and deployment augments assembling sets of genetic words in that clone's DNA library through classical breeding. By choosing a maturation state as well as a genotype, one can then refine which DNA chapters, paragraphs, sentences, and words of that genotype's DNA library are being read at the time of propagation and subsequent development. The kinds and amounts of differences between seedlings and vegetative propagules (or between propagules of different maturation states), and the biologic and economic values of such differences, should be well understood before a program commits to vegetative propagules rather than seedlings. This requires testing, and Chapter 9 presents some principles for designing, executing, and evaluating tests to compare vegetative propagules of various kinds and maturation states to each other, and to seedlings of similar or identical genotype.

The molecular level leads us to biotechnology.

3 Applications of Biotechnology and Beyond

Recent developments in biotechnology and their application to forest-tree species have added present and potential dimensions to clonal forestry. These new technologies are contributing not only to rapid and reliable multiplication of selected genotypes (Chaps. 7, 10, 11, 12 and 13, Vol. 1), but they are also providing tools for the genetic modification of forest tree genomes (Bonga and Durzan 1987; Hanover and Keathley 1988; Chaps. 14 and 15, Vol. 1). In this regard, a fundamental understanding of forest tree genomes is a necessary prerequisite for eliciting introduced genetic changes in clones. Relevant to this is also the area of somatic-cell genetics of woody plants (Ahuja 1988, 1991), a field that has contributed considerably to the identification of clones (Chap. 8, Vol. 1), to the clonal conservation of forest species and populations (Chap. 3, Vol. 2), and to an understanding of the genetics of clones (Chap. 2, Vol. 1), particularly to an understanding of genetic transformation (Chaps. 14 and 15, Vol. 1).

References

Ahuja MR (ed.) (1988) Somatic cell genetics of woody plants. Martinus Nijhoff. Dordrecht. 225 pp
Ahuja MR (ed.) (1991) Woody plant biotechnology. Plenum Press, New York, 373 pp

Ahuja MR, Libby WJ (1993) Clonal forestry. Vol. 2. Conservation and application. Springer-Verlag. Heidelberg

Bonga JM, Durzan DJ (eds.) (1987) Cell and tissue culture in forestry. Vols 1–3. Martinus Nijhoff, Dordrecht. 422, 447, 416 pp

Hanover J, Keathley D (eds) (1988) Genetic manipulation of woody plants. Plenum Press. 519 pp

Harrison RG, Doyle JJ (1990) Redwoods break the rules. Nature 344: 295–296

Larsen CS (1956) Genetics in silviculture. Oliver & Boyd, London. 224 pp

Matthew P (1831) On naval timber and arboriculture – with critical notes on authors who have recently treated the subject of planting. Adam Black, Edinburgh, Scotland. 391 pp

Matthew P (1860) Nature's law of selection. Gardener's Chron Agric Gaz, Apr 7: 312–313

Zobel BJ, Talbert JT (1984) Applied forest tree improvement. J. Wiley & Sons, NY. 505 pp

2 The Genetics of Clones

W.J. Libby[1] and M.R. Ahuja[2]

1 Introduction

This chapter will focus on the ways in which the genetics of clones differs from the genetics of sexually propagated individuals. We give particular emphasis to those differences that influence differences between clonal forestry and zygotic forestry.

2 What Is a Clone?

There are at least three ways of defining clones.

Historically, clones were simply vegetative propagules. Such organisms as grapes, or willows, or African violets, were clonally propagated by rooting cuttings, or sometimes by grafting. The ramets thus produced were nearly identical to their ortets and to each other. The concept of mutations within clones was recognized long before mutations were well understood. A good example is the semi-evergreen poplar that first appeared prior to 1900 as a sport on a branch of the Lombardy poplar clone (*Populus nigra italica*) in Chile (Pryor and Willing 1965). We now know that somatic mutations occur at nontrivial rates, and that each individual tree is therefore a genetic mosaic for many loci. Clonal propagules derived from different parts of such a genetically heterogeneous ortet (or from later donor ramets) will differ from each other at many of these loci, and such differences will continue to occur and accumulate within and between ramets of such organism-level clones over time and through subsequent generations. Yet, it is general experience that genetic differences among members of a clone are much less than are genetic differences among self-sibs, full-sibs, or other kinds of sexually produced relatives in most organisms that have substantial within-population genetic variation, the common case for forest trees.

[1] Department of Forestry and Resource Management, Mulford Hall, University of California, Berkeley, CA 94720, USA
[2] Institute of Forest Genetics, Federal Research Centre for Forestry and Forest Products, Sieker Landstrasse 2, W-2070 Grosshansdorf, FRG

Clonal Forestry I, Genetics and Biotechnology
Ed. by M.R. Ahuja and W.J. Libby

As cell culture has become available technology in biology, it has become fashionable to call cell lines that originated from different single cells of the same organism different "clones". Yet, even these cell lines accumulate mutations, and sublines become slightly but nontrivially different from each other. Furthermore, there is sometimes a tendency for "somaclonal variation" to occur at relatively high rates in some kinds of cell- and -tissue culture, such that some in-vitro clones develop genetic heterogeneity even faster than do the traditional clones produced by rooting or grafting shoots that have developed on intact donor plants (see Chen and Ahuja, Chap. 7, Vol. 1).

Finally, as DNA technology developed, it became possible to clone bits of DNA and even single genes, and to check for the fidelity of such "clones" by sequencing their bases. This book will discuss all three kinds of clones, but the focus will be on the first, traditional kind of clone. Such clones are composed of the zygote (the ortet) and all propagules from it (called ramets). The word "genet" is also sometimes used to indicate a genetically (nearly) identical offspring in such a clone, but we will stay with "ortet" and "ramet", and use primary ramet (from the ortet), secondary ramet (from a primary ramet) etc. if clonal generation is important.

3 Why Aren't More Organisms Clonal?

We have been much helped in this section by J.A. Silander's (1985) chapter in *Population Biology and Evolution of Clonal Organisms* and G. Bell's (1987) chapter in *Evolution of Sex and its Consequences.* They review the conundrum that, when meiosis produces gametes and thereby recombines the genetic variation from two well-adapted parents, many and often most of the offspring have genetic combinations that are less favorable than those of the parents. Since cloning is clearly an option that is used by some species, it seems strange that more of them do not use it, thus staying with proven favorable genetic combinations.

3.1 Genetic Diversity

One explanation for most species' preference for sexual propagation is that genetic diversity is thereby maintained, and new genetic combinations are presented to a spatially or temporally varying environment. This makes sense to thinking organisms such as human scientists. It is harder to understand how nonthinking organisms, such as trees, obeying only the unwritten laws of evolution, manage to opt for this. Natural selection focuses on current-generation selection of individuals rather than on group selection, the latter being perhaps better for some long-term carefully thought-out strategy. The recombination of successful genotypes to produce, on average, less favorable genotypes seems to sacrifice short-term biological advantage for a longer-term course of what seems to be a sort of mystical wisdom.

One possible, although not general, explanation may hold in very fine-grained environments, where diverse offspring may be able to exploit such variable niches better than could a uniform set of offspring. Furthermore, diverse offspring may compete with each other less than would genetically identical offspring.

3.2 Escape from Parasites

It has been observed that viruses, nematodes, and other deleterious parasites are commonly purged from the reproductive line near to or by meiosis, but are passed on and continue to accumulate in vegetatively propagated members of a clone. Thus, no matter how favorable or successful its genotype, a clone slowly becomes less fit due to these parasites, and sexual propagation restores freedom from them and thus renewed health and vigor to zygotic offspring. One wonders whether this might be a cause of the preponderance of sexual propagation over the otherwise-more-advantageous clonal propagation, i.e., whether it has somehow evolved as a method of enforcing sexual propagation for the long-term fitness of a population, overriding the short-term advantages of clonal propagation by conferring greater health on the otherwise genetically less fit offspring of the successful parents. Put another way, is such purging coupled with sexual recombination a necessity because the purging could only be accomplished by meiosis, or was meiosis not necessary for the purging per se, but was it in some way during evolution coupled with the purging to make meiosis and sexual propagation of greater short-term advantage than asexual propagation?

Although the dynamic of this relationship of purging and meiosis is not well understood, it seems likely that the advantages of presenting a minority of better-adapted genotypes to the environment (particularly if the environment is different or changing from that of the parents), buttressed by a meiotic purging of parasites, may well explain why and how most organisms reproduce sexually rather than clonally in nature. As we move into clonal forestry, it will be wise to back clonal deployment with a continuing breeding program, to give the continuing health of the deployed clones some attention, and to consider deployment strategies carefully with respect to such options as monoclonal blocks or intimate mixtures, respectively, on coarse-grained or fine-grained sites.

4 Transmission of Genetic Covariances

Differences in transmission genetics between clonally and sexually reproducing organisms are substantial, and these differences are generally well understood.

In the normal case, half the genes of each parent (and thus half of the additive genetic variance) covary between parent and offspring during sexual propagation; all of the additive genetic component of the ortet is transmitted to each ramet during clonal propagation. Put another way, selecting the best full-sib families from a breeding orchard or seed orchard for use in family forestry captures the best of only half of the additive genetic variation available in that generation (the remainder segregating within the full-sib families), while selecting the best clones from that same population captures the best of almost all of the additive genetic variance (only new mutations may cause the ramets within clones to differ genetically).

While various breeding designs may allow the breeder to *identify* and *estimate* various kinds of nonadditive genetic variation, sexual propagation is very poor at *capturing* it. For example, if a parent is selected as outstanding in part due to particular within-locus genetic combinations (the dominance component of genetic variation), only one gene at each locus (in the case of normal diploids) is passed on to each offspring, and thus none of the dominance component of genetic variance covaries between parent and offspring. A parent may pass one quarter of its additive-by-additive component of epistatic variation to its offspring, on average, but that is the highest fraction of any component of nonadditive genetic variation that can be captured by line-breeding in an outcrossing population. (One can capture higher percentages using inbreeding schemes, but these are rarely employed in forestry.) By contrast, with the exception of mutations, ramets receive all of the nonadditive combinations that contributed to selection of outstanding ortets.

We are now only beginning to study variation among mitochondrial and chloroplast DNAs. Because of their peculiar inheritance (Harrison and Doyle 1990), it seems likely that important genetic variation, if any, will be better captured by clonal forestry than by zygotic forestry. In other words, for those plants with matriclinal inheritance of mitochondria and chloroplasts, only the organelle genes of the female parent are sexually transmitted to the offspring. For those conifers with patriclinal chloroplast but matriclinal mitochondrial inheritance, except for selfing, only one kind or the other of organelle genes will be sexually transmitted from each parent to the offspring, depending on whether the selected tree is used as a male or a female; but clonal offspring of a selected tree receive the full complement of its organelle genes.

5 Clonal Selection

It has long been understood that the accuracy of identification of outstanding clones is more efficient than that of full-sib, polycross, or open-pollinated families. It is perhaps less appreciated that breeding advances for single traits (or simple trait indexes) are a combination of accurately and efficiently identifying outstanding genotypes (clonal selection is better) and of selection differential (family-index selection is better). Thus, clonal selection is attractive for choosing a relatively few clones for immediate deployment, but it is theoretically not as good for breeding, particularly when the number of genotypes that is to be maintained in a breeding line is substantially larger than the numbers to be deployed either clonally or as parents of the next zygotic production line. See Carson (1986), Foster (Chap. 5, Vol. 1), Libby (1969, 1990), and Russell and Libby (1986) for more complete development of these trade-offs.

A topic that is less understood is the accumulation of outstanding performance for several traits in a selected clone, as contrasted to breeding for equally outstanding performance in that set of traits in a zygotic breeding line. Being able to find a few rare clones with unusual combinations of favorable expression for several traits, and then being able to rapidly propagate and deploy such unusual clones, are sometimes given as two important advantages of clonal forestry (Libby 1990). But the theoretical likelihood of finding such clones is not reliably worked out, and this needs some attention.

6 Maturation

A major genetic difference between clonal forestry and zygotic forestry is that zygotic forestry plantations usually (but not invariably, see Libby and Ahuja, Chap. 1, Vol. 2) begin with trees in a juvenile maturation state, while clonal forestry plantations are likely to be planted with propagules at states other than fully juvenile. Some of the reasons for and consequences of this difference are presented in Greenwood and Hutchison (Chap. 3, Vol. 1) and Bonga and von Aderkas (Chap. 12, Vol. 1). Suffice it to say here that there is general agreement that we are increasing our knowledge and understanding of maturation state, but its manipulation and control are not yet reliable technology. Until they are, the problems associated with maturation of clones will remain a major concern when opting for clonal forestry.

6.1 Maturation as a Problem

At least three elements of maturation present problems. The first is operational. In general, as clones mature, they become more difficult to propagate. This raises costs, often to the point of being prohibitive.

The second is that, in general, if propagules from fully mature or nearly mature donors retain that degree of maturation, they grow in ways that are less desirable than those of propagules of the same genotype from more juvenile donors.

The third is that, if test results and management experience are accumulated for one maturation state and the clone then matures to another, the test results will no longer be reliable and management experience may be misleading, even if the new properties of the clone are of positive value.

6.2 Maturation as a Form of Genetic Leverage

It seems likely, although not proven, that maturation is largely or wholly a process of developmental genetics. Thus, the same basic genetic library of a given genotype is being read differently (quantitatively and perhaps qualitatively) as the tree or clone matures.

In many cases, the quality of a particular trait improves with maturation, as for example resistance to some diseases (see Zagory and Libby 1985; Frampton and Foster, Chap. 9, Vol. 1). In some cases, the overall trade-off between traits that become more valuable and those that become less valuable maximizes at some propagule maturation state other than juvenile. In such cases, one can use this additional form of genetic leverage to increase the value of production plantations, by deploying clones at a maturation state at or near where these trade-offs produce maximum value. For radiata pine, for example, such an option is attractive if clones are deployed at a late-juvenile or early-adolescent maturation state (Gleed, Chap. 8, Vol. 2).

6.3 Senescence

Are clones immortal? We do not know. We know of some very old clones. The Thompson seedless grape clone is said to trace its origins to Greece over 3 millenia ago. Some natural redwood and aspen clones may be much older, perhaps approaching or even exceeding 10000 years.

If senescence is merely or mostly the accumulation of pathogens, there are techniques now available for purging meristems of those pathogens without recourse to meiosis. If senescence is really the end-point of a programmed developmental process, we may be able to reverse it (Bonga and von Aderkas, Chap. 12, Vol. 1). If so, then perhaps clones, or sublines of clones, can live without meiotic recombination for very long periods of time.

7 Breeding

This has not been sufficiently emphasized in many previous discussions of clonal forestry, so let us emphasize it here. *A long-term clonal forestry program must be supported by a long-term breeding program.* Clonal forestry should not be done instead of breeding. Clonal selection is only capable of picking the best genotypes produced or available in a particular generation, and clonal forestry systems can then deploy them, first quickly, and later efficiently.

A question that will need attention is whether the clonal forestry program will just opportunistically skim the cream off an ongoing breeding program, or whether the breeding program will be designed specifically to serve the clonal forestry program. The answer will vary with the situation, and with the level of commitment to clonal forestry. One can imagine that substantial differences will exist between these two strategies. For example, in a breeding program committed to serve clonal forestry, parents of the next generation might be heavily weighted to those that have performed well *as clones*, which could include characteristics such as ease of vegetative propagation, low levels of allocation to sexual organs, etc. The first cycle of testing might be clonal, with sib-testing for good families following if there was still some demand for zygotic forestry. The opposite might be true if the commitment were to family forestry, with clonal forestry serving only minor speciality functions. Perhaps there should be one or more clonal forestry multilines among the several or many multilines serving a general breeding program.

The outcome of the contest between early selection and breeding for zygotic forestry vs breeding plus clonal selection for clonal forestry is crucial as to whether full clonal forestry will be achieved. In brief, if the average performance of a zygotic breed is better than the performance of deployed clones, then the zygotic breed will replace the clones. This is a relatively new concern, and it is perhaps the most telling point made in Carson's (1986) paper. With respect to the questions of how many generations and how many multilines it will take for line-breeding to exceed, on average, selected outstanding clones, that seems a tractable problem once we are confident that we can describe the nature of outstanding clones and be surer of the approximate probabilities of their occurrence and of finding them. Calculation of breeding times to specified goals will be complicated by genetic correlations among the important traits, by age at which the set of desired traits can be effectively evaluated, and by different heritabilities for different traits. To a perhaps somewhat lesser degree, these complications also apply to calculating the probabilities of occurrence of outstanding multi-trait clones and of identifying them.

8 Clonal Testing

This topic is well covered in Libby (1987), in Frampton and Foster (Chap. 9, Vol. 1), and in citations therein.

A topic that seems to need further attention is the duration and nature of testing that is needed to supply management with the information to confidently and effectively deploy and manage well-known clones.

9 Clonal Deployment

One of the powerful tools made available by clonal forestry is the ability to effectively deploy well-known clones. Lindgren (Chap. 4, Vol. 1) presents some of the principles and concerns that go with this new tool.

9.1 Diversity

A concern about loss of diversity is often one of the first responses to proposals or conversations about clonal forestry. Surely, there is an opportunity for error here. On the one hand, managers thinking too locally, or with too short a time dimension, are tempted to opt for deploying the one or two very best clones, hoping that they can get away with it. On the other hand, well-meaning legislators, regulators, or others may opt for far too large a required minimum number of clones, thereby losing many of the advantages of clonal forestry (see

Muhs, Chap. 13, Vol. 2; Zobel, 1992). But, as in the Goldilocks story, there may be numbers that are "just right", serving the clonal forestry features of management efficiency and genetic gain, and providing greater effective deployed diversity than could zygotic forestry (Libby 1982).

9.2 Longevity

How long the best clones will be deployed is more of a question than a concern, although one can find concern in some quarters. The answer will depend on the answers to questions posed in this chapter's sections 6.1, 6.3 and 7, above. With respect to management effectiveness, we think that the answer is "the longer, the better". We offer the Lombardy poplar as a case in point.

10 Propagation

Unlike zygotic forestry, which pretty much depends on seeds, clonal forestry may be able to choose among grafting, rooting cuttings, inducing plantlets from tissue culture, and somatic embryogenesis. These topics are described and discussed in Chapters 7, 10, 11 and 13, Vol. 1.

The high cost of such propagation is often given as an argument against clonal forestry. However, we can expect various efficiencies to develop as dedicated creative workers gain experience with these techniques. Zobel's (1992) experience in South America produced the following quote in less than two decades: "The cost differential between rooted cuttings and seedlings will probably diminish rapidly until, as has happened for some *Eucalyptus* programs, rooted cuttings are produced for the same cost as seedlings."

11 Identification

The ability to verify the stated identity and fidelity of trees for a variety of purposes has long been desired. It is generally easier or at least more certain with clones than with genetically segregating heterogeneous lines or families. Cheliak (Chap. 8, Vol. 1) describes some of the techniques becoming available.

12 Genetic Engineering

While genes inserted or modified by techniques of genetic engineering may be spread through a breeding population by sexual recombination, there will be cases where it will be advantageous to replicate them (for testing) and ultimately use them in the constant genetic backgrounds provided by clones. This is particularly true if clones exist with generally excellent qualities, but with one or two disqualifying flaws. By using genetic engineering to correct the flaws, the excellent qualities already present won't be sacrificed to recombination, and good clones can thus be made even better.

References

Bell G (1987) Two theories of sex and variation. In: Stearns SC (ed) Evolution of sex and its consequences. Birkhäuser, Basel Boston, pp 117–134

Carson MJ (1986) Advantages of clonal forestry for *Pinus radiata* – real or imagined? NZ J For Sci 16:403–415

Harrison RG, Doyle JJ (1990) Redwoods break the rules. Nature (London) 344:295–296

Libby WJ (1969) Some possibilities of the clone in forest genetics research. In: Bogart R (ed) Genetics lectures, vol 1. Oregon State Univ Press, Corvallis, OR, pp 121–136

Libby WJ (1982) What is a safe number of clones per plantation? In: Heybroek HM, Stephen BR, von Weissenberg K (eds) Resistance to diseases and pests in forest trees. Pudoc, Wageningen, pp 342–360

Libby WJ (1987) Testing for clonal forestry. Ann For 13:69–75

Libby WJ (1990) Advantages of clonal forestry revisited. In: Proc Joint Meet WFGA & IUFRO Working parties S2.02–05, –06, –12 and –14. Olympia WA. 20–24 Aug 1990, Weyerhaeuser, Tacoma, WA, sec 5.13, pp 13

Pryor LD, Willing RR (1965) The development of poplar clones suited to low latitudes. Silv Genet 14:123–127

Russell JH, Libby WJ (1986) Clonal testing efficiency: The trade-offs between clones tested and ramets per clone. Can J For Res 16:925–930

Silander JA Jr (1985) Microevolution in clonal plants. In: Jackson JBC, Buss LW, Cook RE (eds) Population biology and evolution of clonal organisms. Yale Univ Press, New Haven, pp 107–152

Zagory D, Libby WJ (1985) Maturation-related resistance of radiata pine to western gall rust. Phytopathology 75:1443–1447

Zobel, B (1992) Vegetative propagation in production forestry. J For 90(4):29–34

3 Maturation as a Developmental Process

M.S. Greenwood[1] and K.W. Hutchison[2]

1 Introduction

Maturation, cyclophysis, ontogenetic aging, phase change, and phasic development are terms that have all been used interchangeably to refer to some of the changes in woody plant developmental behavior that occur with increasing age. The maturational process begins even during embryogenesis and early seedling development, and involves a complex array of morphological traits and developmental phenomena (Greenwood 1987). These phenomena have been comprehensively reviewed (Bonga 1982; Hackett 1985; Zimmerman et al. 1985; Greenwood 1987), and rejuvenation is discussed elsewhere in this Volume (Bonga and von Aderkas, Chap. 12, Vol. 1). The morphological effects of maturation described here are closely similar to those in a wide variety of plants (Hartmann and Kester 1983). We will deal here primarily with maturation phenomena in conifers beginning with 1-year-old seedlings.

While the response of a plant to its environment is clearly a function of its genetic make-up, its apparent ability to exhibit altered responses with increasing age (that are not just due to increased size and complexity of the plant) can justifiably be considered to be a form of maturation if the continuing differentiation of its cells, tissues, and organs changes in nature with age. Any differentiation of cell lines or organs that occurs must begin with apical meristems, which by definition are sources of undifferentiated cells. We are then confronted with the paradox that a meristem from a mature tree must have differentiated in some way in comparison to a similar meristem from a juvenile tree, even though both are sources of apparently undifferentiated cells. Whatever changes have occured in these apparently undifferentiated apical cells must be distinct from, and precede, other forms of cellular differentiation (such as that involved in the formation of xylem, phloem, or leaf tissues) that continuously occur in shoots of any age. The paradox of the differentiation of undifferentiated cells can be avoided if maturational behavior results solely from the effects of an increase in the size and complexity of the plant (Borchert 1976a). However, size-related developmental changes appear to be distinct from those due to maturation, and are more appropriately referred to as aging (Wareing 1959).

[1]Department of Forest Biology, 104 Nutting Hall, University of Maine, Orono, ME 04469, USA
[2]Department of Biochemistry, 275 Hitchner Hall, University of Maine, Orono, ME 04469, USA

Clonal Forestry I, Genetics and Biotechnology
Ed. by M.R. Ahuja and W.J. Libby

The purpose of this chapter is to critically examine specific examples of developmental changes that appear to be associated with maturation, with emphasis on experimental demonstration of the nature and relationship of changes in growth potential, foliar characteristics, root regenerating ability, and reproductive competence. The observation that all maturational traits do not develop at the same rate raises the key question of how maturation occurs. Does it have a single cause, or do maturational traits develop independently of one another (Borchert 1976b)? The developmental effects of maturation can be observed on intact trees of different ages, but the critical observation is that mature forms of traits are continued by grafted scions or rooted cuttings from these trees, where the effects of increased size and complexity are removed. Observations and experimental results from both situations will be compared here, in hopes of clarifying how the mature state originates. The wide variety of terms used for these phenomena probably reflects the state of confusion about the developmental basis for them.

2 Effect of Maturation on Long Shoot Growth Behavior

2.1 Height and Diameter

Forward and Nolan (1964) observed that the shoots in the upper two thirds of the live crown of red pine trees aged 20, 27, and 50 years showed less elongation growth with increasing age. Since the trees were growing on the same site, they attributed the decrease to age, but noted that environmental variation would easily override such differences. The commonly observed sigmoid curve for total height growth of forest trees may also be a maturation phenomenon. For example, the annual height increment for Norway spruce reaches a maximum at about age 20 on good sites, but not until age 35 on poor sites, although more total height growth occurs on the former (Assman 1970). Faster growth due to good site may result in earlier maturation if its attainment is a function of size (or, more specifically, number of cell divisions) rather than chronological age.

Reduced capacity for height and diameter growth with increasing age can be readily demonstrated by rooting cuttings or grafting scions from trees of different ages. Studies carried out on radiata pine (Sweet 1973), slash pine (Franklin 1969), radiata pine (Hood and Libby 1978), loblolly pine (Greenwood 1984), giant sequoia (W.J. Libby, unpubl.), eastern larch (Greenwood et al. 1989) have all yielded similar results. Juvenile grafts and cuttings not only produce more stem growth, but, as would be expected, also produce more needles and more total shoot biomass (Hood and Libby 1978; Greenwood 1984). Since rooting ability declines with age, decreases in height and diameter could be a function of less vigorous root systems on cuttings from older ortets. However, parallel experiments done with grafted scions yielded similar results, where root systems across all age classes were initially identical. In loblolly pine and larch, grafting success was not affected by ortet age.

Grafting success of scions from eastern larch trees ranging from 1 to 74 years in age did not exhibit a consistent trend with age, but was uniformly high

(90–95%, see Greenwood et al. 1989). In addition, hydraulic conductivity across the graft union does not appear to account for the reduced growth of the mature grafts (Greenwood, unpubl.). One year after grafting, conductivity through the graft union was the same as the root stock on vigorous grafts. Also, mature grafts which had exhibited very slow growth for three growing seasons did not exhibit reduced hydraulic conductivity through either the graft union or the rootstock below.

The effect of age on height and diameter growth of scions of eastern larch are shown in Figs. 1a and b. The time course for the effect of age on height and diameter is curvilinear for both, with the rate of change decreasing with age. The increased diameter of the juvenile scions appears to be due to increased radial tracheid diameter. Over 4 years, the number of cells in radial files is almost exactly the same in juvenile and mature stems, but radial diameters are consistently greater each year in juvenile stems. Tracheid length is roughly the

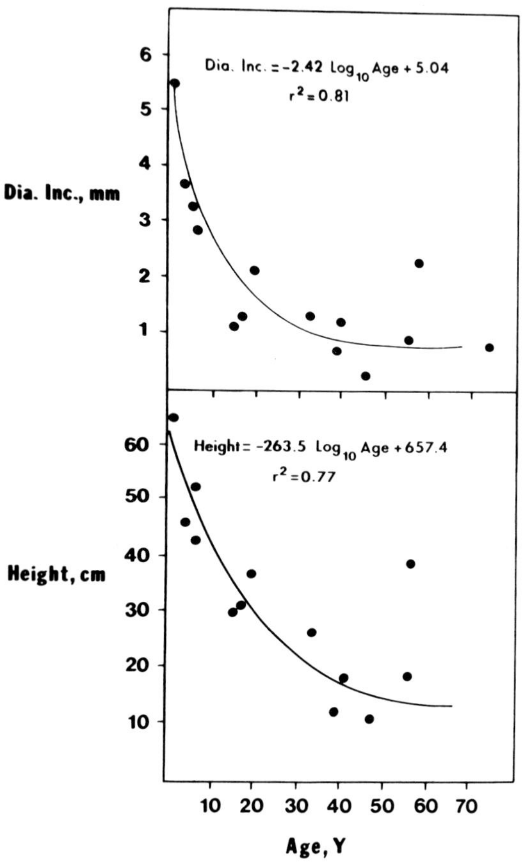

Fig. 1. Diameter and height growth of grafted scions of eastern larch ortets of different ages after 1 year. Each *point* represents the mean from 5 to 29 scions in each of 14 different age groups. Each scion was pruned to a length of 20 cm by removing the terminal bud, and diameters were similar across all ortet ages

same, and tangential diameters are slightly greater only in the first year after grafting (Takemoto 1991). Unfortunately, there are very few demonstrations of the time course of reduced growth potential as a function of age in other species. However, limited data available for radiata and loblolly pines indicate a pattern similar to that for eastern larch (Greenwood 1987).

Hackett (1985) and Zimmerman et al. (1985) appear to have attributed reduced height and diameter growth to increased complexity and size of the tree, i.e., to aging, not maturation (as defined by Wareing 1959). However, the fact that scions of the same size, from trees of different ages, exhibit growth rate differences, indicates that decreased growth potential is also a maturational characteristic. While grafting onto juvenile root stock does appear to reinvigorate scions from older trees, they cannot initially grow as fast as scions from younger trees, even though the rate of graft take and the final success rate do not appear to differ. Thus, in our opinion, these results demonstrate that loss of vigor can be due to aging, maturation, or both.

2.2 Branchiness and Shoot Growth Habit

Trees originating as grafted scions or rooted cuttings from juvenile (about 1 year old) ortets of eastern larch and loblolly pine exhibit greater branchiness (branches per unit stem length) and/or a greater tendency towards orthotropic growth than those from more mature ortets (Fig. 2). The transition to the mature form for branchiness appears to be abrupt for both loblolly pine and larch instead of the curvilinear pattern described above for height and diameter (Greenwood 1984; Greenwood et al. 1989). Grafted scions of larch only exhibit orthotropic growth if they come from 1-year-old trees, while loblolly pine scions from all ages sampled exhibited orthotropic growth. Scions from both juvenile and mature larch trees were decapitated so that the main shoot developed from a lateral bud, to minimize topophytic effects. However, loblolly pine rooted cuttings exhibited some plagiotropic growth only when taken from ortets older than 1 year (Greenwood and Nussbaum 1981). In general, vegetative propagules of *Pinus* sp. show less plagiotropism than other genera of the Pinaceae. Also, the expression of plagiotropism in grafted Douglas fir can be greatly decreased by grafting scions onto older rootstock (Copes 1987).

The manifestation of orthotropism is not solely a maturational (cyclophytic) characteristic (Roulund 1979; Hartmann and Kester 1983; Power et al. 1988). Redwood cuttings from progressively higher branch orders from seedlings of the same age show an increased tendency towards plagiotropic growth, which the authors attribute to a positional (topophytic) effect. This could also be a consequence of the maturational process occurring at a different rate in lateral meristems. Also, since plagiotropism can involve the formation of compression wood on the upper side of the branch (Starbuck and Roberts 1983), the maturational process may directly or indirectly affect the cambium as well. In addition, cuttings from similar branch orders from older plants show even more plagiotropic tendencies for both redwood and Norway spruce. Also, the abrupt decrease in branchiness and orthotropic growth are distinct from the gradual change observed for other maturational characteristics (Greenwood et al. 1989).

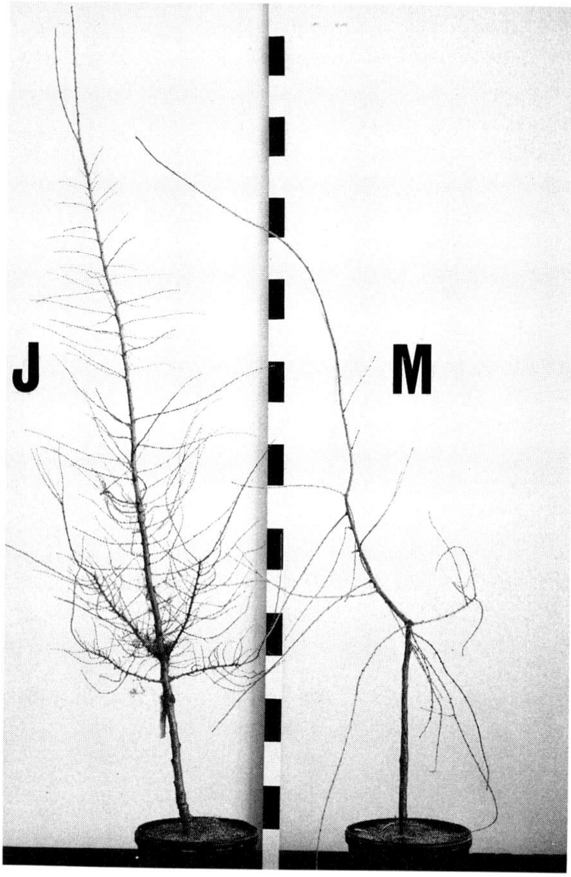

Fig. 2. Scions from a 1-year-old ortet (*J*) and a 25-year-old ortet (*M*) after two growing seasons. The 1-year-old ortet was grown from open pollinated seed from the 25-year-old ortet. Note plagiotropism, reduced diameter, and branching of the mature graft

3 Foliar Characteristics

3.1 Morphology and Physiology

3.1.1 Morphology

Foliar characteristics change markedly as maturation progresses in a wide variety of conifers. Changes in needle form, dimensions, pigment (chlorophyll and anthocyanin) content, surface waxes, and CO_2 fixation have all been

Table 1. Foliar characteristics of selected conifers and *Hedera helix*

	Species (reference)	Juvenile	Mature
1.	*Cupressus* (Lagner and Melchior 1968)	Needles relatively long and spine-like, with internode between	Needles closely spaced, scale-like, tightly appressed to branch. Transition occurs abruptly in approximately 2 years
2.	*Larix laricina* (Greenwood et al. 1989)	Relatively less chlorophyll (mg g^{-1} dry wt.) in both long and short shoot needles; juvenile long shoot needles longer and thinner	30–50% more chlorophyll in both long and short shoot needles, which is associated with a significant increase in net photosynthesis (gCO_2 m^{-2} S^{-1}). Transition is gradual
3.	*Picea* sp. (Steele et al. 1989b; Hutchison et al. 1990b)	Juvenile needles spine-like, square in cross section, much smaller than mature needles	Mature needles blunt, flattened, massive in comparison to juvenile, 3x wider, specific leaf area (mm^2g^{-1}) about 25% that of juvenile. Less stomatal conductance (50%), which is associated with 30% less photosynthesis (gCO_2 m^{-2} S^1). Transition is gradual
4.	*Pinus taeda* (Greenwood 1984)	Fascicular needles shorter, thinner, with less dry weight per needle than mature. Yellower in appearance	Fascicular needles longer, thicker, more dry weight, darker in green (more chlorophyll?). More resistant to infection by fusiform rust. Transition is gradual
5.	*Hedera helix* (Bauer and Bauer 1980)	Juvenile leaves lobed, thinner, less stomatal conductance which results in less photo synthesis (mg CO_2 dm^{-1} h^{-1}). Juvenile form has more chlorophyll (mgg^{-1})	Mature leaves rounded, thicker, more stomatal conductance, more photosynthesis (mg CO_2 dm^{-2} h^{-1}), no anthocyanin in petioles. Transition is gradual

reported (Table 1). Following germination, most conifers exhibit indeterminate growth resulting in acicular needles throughout most of the first growing season. In many cases these simple needles are considerably smaller and distinctly different from the foliage produced after the first bud is set. However, the post-juvenile foliage that forms thereafter also exhibits gradual change with increasing age. For example, the fascicular needles of loblolly pine are thicker and slightly longer with increasing age, while the more mature needles of Sitka spruce and eastern larch are considerably thicker (Greenwood 1984; Steele et al. 1989; Sherman 1989). Mature needles of red spruce are also considerably thicker and heavier than juvenile needles. The length-to-width ratio is smaller with increasing age in all four species; i.e., needles become more massive with age, although the change is relatively less in larch than in the other three species. In addition, the amount of cuticular wax per unit surface area appears to increase with age in Sitka spruce (Steele et al. 1989). In general, the changes in foliar morphology are curvilinear with increasing age, changing most rapidly in the early years.

Foliar chlorophyll content also appears to change with age in some species. In eastern larch, chlorophyll content on a needle dry-weight basis increases almost 50% between 1 and 74 years (Greenwood et al. 1989). The time course for this change follows the same curve shape as height and diameter (see Fig. 1) but with opposite slope. While juvenile foliage of both radiata and loblolly pine appears yellower than mature, chlorophyll content has not been measured. In red spruce, there appears to be little difference in chlorophyll content between the foliage on grafted scions from juvenile and mature trees (Hutchison et al. 1990a), but the juvenile foliage is more photosynthetically active, due to greater stomatal conductance (see Table 1). However, analysis of juvenile and mature foliage of English ivy indicates that there is more chlorophyll present in the former.

3.1.2 Physiology

While mature needles appear to be somewhat more massive, the effects of maturation on foliar morphology appear to be quite variable, and the consequences of these changes for photosynthetic activity are very different for eastern larch and red spruce. The mature foliage of larch exhibits significantly more net photosynthesis per m^2 of foliage, and photosynthesis is significantly correlated with chlorophyll content (Sherman 1989). In contrast, mature foliage of red spruce grafted scions exhibits significantly *less* net photosynthesis than juvenile foliage, which is probably due to an even greater decrease in stomatal conductance (Hutchison et al. 1990a).

3.1.3 Disease Resistance, Response to Atmospheric Pollution

Striking effects of maturation on disease resistance could be due to changes in foliar morphology. Mature propagules of radiata pine are more resistant to western gall rust [*Endocronartium harknessii* (Moore) Hiratsuka] (Zagory and Libby 1985), and the authors cite similar maturation-related resistance to fungal diseases for *Thuja plicata* Lamb. and *Pinus strobus* L. Ohba (Chap. 4, Vol. 2) reports that rooted cuttings (presumably from mature trees) from sugi are more resistant to needle blight (*Cercospora sequoiae*) than seedlings. Amerson et al. (1988) report that loblolly pine tissue culture plantlets, which exhibit accelerated maturation, are more resistant to fusiform rust (*Cronartium quercum* F. ap. *Fusiforme*). In addition, even after having been grafted onto 2-year-old rootstock, mature red spruce scions appear to be more resistant to ozone damage than juvenile scions (Rebbeck, unpubl.). We speculate that the size of the more massive mature needles, accompanied by possible difference in cell wall constituents (see Table 1), may make them more resistant to fungal penetration. Reduced ozone damage may be due to reduced stomatal conductance exhibited by mature red spruce foliage.

4 Reproductive Competence

4.1 Demonstrating the Effects of Maturation

Reproductive competence, defined as the ability to form reproductive structures whether under natural, permissive conditions, or in response to a variety of treatments, has traditionally been the chief criterion for maturation in woody plants. But reliance on ability to flower as an indicator of the mature state has recently been questioned (Zimmerman et al. 1985). Failure of trees to flower when they are small does not necessarily indicate reproductive incompetence since environmental or hormonal manipulation can repeatably induce flowering on such trees. But induced flowering on small trees does not mean that they have become totally mature, since other indicators of the mature state are unaffected, and flowering does not continue without treatment (see Pharis 1974). While there is little doubt that repeated flowering by a tree in its natural environment is a good indicator that the mature state has been achieved, reproductive competence as defined above is not a sufficient indicator of the mature state. Nonetheless, the best way to obtain flowering on small trees is to achieve a species- or genotype-specific minimum size as rapidly as possible, then apply flower stimulation treatments (Hackett 1985). This rule appears to apply to whole and grafted trees, even when the latter are made with mature scions.

Grafting scions from trees of different ages onto uniform rootstock permits assessment of the maturation state of their apical meristems, without rootstock maturation, overall size, or environment confounding the comparison. When scions from mature flowering trees are grafted onto juvenile rootstock, they do not flower well initially; but they do exhibit very vigorous growth compared to their recent growth on the mature donor tree. For example, although eastern larch is reproductively precocious (Fowler 1986), in our research neither greenhouse-grown scions from juvenile nor from mature trees flowered much at all in the first 2 years following grafting, despite putting on about 2 m of growth. Only 1 of 20 mature grafts produced a few male strobili, while 3 of 20 juvenile grafts produced a few female strobili in the spring of the third growing season after grafting (unpubl. data). But in the spring of the fourth growing season over 60% of the grafts flowered, with the juvenile grafts flowering more heavily than the mature (Greenwood et al. 1989, and see Table 2). The juvenile scions of larch exhibited a significantly greater total reproductive competence than mature, but produced mainly female strobili. Similar results have been obtained in subsequent experiments (Eysteinsson and Greenwood 1990). These data are compared with the results from a similar experiment with loblolly pine (see also Table 2), which is not considered to be reproductively precocious (Dorman and Zobel 1973). Like larch, very little flowering occurred in the first 2 years after grafting, but unlike larch, the juvenile scions of loblolly pine flowered significantly less than did the mature scions. Juvenile scions from both larch and loblolly pine exhibit more vegetative growth, including more branches per unit length of stem. In both cases there are more potential flowering sites on the juvenile grafts, but the loblolly pine juvenile grafts produced very few strobili, in contrast to larch. Juvenile loblolly pine clearly exhibits less reproductive competence than larch.

Table 2. Reproductive competence of grafted scions from 1-year-old tree (juvenile) compared with scions from older trees (mature) for both eastern larch and loblolly pine.

	Juvenile			Mature		
	♀/tree	♂/tree	♂/♀	♀/tree	♂/tree	♂/♀
Loblolly pine	0.41	0.02	0.05	2.32	2.58	1.10
Eastern larch	64.20	5.85	0.09	3.95	13.20	3.34

Flowering was scored after 3 growing seasons on greenhouse-grown grafts for both species (data from Greenwood 1984 and Greenwood et al. 1989). The juvenile and mature trees are related at the half-sib level (the juvenile scions came from seedlings grown from open pollinated seed from the mature trees). Individual female strobili were scored for all trees; individual males for larch, but male clusters for loblolly pine.

4.2 Sex Ratio as an Indicator of Maturation

Both juvenile and mature scions of both species produce relatively few male strobili, which is reflected in a male/female ratio of about 1 or less (see Table 2) compared with ratios of 6 to 10 or more on older trees (Schmidtling 1983). Although male flowering is observed on small, potted trees grown in a greenhouse or lath house (W.J. Libby, unpubl.), field-grown trees of both species generally produce only females when flowering begins, which is commonly observed in other tree species (Ross and Pharis 1987). For example, an examination of over 6000 4-year-old, field-grown loblolly pine showed that about 25% of the trees produced female strobili, while less than 0.1% produced males (R.C. Schmidtling, unpubl.). Also, Caron (1986) counted male and female strobili on field-grown black spruce ranging from 8 to 18 years old from seedling establishment. Almost no strobili were produced until age 9, and the amount of flowering and the male/female ratio increased steadily thereafter (Fig. 3).

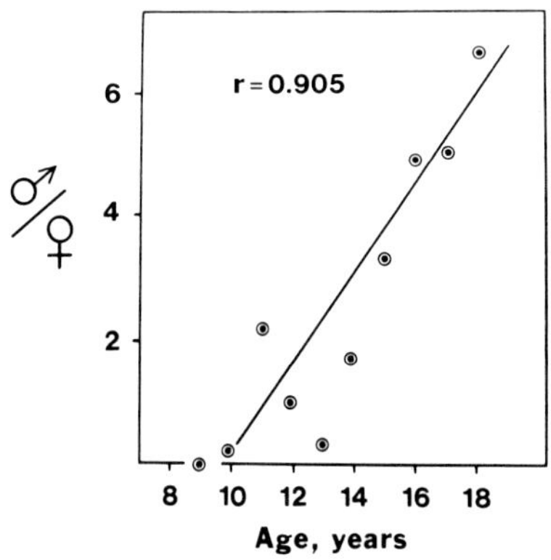

Fig. 3. Effect of maturation on sex ratio of black spruce. Female and male strobili were counted on trees in plantations 8, 10, 12, 13, and 16 years from seed. (After Caron 1986)

Therefore, we propose that sex ratio may be a better indicator of maturation state than overall reproductive competence, since female strobili predominate on juvenile material in most conifers, even under a variety of environmental conditions or in response to flower stimulation treatments (Ross and Pharis 1987).

4.3 The Relation Between Size and Maturation: Effects of Topworking

We can, therefore, conclude from the above that both a minimum size, as well as a minimum maturation state, are apparently required for flowering to occur. We would then predict that if juvenile scions were grafted onto large, sexually mature trees (topworking), that the scions would flower better than the original ortet, if size per se and maturation state were the main determinants for expression of reproductive competence.

Topworking juvenile (1-year-old) scions of loblolly pine throughout the crowns of 15-m-tall, 17-year-old (from grafting) seed orchard trees representing four clones (which were the maternal parents of the scion material) resulted in more frequent flowering of the scions, especially for male strobili, when compared with field-grown trees from the same open pollinated families (Greenwood and Gladstone 1978). Flowering was assessed on 5-year-old, field-grown trees, and on the scions 4–5 years after grafting (see Table 3). At the time flower counts were done, the grafts were from 30 to 183 cm in length, while field-grown trees averaged 270 cm in height. Although the grafts grew less than the field-grown trees, the frequency of occurrence of male flowering was about 20-fold greater for the grafts, while female flowering was about seven-fold greater. The preponderance of male flowering may have resulted not only from the large size of the understock, but from the grafts having been made on secondary branches so that they did not develop into vigorous primary branches where females are usually borne. Also, after 5 years, most of the grafts were in the lower half of the crown, where male strobili predominate. Nonetheless, examination of photographs of the grafted scion and adjacent understock branches showed that the mature understock produced male clusters with many catkins on virtually every

Table 3. Flowering (% of grafts or trees with ♀ or ♂) by grafted scions (topworked into reproductively mature trees) and field-grown trees representing four half-sib loblolly pine families after about 5 years

	Topworked grafts		Field-grown trees	
Family #	% with ♀	% with ♂	% with ♀	% with ♂
1	23	69	4	13
2	20	60	17	0
3	13	67	8	0
4	23	38	8	0
$\bar{X}$	20	59	9	3

available branch tip, compared with male clusters with few catkins on a minority of the available branch tips of the nearby juvenile scions. Overall, flowering on the topworked juvenile grafts was not nearly as frequent as on the adjacent branches of the rootstock. Virtually all primary and secondary branches of the rootstock adjacent to the grafts bore male and/or female strobili, and their foliage was darker green and coarser than that of the grafts. The scions clearly had retained juvenile foliar characteristics even after 5 years, and were not as reproductively competent as the mature upper crown rootstock branches. Although the rootstock promoted flowering by the scions, the juvenile characteristics of the scions, including flowering, were very persistent. There are other reports of attempts to stimulate flowering in *Pinus* by topworking, some of which were successful (Mirov 1951; McDaniel and Einert 1976), while others were not (Barnes and Bingham 1963; Simak 1979).

4.4 Conclusions

In conclusion, the maturation state of grafted conifer scions strongly influence their reproductive behavior, regardless of the type of rootstock used. Even in nonprecocious species, scions from young (1-year-old) plants will flower in response to various treatments, and such juvenile scions will have a greater tendency to form female strobili. $GA_{4/7}$ applications and cultural manipulations cannot completely reverse the failure of juvenile material to flower, and $GA_{4/7}$ preferentially stimulates female strobilus production (e.g., Greenwood 1981; Ross and Pharis 1987). However, even though the relative proportion of male strobili can be slightly increased by topworking, girdling, and a number of other treatments (Greenwood and Schmidtling 1981; Ross and Pharis 1987), the effects of maturation on sex ratio are persistent. The ratio of numbers of male to female strobili may well be a more precise indicator of maturation state in conifers than overall reproductive competence alone, but should be assessed using grafted material on uniform rootstock, under uniform conditions. Although increased size promotes flowering, the role of size on the expression of reproductive competence is, in our opinion, an aging phenomenon. The physiological effects of increased size and complexity on flowering are not well understood, but continuing cell division of an apical meristem may be the actual mechanism that drives the maturation process (Robinson and Wareing 1969; Sussex 1976; Hartmann and Kester 1983). The experiments with grafted scions do not, in our opinion, indicate that the ability to flower is solely a function of increased size and complexity of the plant.

5 Rooting Ability

Next to reproductive competence, the ability of cuttings to form roots is the commonest and most widely studied maturational characteristic (e.g., Gardner 1929; Thimann and Delisle 1939). In addition, studies of the rooting ability of cuttings taken from different levels of the crown provide evidence that maturational gradients exist in individual trees (Tufuor 1973; Foster and Adams 1984; Hartmann and Kester 1983), with maturation increasing towards the top of the tree. For this reason, cuttings should be taken from the uppermost part of the crowns of trees of different ages in order to compare the effects of maturation on rooting.

The effects of maturation on rooting by cuttings taken from eastern larch and Sitka spruce are shown in Fig. 4. The Sitka spruce cuttings were taken directly from the uppermost whorl of lateral branches of Sitka spruce trees of different ages. In contrast, the larch cuttings were taken about 4 months after grafting from scions, which were taken from the terminals of primary branches in the upper 25% of the live crown (Steele et al. 1989b; Greenwood et al. 1989). As in the spruce experiment, the larch cuttings were taken from lateral branches of the developing scions, so that the effects of varying size and topophytic position could be reduced. Morgenstern et al. (1984) report strikingly similar effects of age on rooting of eastern larch, even when the cuttings were taken from field-grown trees of different ages. In addition to positional effects within a tree, the effect of maturation on rooting is easily confounded by the large amount of variability in rooting ability among ortets of the same apparent maturation state (e.g., Tufuor 1973; Foster et al. 1981). Consequently, large numbers of cuttings from a wide variety of individuals are included in the results presented here. For both species, rooting ability declines curvilinearly with age. While rooting decline in larch is linear if plotted against the $\log_{10}$ age, the more pronounced

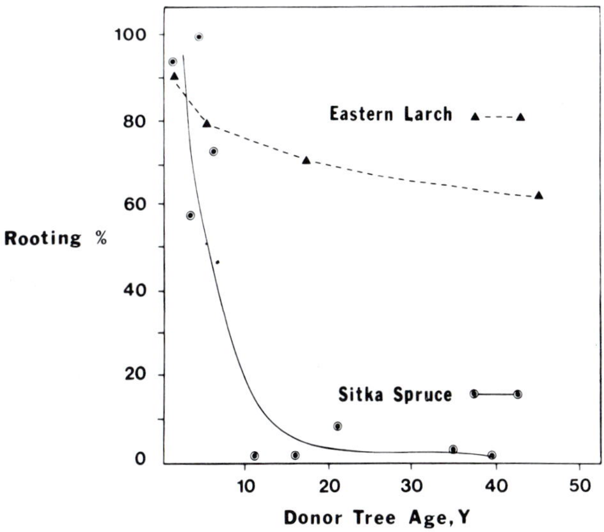

Fig. 4. Effect of maturation on rooting of eastern larch and Sitka spruce. All cuttings were taken from lateral shoots, but larch cuttings were taken from branches that elongated on scions about 4 months after grafting. In the latter case, effects of environment on the vigor of the cutting prior to rooting would be minimized. (Based on data from Steele et al. 1989b; Greenwood et al. 1989)

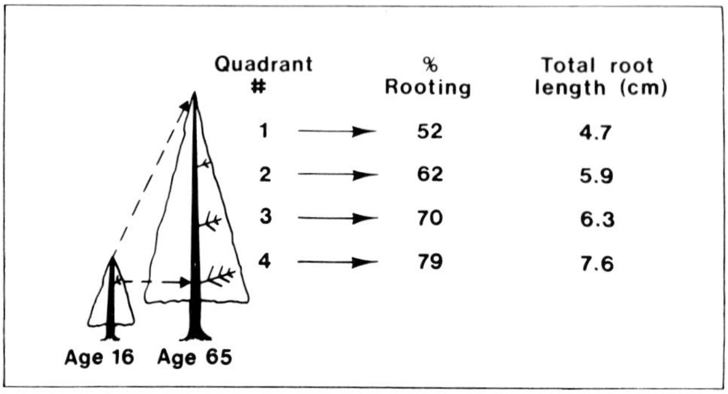

Fig. 5. Demonstration of a gradient of maturation state persisting in 65-year-old western hemlock trees. Rooting percentage and the total length of roots produced by the cuttings decreases with tree height. The relative amount of growth produced by the leader and a lateral branch between ages 16 and 65 is compared schematically. (Based on data from Foster et al. 1984)

decline in spruce can be linearized with a Gompertz function (Steele et al. 1989b). In both experiments, ANOVA showed that the effects of age on rooting were significant at $p < 0.001$.

Rooting ability declines with maturation state for many other species, and the time courses appear to vary widely between species (Libby, unpubl.). In species where adult trees are relatively difficult to root, such as sugar pine and loblolly pine, the decrease may be very abrupt (Gardner 1929; Libby, unpubl.). In loblolly pine, only cuttings from seedlings in their first year root well.

The existence of a gradient in rooting ability within single trees is evident from the results of Foster and Adams (1984), where rooting was compared between four levels of the live crowns of several trees (see Fig. 5). The authors attribute these differences to the existence of a maturational gradient within the tree. The same phenomenon may be evident in a number of other species (Tufuor 1973; Hartmann and Kester 1983). While the base of the tree has existed longer than the top, the terminal shoot is the developmentally most mature part of the plant. The cells of this terminal apical meristem have undergone many more cell divisions than those of a lateral branch near the base of the tree. Such results have been used as evidence to support the hypothesis that maturation has a cellular basis (Hartmann and Kester 1983), and that maturation state is a function of cumulative cell divisions (Robinson and Wareing 1969; Greenwood 1984).

6 Gene Expression and Maturation

Currently there is little information on the relationship of gene expression to phase change in woody plants. Therefore, models for the role and the regulation of gene expression in maturation are more speculative than predictive. In the absence of large amounts of data, we will focus this discussion primarily on two arguments. First, we will discuss whether it seems reasonable and/or feasible to investigate maturation from a genetic or molecular genetic approach. Second, we will discuss what are some of the potential genetic mechanisms which a plant may use to control this complex developmental process.

There are data that do suggest that genes play a role in the regulation of maturation. The observations of Greenwood (1984) in loblolly pine, and Greenwood et al. (1989) in larch, suggest that a stable genetic change has taken place in the meristematic tissue of the plants. In both cases, scions from mature trees grafted onto juvenile rootstock maintain their mature characteristics, in spite of their reduced complexity and size, their increased vigor, and the presence of whatever signals may be coming from the juvenile rootstock. Perhaps more direct examples are the observations that in loblolly pine both precocity and fecundity are heritable characteristics that vary independently of one another (Schmidtling 1981, 1982), and that there is considerable clonal variation within a species in the ability to form roots (Foster et al. 1981). These data all support the proposition that at least some of the traits associated with maturation are under genetic control. A second conclusion one might draw is that the control of any single trait is either independent of, or can be genetically divorced from control of the other traits.

Each of the maturational traits described above is likely to be affected by a number of different genetic loci, which can be classified into quantitative trait groups. The number of loci involved in each trait is, of course, unknown. There is, also, no reason to exclude the possibility that individual members of a quantitative trait group can be part of more than one group. Variability in the expression of the traits could be due to differential expression of individual genes within these groups. The possibilities become more complex when one considers that an individual gene may be part of a multigene family. Additional variability in the expression of maturation characteristics may result from differential expression within such a family, just as individual members of the multi-gene family encoding the small subunit of ribulose-1,5-bisphosphate carboxylase/oxygenase (rbcS) are differentially expressed (for example, see Dean et al. 1985).

This discussion of the intricacy of the system is not to argue that maturation is too complex a problem to approach on a genetic or molecular genetic level. Indeed, the ability to work with specific genes will reduce the complexity of the problem. For example, Paterson et al. (1988) found in the tomato that fruit mass, fruit solids, and fruit pH could each be affected by four to six different loci. What is relevant to this discussion is the fact that the effect of each locus within a quantitative trait group could be separated genetically from the other loci in the same group. In other words, quantitative trait loci appear to operate in an additive, as opposed to cooperative manner. This suggests that the study of the expression of any one gene can be carried out independently of consideration of the rest of the haplotype for any given trait. Initial studies can focus on one gene affecting a specific maturation characteristic, e.g., root or cone formation, without having to consider other genes which affect either the same, or different traits. It would seem likely that once the regulation of one maturation-associated gene is understood, the results could be extended to other members of the quantitative trait group, if not to the process of maturation as a whole.

If such differential gene expression does occur in association with phase change, there are a number of questions to be answered. Is (are) the regulatory sequence(s) for each gene affected, and if so, how? Is the controlling mechanism epigenetic in the classical developmental sense, or is the mechanism within the genome itself? Are the regulatory sequences affected independently of one another, or are there maturation-specific genes which, in turn, affect all of the multigene families?

Whatever the mechanism(s) involved, they must meet at least two criteria. First, they must be relatively stable once imposed, so that maturation is apparently not reversible by environmental changes. This restriction would seem to preclude classic genetic control models utilizing small effector molecules, DNA binding proteins, and target sequences surrounding the promoters for the requisite genes. Regulation via changes in DNA methylation, alteration in chromatin structure, and genetic rearrangements are all possible mechanisms that come to mind, and that fit this criterion. Interestingly, such a requirement would not exclude a cellular (i.e., nongenetic) mechanism involving the internalization and turnover of membrane-bound receptors, making them unavailable for use.

Second, they must ultimately be reversible so that the next generation can proceed normally through the juvenile phase. This provision would seem to except DNA rearrangements. However, a simple rearrangement, such as seen with phase variation in *Salmonella* (Zieg et al. 1978), would still be possible. In this system, the control region is involved in a specific and reversible inversion. The phenotype of the bacterium is dependent upon the orientation of the control region. An alternative model relying on DNA rearrangements would involve genetic rearrangement or loss of genes associated with the onset of maturation. Reversal would involve gene conversion, using cassettes of information stored elsewhere in the genome, similar to what is seen with mating-type switching in yeast (see Nasmyth 1982). It should be pointed out that this second requirement is actually not necessary if some population of the cells in the plant never mature and are recruited as the precursors for gametogenesis. Cellular control mechanisms such as the receptor internalization mentioned above would also fulfill this stipulation.

DNA methylation is an alternative mechanism which meets both criteria of stability and reversibility (Hood and Libby 1978). We have measured methylation of DNA in juvenile and mature larch, and see no significant difference (Greenwood et al. 1989). However, changes in the level of DNA methylation at specific sequences would be undetected by such techniques. Recent work in corn suggests that DNA methylation may be at least one of the mechanisms of control during maturation.

Maturation in corn proceeds upward along the stalk, with the base of the plant most juvenile and the top of the plant most mature. In this regard, the gradient is similar to that reported in the crowns of conifers (e.g., Foster et al. 1981). The maize juvenile form has short internodes, similar to what we have reported for the juvenile form of larch. Cuttings from the juvenile portion of the stalk root more readily and rejuvenate to produce a new plant. This is, in fact, the basis of the normal production of tillers by the plant. The upper, or mature, portions of the plant are associated with the onset of flowering. And maturation is essentially irreversible at the somatic level. Rooted cuttings from the top of the plant will not produce a new stalk, but a rooted ear. Rejuvenation occurs only by passage of the genome through the gametophyte generation.

In some recent studies, Fedoroff and co-workers (Fedoroff and Banks 1988; Banks et al. 1988) have investigated the expression and genetic activity of the maize transposable element, *Spm*. They have found that the elements are more actively expressed in the juvenile portion of the plant than in the mature portion of the plant. Significantly, for this discussion, the level of expression was inversely correlated with level of methylation in the sequences surrounding the transposable element control regions. Extrapolation of these observations to maturation in woody plants is a significant but intriguing jump.

It is our working hypothesis that maturation is the result of genetic changes in the meristem of the tree, and that the different phenotypes of the mature and juvenile phases will be reflected in the RNA transcripts found in the respective plants. We recognize that cellular mechanism such as mentioned above might be active, but most of the data on gene regulation during development suggest that the primary mode of control is at the level of transcription. We have initiated a study to detect genes which are differentially expressed in juvenile and mature trees, using larch as a model system.

We approached this problem in two ways. Our initial approach was to screen cDNA libraries representing sequences expressed in either juvenile or mature trees, using differential and subtractive hybridization (see Hutchison et al. 1991). After an extensive search, we have been unable to identify any sequences that are uniquely expressed in either the juvenile or mature phase (Hutchison et al. 1991). Our conclusion from this effort is that there are few, if any, transcripts that are unique to a specific phase of growth, and that maturation may be a consequence of quantitative changes rather than qualitative changes in gene expression. Hackett and his co-workers have reported similar results in the study of maturation in *Hedera*. Using juvenile and mature plants from a single genetic background, a screen of approximately 10 000 clones yielded two clones that are differentially expressed. One of these encodes a cell-wall protein and one encodes one member of the chlorophyll a/b-binding protein multigene family (cab) (Hackett et al. 1991).

As an alternative approach, we have been using known DNA clones from other plants that may be developmentally regulated and differentially expressed between juvenile and mature plants. Using this approach, we determined that the cab multigene family is expressed at higher levels in newly expanding short shoots from juvenile larch trees than in the comparable tissue from mature trees (Hutchison et al. 1990b). As we have found for other maturation-related changes, most of the decline in cab expression occurs within the first 5 years (Greenwood et al. 1989; Hutchison et al. 1990b). We do not know if differences in cab expression are due to differential transcription rates or changes in the rate of cab mRNA turnover. In some preliminary experiments we have determined that some of the differences in cab mRNA levels may be due to specific changes in the expression of individual members of the cab multigene family (Hutchison et al. 1990c). Expression of cab genes in late season foliage was highly variable (Hutchison et al. 1991). This may be due, at least in part, to difficulty in extracting RNA from such tissue samples. We have also found that expression of another multigene family encoding part of the photosynthetic apparatus, namely the genes encoding the small subunit of ribulose 1,5-bisphosphate carboxylase/oxygenase are not differentially expressed between juvenile and mature plants.

Our results have shown that though qualitative changes in gene expression may not occur during phase change in conifers, quantitative changes do occur. We can use the expression of the cab multigene family as a model to lead us ultimately to the underlying mechanisms responsible.

As we have attempted to show in the previous sections, the issue of phase change is sufficiently complex to generate disagreement as to what the phenotypes can be specifically correlated with the juvenile or mature phases. As these questions are examined more deeply, one can be left with the impression that maturation is a form of confusion with no consistent cause or effect. This hypothesis is not experimentally approachable. Much of the confusion, though, comes from comparing the maturation-associated changes across species, genera, families, etc. As an alternative approach, we can start with the assumption that the underlying mechanisms are the same in all or most plants, and it is simply the resulting phenotype that is variable. What this suggests is that the phenomenon we study should be consistently associated with phase change within the species one is studying, but that it need not necessarily be associated with maturation in all systems.

A corollary to this statement is that maturation-related phenomenon need not necessarily be irreversible in all plant systems. If the studies in animal systems tell us anything, they tell us that one can get a wide range of phenotypes associated with growth and development, but that in many, if not most cases, these have an underlying common mechanism of control. The phenomenon of maturation is a complex developmental problem and the genetic mechanisms of control are likely to be similarly complex at the molecular level. We believe, however, that the genetic mechanisms that ultimately control phase change are likely to be simpler, at the conceptual level, than might be suggested by the variety of phenotypic changes associated with maturation.

Acknowledgments We wish to thank Wes Hackett for illuminating discussions and sharing his recent results with us, and Bill Libby for his useful contributions and insights. We are also grateful to Mary Lou Hodge and Helen Belyea for their labors in editing and preparing this chapter.

References

Amerson HV Jr, Frampton LJ, Mott RL, Spaine PC (1988) Tissue culture of conifers using loblolly pine as a model. In: Hanover JW, Keithley DE (eds) Genetic manipulation of woody plants. Plenum, New York, pp 117–137

Assman E (1970) The principles of forest yield study. Pergamon, New York

Banks JA, Masson P, Fedoroff N (1988) Molecular mechanisms in the developmental regulation of the maize *Suppressor-mutator* transposable element. Genes Dev 2: 1364–1380

Barnes BV, Bingham RT (1963) Flower induction and stimulation in western white pine. US For Serv Res Pap INT-2

Bauer H, Bauer U (1980) Photosynthesis in leaves of the juvenile and adult phase of ivy (*Hedera helix*). Physiol Plant 49: 366–372

Bolstad PV, Libby WJ (1982) Comparison of radiata pine cuttings of hedge and tree-form origin after seven growing seasons. Silv Genet 31:9–13

Bonga JM (1982) Vegatative propagation in relation to juvenility, maturity, and rejuvenation. In: Bonga JM, Durzan DJ (eds) Tissue culture in forestry. Nijhoff/Junk, The Hague, pp 387–412

Bonga JM, von Aderkas P (1993) Rejuvenation of tissues from mature conifers and its implications for clonal propagion in vitro. In: Ahuja MR, Libby WJ (eds) Clonal forestry, vol 1: genetics, and biotechnology. Springer, Berlin Heidelberg New York, pp 182–199

Borchert R (1976a) The concept of juvenility in woody plants. Acta Hortic 56:21–33

Borchert R (1976b) Differences in shoot growth patterns between juvenile and adult trees and their interpretation based on systems analysis of trees. Acta Hortic 56:123–130

Caron GE (1986) Development of branch patterns and seed production in young black spruce [Picea marina (Mill.1) B.S.P.]. PhD Thesis, Univ New Brunswick, Fredericton NB

Cecich RA (1981) Applied gibberellin $A_{4/7}$ increases ovulate strobili production in accelerated growth jack pine seedlings. Can J For Res 11:580–585

Copes DL (1987) Effects of rootstock age on leader growth, plagiotropism, and union formation in Douglas-fir grafts. Tree Plant Notes 38:14–18

Dean C, van den Elzen P, Tamaki S, Dunsmuir P, Bedbrook J (1985) Differential expression of the eight genes for the small subunit of petunia ribulose-1,5-bisphosphate carboxylase into three gene families. EMBO J 4:3055–3061

Dorman KW, Zobel BJ (1973) Genetics of loblolly pine. USDA For Serv Res Pap WO-19

Eysteinsson T, Greenwood MS (1990) Promotion of flowering in young *Larix laricina* grafts by gibberellin $A_{4/7}$ and root pruning. Can J For Res 20:1448–1452

Fedoroff NV, Banks JA (1988) Is the *Suppressor-mutator* element controlled by a basic developmental regulatory mechanism? Genetic 120:559–578

Forward DF, Nolan NJ (1964) Growth and morphogenesis in the Canadian forest species. VII. Progress and control of longitudinal growth of branches in *Pinus resinosa* Ait. Can J Bot 42:923–950

Foster GS, Adams WT (1984) Heritability, gain and C effects in rooting western hemlock cuttings. Can J For Res 14:628–638

Foster GS, Martin JA, Caldwell TD (1981) Genetic variation in rooting of western hemlock cuttings. Crown Zellerbach Corp, Res Ms 32

Fowler DP (1986) Strategies for the genetic improvement of important tree species in the maritimes. CFS-Maritimes, Fredericton NB, Can, re. m-x:156

Franklin EC (1969) Ortet age has strong influence on growth of vegetative propagules of *Pinus-elliottii* In: 2nd World Consult forest tree breeding, IUFRO, Washington, Aug 7–16, Pap FO-FTP-69–718

Gardner FE (1929) The relationship between tree age and the rooting of cuttings. Proc Am Soc Hortic Sci 26:101–104

Greenwood MS (1981) Reproductive development in loblolly pine II. The effect of age, gibberellin plus water stress and out-of-phase dormancy on long shoot growth behavior. Am J Bot 68:1184–1190

Greenwood MS (1984) Phase change in loblolly pine: shoot developments as a function of age. Physiol Plant 61:518–522

Greenwood MS (1987) Rejuvenation of forest trees. Plant Growth Reg 6:1–12

Greenwood MS, Gladstone WT (1978) Topworking loblolly pine for precocious flowering, Tech Rep 042-3004/78/80. Weyerhaeuser, Hot Springs, AR

Greenwood MS, Nussbaum ES (1981) Rooting ability and vegetative growth performance of stem cuttings from one and five-year-old ortets of loblolly pine. In: Proc 16th South for Tree Improv Conf, Blacksburg VA, May 26–29, pp 176–183

Greenwood MS, Schmidtling RC (1981) Regulation of catkin production. In: Franklin EC (ed) Pollen management handbook. USDA, For Serv Agric Handb, pp 20–25

Greenwood MS, Hopper CA, Hutchison KW (1989) Maturation in larch. I. Effect of age on shoot growth, foliar characteristics, and DNA methylation. Plant Physiol 90:406–412

Hackett WP (1985) Juvenility, maturation, and rejuvenation in woody plants. Hortic Rev 7:109–155

Hackett WP, Murray J, Woo H (1991) Biochemical and molecular analysis of maturation related characteristics in *Hedera helix*. In: Woody plant biotechnology, Ahuja MR, ed., Plenum NY

Hartmann HT, Kester DE (1983) Plant propagation: principles and practices. Prentice Hall, Englewood Cliffs, 727pp

Hood JV, Libby WJ (1978) Continuing effects of maturation state in radiata pine and a general maturation model. In: Hughes KW, Henke R, Constantin M (eds) Propagation of higher plants through tissue culture. Proc Symp Univ Tenn, Knoxville, pp. 220–232
Hutchison KW, Singer PB, Greenwood MS (1988) Gene expression during maturation in Eastern larch. In: Hälgren J-E (ed) Proc Trans Kempe Symp 8: 101–114
Hutchison KW, Greenwood MS, Sherman C, Rebbeck J, Singer P (1990a) The molecular genetics of maturation in eastern larch [*Larix laricina* (Du Roi) K. Koch]. In: Rodrigues R (ed) Plant aging, basic and applied approaches. Plenum, New York, pp 141–145
Hutchison KW, Sherman CD, Weber J, Smith SS, Singer PB, Greenwood MS (1990b) Maturation in larch. II. Effects of age on photosynthesis and gene expression in developing foliage. Plant Physiol 94: 1308–1315
Hutchison KW, Singer PB, Greenwood MS (1991) Gene expression during growth and maturation. In: Ahuja MR (ed) Woody Plant biotechnology. Plenum NY, pp 69–75
Hutchison KW, Singer PB, Greenwood MS (1990C) Structure and expression of the *cab* gene family during maturation in conifers. Abst. IUFRO Molecular Genetics Working Party, Fallen Leaf Lake CA
Langner W, Melchior GH (1968) Morphologische merkmale und physiologisches verhalten in eirner einzelbaumnachkommenschaft von Chamaecyparis pisifera-plumosa aurea. Silv Genet 17:82–88
McDaniel AT, Einert AE (1976) A new technique for budding Scotch Pine. Plant Propagator 22: 10
Mirov NT (1951) Inducing early production of pine pollen. For Res Note #80, Ca For Range Exp Stn, Berkley, For Res Note 80
Morgenstern EK, Nicholson JM, Parks YS (1984) Clonal selection in *Larix laricina*. I. Effects of age, clone and season on rooting of cuttings. Silv Genet 33: 155–160
Nasmyth KA (1982) Molecular genetics of yeast mating type. Annu Rev Genet 16: 439–500
Paterson AJ, Lander ES, Hewitt JD, Peterson S, Lincoln SE, Tanksley SD (1988) Resolution of quantitative traits into Mendelian factors by using a complete linkage map of restriction fragment length polymorphisms. Nature (London) 335: 721–726
Pharis RP (1974) Precocious flowering in conifers: the role of plant hormones. In: Ledig FT (ed) Toward the future forest: applying physiology and genetics to the domestication of trees. Yale Univ School For Environ Stud, New Haven, pp 51–80
Power AB, Dodd RS, Libby WJ (1988) Cyclophysis and topophysis in coast redwood stecklings. Silv Genet 37: 8–14
Robinson LW, Wareing PF (1969) Experiments on the juvenile-adult phase change in some woody species, New Phytol 68: 67–78
Ross SD, Pharis RP (1987) Control of sex expression in conifers. Plant Growth Reg 6: 37–60
Roulund H (1979) Stem form of cuttings related to age and position of scions [*Picea abies* (L) Karst.] For Tree Improv 13: 4–26
Schmidtling RC (1982) Genetic variation in fruitfulness in a loblolly pine (*Pinus taeda* L.) seed orchard. Silv Genet 32: 76–80
Schmidtling RC (1983) Rootstock influences flowering, growth, and survival of loblolly pine grafts. For Sci 29:117–124
Schmidtling RC (1981) The inheritance of precocity and its relationship with growth in loblolly pines. Silv Genet 30: 188–192
Sherman C (1989) The effect of maturation on net photosynthesis in eastern larch. MS Thesis, Dep Bot Plant Pathol, Univ Maine, Orono
Simak M (1979) Precocious flower induction in *Pinus sylvestris* by grafting. In: Bonner F (ed) Proc symp flowering and seed development. Mississippi State Univ, Starkville, pp 63–76
Starbuck CJ, Roberts AN (1983) Compression wood in rooted cuttings of Douglas fir. Physiol Plant 57: 371–374
Steele MJ, Coutts MP, Yeoman MM (1989a) Developmental changes in Sitka spruce as indices of physiological age. I. Changes in needle morphology. New Phytol 113: 367–375
Steele MJ, Coutts MP, Yeoman MM (1989b) Developmental changes in Sitka spruce as indices of physiological age. II. Rooting of cuttings and callusing of needle explants. New Phytol 114: 111–120
Sussex I (1976) Phase change: physiological and genetic aspects. Acta Hortic 56: 275–280
Sweet GB (1973) The effect of maturation on the growth and form of vegetative propagules of radiata pine. NZ J For Sci 3: 191–210
Takemoto Y (1991) Maturation in tamarack: its effect on needle and stem wood morphology. MS Thesis, Dep For Biol, Univ Maine, Orono

Thimann KV, Delisle AL (1939) The vegetative propagation of difficult plants. Arnold Abor 20:116–142

Tufuor K (1973) Comparative growth performance of seedlings and vegetative propagules of *Pinus radiata* (D.Don) and *Sequoia sempivirens* (D.Don) Endl. PhD Thesis, Univ Cal, Berkeley

Wareing PF (1959) Problems of juvenility and flowering in trees. J Linn Soc London Bot 56:282–289

Zagory D, Libby WJ(1985) Maturation-related resistance of *Pinus radiata* to western gall rust. Phytopathol 75:1443–1447

Zieg J, Hilmen M, Simon M (1978) Regulation of gene expression by site-specific inversion. Cell 15:237–244

Zimmerman RH, Hackett WP, Pharis RP (1985) Hormonal aspects of phase change and precocious flowering. In: Pharis RP, Reid DM (eds) Encyclopedia of plant Physiol, vol 11. Springer, Berlin Heidelberg New York, pp 79–115

4 The Population Biology of Clonal Deployment

D. Lindgren[1]

1 Introduction

Clonal deployment can be discussed based on experimental information, practical experience and theoretical deduction. Reasonably safe conclusions require unrealistically large experiments with an experimental setup relevant to the suggested practice, and followed over most of a rotation time. This is seldom or never achieved with forest crops. Even if an understanding is possible in a few specific situations, proper actions depend greatly on the specific context and are difficult to generalize. Optimal clonal deployment depends on the characteristics of unspecified pests or diseases and on the future context of forestry, and these factors cannot be safely predicted. What can be done is to discuss some factors, facts, principles, and ideas of possible relevance.

2 Does it Matter What Trees Grow Together?

Clonal forestry provides more options to the forester than does the planting of unidentified mixtures of "improved" genotypes. Clonal trees of different genotypes can be deployed in different ways, forming forest populations with characteristics that may be different from conventional forests on both the individual and population level.

2.1 Competition

Trees compete for limited resources. Trees of the same genotype are likely to make similar demands at the same time. Thus, plantations where trees of the same genotype often compete may be less able to fully utilize the ecological space than plantations of genotype mixtures.

[1]Department of Forest Genetics and Plant Physiology, Swedish University of Agricultural Sciences, S 901 83 UMEÅ, Sweden

Clonal Forestry I, Genetics and Biotechnology
Ed. by M.R. Ahuja and W.J. Libby

2.2 Evolution Makes Trees Selfish

A successful genotype, in an evolutionary sense, is one that transmits its genes to successful progeny in competition with other genotypes. Therefore a certain degree of selfishness is an unavoidable consequence of natural selection. By an intentional selection and deployment strategy, it ought to be possible to obtain a crop where trees interact with each other in a more positive (i.e., productive, in the opinion of mankind) manner than in nature. Clones that perform well in monoculture, or jointly with other selected productive clones, may be generally more cooperative than random genotypes. Theoretically, a population of clones could be constructed that coworks well as a group.

2.3 The Most Valuable Clones May Vary with the Situation

In a multiclonal plantation, the better clones will produce more value per plant. Thus it may be more important that some upper percentile of a given mixture in a particular locale is good, than that the average is good. Furthermore, under various situations, different genotypes of that mixture may do well. If it cannot be accurately predicted which genotypes perform best in a particular situation, and if the available clones are likely to respond differently to different situations, then it would be advisable to plant a clonal mixture. Lack of stability in individual clones is a good argument for the use of clonal mixtures. In practice, a reliable identification of a clone that is best under all circumstances can not be made, given realistic constraints of time and resources for testing. Therefore it is theoretically possible to obtain higher production on average by using mixtures than by using the same single topranking clone over sites that vary in a patchy and difficult-to-characterize manner.

2.4 Production Versus Diversity

In agriculture, genetic variation within a plantation is usually minimal. Achieving a uniform crop, harvestable at a specific time, in a technically simple way, is an economically driven objective. A small loss in production may be acceptable in order to achieve this uniformity. A high degree of pest control is both essential and usually socially tolerable. Uniformity is often desirable from administrative and legal points of view. In agricultural crops, it has been found that performance and yield are sometimes improved by mixing appropriate "pure lines" (Harper 1977). Such pure lines are analogous to clones and, in this way, this body of agricultural literature is relevant to clonal forestry. A substantial variety of results has been observed and reported, from mixtures performing more poorly than the worst of those grown pure, to mixtures growing better than the best of those grown pure. On average, mixtures performed 4–10% better than the average of their components grown pure, but only a minority of the mixtures performed better than the best of their components grown pure. (There is a risk here of a bias, as results indicating an effect of mixing may be published more frequently than results not indicating differences.) It should be noted that most of these experiments have been done with crops which have been selected for many generations to perform well in monoline plantings. In a few experiments, lines or clones have been drawn from nature, where evolution has proceeded in a biotically heterogeneous environment. The relative advantage of the mixtures has usually been substantially greater in such experiments, which are more analogous to the forest situation. Simmonds (1979) summarized that "there is evidence that different genotypes in a population often interact, in ways yet far from understood, to produce slightly higher means and lower variances: that is, heterogeneous populations tend to be more productive and more stable." Most experiments thus far mix just two components, and thus the available data from these experiments contribute little guidance for discussions about the possible need for ten or more clones in a mixture.

If genetic diversity is advantageous for production, one would expect species mixtures to perform better than pure species. There are indications that this sometimes happens, but the evidence is not very clear (Trenbath 1974; Agestam 1985).

2.5 Pests and Pathogens Versus Diversity

A discussion of biotic risks and consequences is key to the issue of clonal deployment. There are many different parasite-host systems. Biotic systems are able to adapt and change the rules of the game during the game. Therefore, model building is difficult, and predictions are unreliable. Relevant experiments with conifers are practically nonexistent and the existing practical experience is difficult to generalize. Thus, analogies to experiments with short-rotation crops are justified.

In line with theoretical expectations, reductions in the size of pathogen populations in mixed stands (relative to pure stands) have been observed on numerous occasions in agricultural situations (Burdon 1987). Results were compiled from several studies comprising 103 two-component mixtures and ten different host-pathogen combinations. In the mixtures, disease levels were always less than in the most affected monoculture, and often substantially below the arithmetic mean of the components in pure stands.

It would seem that, with respect to a particular pathogen or virulence strain, resistant varieties often protect susceptible ones by reducing the infection pressure. The build-up of host-specific pathogens is retarded in a mixture of lines and in an annual crop even a small delay in infection build-up may result in a considerable decrease in disease. However, because of the longer time to harvest, the protection effect may be smaller in forests.

There have been cases where the mixture was disadvantageous (e.g., Burdon 1987). Agriculturists sometimes talk about "dirty multilines". A susceptible line or variety may be the entrance for a disease to a stand. There are also other mechanisms by which a mixture may constitute a better niche for a parasite. A reminder that ecological diversity may help pathogens is the existence of alternate host systems in nature.

Domesticated and wild systems are different. Part of the domestication strategy for food and feed crops is to get rid of all unappetizing, unpleasant substances that have evolved for pest protection. Thus, the natural level of protection has been much reduced for most agricultural crops. Protection mechanisms are expected to be different for forest trees compared to the relatively short-lived, even-aged, and crowded agricultural crops with a long history of domestication. Indeed, an individual tree is sufficiently large and long-lived to witness the evolution of some organisms.

Breeding may be regarded as an activity that reduces diversity. Zobel (1982) states that it is most fortunate that in forest trees almost all economically important traits are genetically independent from characteristics of pest resistance. It is therefore possible to breed for better economic characteristics, while at the same time breeding for or at least maintaining pest resistance.

There are differences among genotypes of forest trees concerning their interactions with fungi, insects, and animals. In experiments involving clonal material, significant clonal differences are frequently observed. In most situations this may simply mean that some clones are harmed more, and others less, although on average the harm will be the same as in a seedling population. The biotic damage could be reduced by selecting the most resistant clones. (It is likely that the quantitative effect of such improvements will be exaggerated due to differences between the "test" and "real exposure" situations.)

A situation that may make clones more vulnerable than seedlings is when a clone constitutes an ecological niche large enough for a parasite to become adapted. Thus, if a clone is planted over a large enough area and time, it may become more susceptible as the parasite adapts. There may be a lower and an upper limit on the size of niche which is compatible with an increased vulnerability of clones. Clones are not expected to be relatively more vulnerable if an individual tree is large enough to function as a niche. If a clone is only a small share of the niche to which a parasite becomes adapted, the parasite will anyway confront many genotypes and is unlikely to become specially adapted to a particular clone. In practice, economically significant pests and pathogens may often have niches larger than a species. For example, the important fungal diseases for *Picea abies* (Norway spruce) are *Armillaria mellea* and *Heterobasidio annosum*, and the important harmful insects are *Hylobius abietes*, *Dendroctonus micans*, *Neomyzaphus abietina*, and *Lymantria monacha* (Roulund 1983). None of these species is constrained to a single conifer species as a host. This suggests that genetic diversity is of limited importance for parasite adaptation and spreading in practical forestry.

An argument for using only a few (distinctly different) clones is that a pest or pathogen may be stopped or slowed by clonal boundaries, while a greater number of clones would provide more genetic continuity and a smoother travel corridor (Libby 1982). However, it may be questioned if this mechanism is typical of forest systems.

Genetic diversity is often regarded as a preventative measure against disease (Scmit 1978; Umearus 1982; Zobel 1982). However, this view is controversial. Heybroek (1982) concludes that "Broad generalizations on the effect of mixing of genotypes on the health of a stand or its components are dangerous. The effect may be different for each disease, site, host and case. Mixing may even be detrimental. A way to avoid possible detrimental effects of clonal mixing may be to restrict mixes to origins from the same provenances when the mixture is not likely to be more diseased than the natural forest."

2.6 Is it Necessary to Maintain the Natural Diversity?

Economically oriented foresters like uniform crops. Much of the diversity in a natural population is there for evolutionary reasons and does not contribute to production. The environment during the life of a domesticated tree is uniform and predictable compared to a wild one. In the initial phase of forest-tree improvement, genetic variation among parents is likely to be greater than among parents in a natural stand. This occurs because the parents are drawn from different stands. Thus, between-stand variation, arising mostly from drift and selection for somewhat different environments, is added to the within-stand variability. The practice of selection reduces variability. However, such reduction will be significant only for selected traits whose phenotypic heritability is high, and which are the primary target of selection (or linked traits). This situation is rare. Thus, probably reduced genetic variation compared to the natural situation is seldom a threat to the productivity of a forest stand comprising a reasonable number of genotypes.

2.7 The Progeny of the Clones

Seeds or pollen from plantations may play a role in subsequent forest generations. They may spread to adjacent areas. There may be deliberate regeneration from clonal forests, such as natural regeneration or seed collection. Seed collections could conceivably be made by mistake, in the case of poorly documented plantations. Others may be intentionally made, especially since the cost of the seed would be low relative to seed orchard seeds, while the genetic quality of the seed crop may be perceived to be compatible with that of clonal seed-orchards. Actually, it is possible that larger areas will be reforested with seedlings raised from seeds harvested in clonal forests than with clones, and thus the progeny of the clonal forests may be the most important aspect of clonal forestry.

In some respects, the considerations for plantation clonal deployment are similar to those of seed-orchard establishment. If there are few clones, selfing will be common, which results in physiologically and genetically poor seeds. This is one argument for using more than a few clones, and for avoiding a situation of contiguous members of a clone. Further, an argument against the exploitation of only one or a few outstanding families in a clonal plantation is the risk that inbred seeds would be formed.

The concerns about inbreeding during regeneration of a clonal forest are probably quite hypothetical. Where the intensity of culture is so high that clonal forestry is used, it seems unlikely that natural regeneration or uncontrolled seed collections will be made. Also, pollen contamination will often contribute to the gene pool, reducing any inbreeding or diversity problems.

In some cases it is desirable to prohibit any reproductive contribution of the current forest to the subsequent one. For example, with the introduction of an exotic species or a nonlocal provenance, it is usually the intention to control spreading. In this case, planting a single clone may be a tool for environmental protection, as production of sound seeds will be minimal and very few progeny will survive to maturity. Clones selected for poor fertility would serve the same purpose.

2.8 Environmental and Political Considerations

Clonal deployment involves the consideration of more than just maximizing yield and minimizing risks. Environmental, ecological, ethical, social, commercial, and political factors may often get more weight than forest production. This may result in more conservative policies than would otherwise be prescribed, e.g., requirements for more clones. The need for policies in itself may arise more from factors other than biological.

Clonal forestry is usually practised within the context of fairly intensive culture. Therefore, how clones are deployed is likely to have limited importance compared to all the other impacts of intensive forestry. Keeping this in mind will help to achieve a more objective perspective when considering the implications of clonal forestry.

3 Interactions Between Deployment and Testing

3.1 The Better the Clone, the More Copies Are Acceptable

Two issues in clonal deployment, in addition to that of the number of clones, is the relative numbers of copies of each clone and how to deploy the individual clones. The most simplistic approach is to deploy all clones in equal amounts. This simplifies and satisfies production requirements and ensures a certain level of diversity. Thus, it is a tempting enough solution to have become the basis of deployment policies. However, there is little biological justification for equal representation. Not only is this situation never found in nature, but there are good reasons for intentionally deploying clones in *unequal* proportions. The benefit of using a good clone evidently increases with the genetic value of the clone and the number of copies. There is some risk or disadvantage associated with making many copies of the same clone. From most points of view this risk is unrelated to the genetic value of the clone. As we are usually willing to accept more risk for more gain, it is logical that the better the clone, the wider its use. Actually, to exploit the better genotypes to a higher extent is the essence of plant breeding.

A useful way of exploring the relative frequency issue is to adopt "effective number of clones" (N_e) defined in the following way:

$$N_e = (\sum f_i)^2/\sum f_i^2 \quad \text{or} \quad N_e = (\sum p_i)^2/\sum p_i^2 = 1/\sum p_i^2,$$

Where f_i = frequency of clone i and $p_i = f_i/\sum f_i$ = proportion of clone i.

Note that if clones are equally represented, the true and the effective numbers of clones are the same. The effective number of clones is related to the entity $\sum p_i^2$. There are good reasons to use this as a quantitative measure of the disadvantage associated with using some clones in high proportions (Lindgren et al. 1989). It measures the probability that two trees belonging to the same clone will be near neighbors in random mixtures, and thus interact with each other by competing for limited resources, by facilitating the spread of other organisms, by mating, etc.

An algorithm for determining the optimal use of a clone has been derived (Lindgren et al. 1989; Bondesson 1989). The gain of using a clone is proportional to its use multiplied by its value. The gain given a certain effective number of clones is maximized when the utilization of a clone is linearly related to its value. In the protocols suggested by Lindgren et al. (1989), compared with a situation when the same effective number of clones is used in equal proportions, diversity is maintained by using more clones, but with many copies from the very best and relatively few from the poorer of those being used.

3.2 The More Well Known a Clone Is, the More Extensive Use Is Acceptable

It seems uncontroversial to state that knowledge reduces risk, and that a more well known and thoroughly tested clone can be used more, both in total and in each plantation. This is reflected in the Swedish prescriptions for clonal deployment (Hedström and Krutzsch 1982).

The ortets of the clones may be more or less well known, and this ought to be reflected in their use. If it is a tested or local provenance, there is information about adaptation. If it is a tested family, much knowledge is already available. The information about the ortets, or the population from which they are drawn, is usually based on trees older than those in the clonal tests. Thus this knowledge can build confidence in performance over a large part of the rotation period, even if the actual clonal tests are young. If something really new appears, as may become available with recombinant DNA technology or somatic-cell fusion, requirements for testing are high.

One advantage of vegetative propagation is the ability to make use of hybrids where the parents are drawn from different species or widely different provenances. If such hybrids do not exist in nature or in previous tests, they may be considered less well known and thus require more extensive testing.

The amount and quality of information will be different for different clones. Members of the currently deployed set of clones are in different trials of different size and design; the age, development, and measured characters may differ; there are various amounts of information from relatives. Many such difficulties caused by lack of symmetry of information, can be handled by predicting genetic values using modern analytical techniques, such as BLUP (Best Linear Unbiased

Predictor). An example of an application, evaluating a clonal trial, is given by Högberg and Danell (1989).

Even in a complex situation, predicted genetic values for "benefit" may be assigned to clones. The certainty and relevance of the information can be built into the prediction algorithms by assigning appropriate weight and scale to the information. When relating the utilization linearly to clonal value, the amount and quality of knowledge about the clones would contribute to their deployment weighting and site assignments.

3.3 Interactive Clones

In experiments, clones often differ in their interaction with test sites (e.g., Kleinschmit 1985), thus the variance component of clone-site interaction may be attributed to some clones more than others. These differences could be exploited. First, interactive clones constitute a powerful research tool for classification of sites. Once the site classification has been decided (by clonal results or in other ways), different strategies could be used.

If a reliable conclusion can be made that certain clones perform better on a well-defined type of site, these clones could be deployed on site types where they perform best.

A common strategy is to cull interactive clones due to the perception of higher associated risk. We may assume that evolution often creates genotypes that are more interactive than desirable for a domesticated forest, thus the risk of reducing interaction may be low. That a clone is characterized as interactive may reflect that it has been tested on the fringe of the environments where it is best adapted. Therefore, interactive clones could be considered for use in an adjacent zone, if their origin makes it likely they would be well adapted there.

However, the easiest and perhaps best strategy is to ignore interaction, and use clones based on their average performance only. Considering interaction means to accept a lower average performance. For reliable information on specific clone-site interactions there are large demands on test size and design, compared with the situation where only general effect information is required. Also, for several reasons, it is unreliable to extrapolate interaction effects from an experimental to a plantation situation. One common reason that large interaction effects are obtained is that the test sites cover too wide an environmental range for the species involved. The appropriate operational response to this situation is to change species, rather than genotypes within a species. Higher stability will probably be obtained implicitly by selection of clones for average performance over different environments.

4 Deployment to a Plantation

Plantation refers to a relatively small, contiguous area that is planted at one time. In this section the focus is mainly on deployment to a single plantation. In the Section 5, below, the focus is switched to the regional level.

4.1 Monoclonal Plantations

There are several good reasons to keep clones together in blocks. This is practised is some forestry programs, and in many agricultural equivalents.

Monoclonal plantations may cost less to establish, tend, and harvest than a more genetically diverse plantation. There are advantages in keeping clones separate when producing and providing plants (cf. Sect. 4.4). They may also require less silvicultural expertise. Practical experience and knowledge of the clones can accumulate. A failed or infected clone could be easily identified and removed. Management regimes could be adapted to specific clones. At harvest, the extracted products would be expected to have greater uniformity than products from an intimately mixed plantation, which is an advantage.

One option in clonal forestry is to select unselfish clones (cf. Sect. 2.2) that function well in stands. To identify compatible clones for mixtures may be an extremely complicated task, but it is relatively easy to test individual clones in monoclonal plantations of modest size (Hall 1982).

Monoclonal plantations may be a particularly interesting option for intensive forest management programs, where the information base is large. Thus, for example, if there are pest problems, a specific prescription can be applied. An advantage of mixtures may be that they make better use of limited resources, but under intensive culture many of these constraints have been removed. For example, nutrient limitations have been reduced by fertilization. Environmental variation is smaller. Test results are more reliable and more reproducible. The heritability and thus gain will be higher, and the greater stability of a mixture becomes less important (Hall 1982).

4.2 Clonal Mixtures

Many of the arguments for mixed culture are simply the inverse of those for monocultures.

The more uncertain the future pest situation, the stronger are the reasons to use mixtures (Sect. 2.5). Many aspects of intensive culture may favor pest establishment and increased susceptibility (Hall 1982; Zobel 1982). Thus, it is safer to use mixtures if the intensive forestry system is not a well-known one. For long-rotation species, where reliable testing takes long time and clonal aging is a problem, clones may never live long enough to be well known. With such species, mixtures are likely to be the preferred option.

4.3 Number of Clones in Clonal Mixtures

As it is uncertain whether one or many clones is best, it is evident that any numerical interpretation of "many" will be highly speculative, and mainly based on theoretical deductions. The explanation as to why it is difficult to give straight answers may be that it does not matter very much. Even when a certain model suggests an optimal number, the penalty for using another number may be small.

Some of the general considerations in determining the number of clones in a mixture are as follows:

a) The rotation age of the species – lower numbers can be used if the species is shorter-lived.
b) The fraction of the initial plants remaining at harvest – lower numbers can be used if the majority remain.
c) The intensity of the system – lower numbers can be accepted if the system is intense.
d) Whether the clones considered are well known and high-ranking (see section 3).

Huehn (1985, 1986a, b, c, 1988) has carried out several theoretical studies on the necessary number of components in mixtures including considerations of phenotypic yield stability (measured by variance), yield, risk (see below), and the juvenile-mature correlation of mixtures. The models are general, and it is not easy to adapt them to specific forest situations and to specify relevant values on needed parameters. He concludes (Huehn 1988) that the appropriate number of clones per mixture is in the order of tens, rather than just a few or hundreds.

Considerations about risk of unacceptable losses may constitute a rationale for decisions of clonal numbers. Fluctuations around a mean will be higher in a small sample than in a large one. The cause of mortality may be unpredictable, for example, air contamination, climatic change, or new diseases. Estimations of risk can be made for varying numbers of clones in a mixture and a fixed minimal acceptable plantation quality (Ohba 1978; Libby 1982; Huehn 1986b). These calculations generally indicate that clone number seldom matters if it exceeds a certain level. In certain situations a mixture of a few clones may be worse than a single clone. Weaknesses of these models are that there is seldom a sharp border between successful stands and failures, and that failure for some stands is normal. Thus, risk must be considered on a higher level than individual stands.

Clonal verification procedures will be increasingly difficult with increasing clonal number.

A tree interacts most with its closest neighbors, and there are only about four to six close neighbors. Thus, the concept of clonal interactions between neighbors may not justify the use of large numbers of clones, practically no more than seven to ten.

4.4 Intimate Mixtures or Clonal Rows?

To receive the predicted advantages of clonal mixtures it seems biologically best to mix the clones as intimately as possible. There may, however, be some reasons to arrange the clones in rows. It is advantageous to grow plants of the same clone together in the nursery and deliver plants clonewise:

a) Undesirable characteristics can be detected and these clones discarded before planting;
b) Nursery management can become clone-specific;
c) It simplifies record-keeping;

d) Physical mixing of clones is difficult and expensive and the plants may suffer from the mixing.

To obtain these advantages, clones can be individually propagated and delivered in clonal sets with a number of plants of the same clone intended for a planting row. Monoclonal rows may constitute a favorable compromise of advantages and disadvantages of monoclonal plantations and mixtures. There may also be some specific merits:

a) Every forest become a demonstration trial.
b) There is sometimes an option to get feed-back of the relative merits of clones, thus plantations may partly function as clonal tests.
c) Rows with inferior clones can easily be removed at thinning.
d) Complementary clones may be deployed in adjacent rows; e.g., each second row may be a clone meant for early harvesting at thinning.

4.5 Mixtures with Seedlings and Other Species

A mixture of cuttings and seedlings has several advantages. The availability of cuttings is usually more constrained by economic, legal, and biological reasons than that of seedlings. If the cuttings grow better, each will occupy more of the productive capacity of the site. Thus, a fixed number of cuttings could conceivably increase production more in mixtures with seedlings than grown pure. The seedlings, then, play the role of somewhat lowering risk, increasing diversity and being a “filler”, when clonal materials are limited in supply or expensive.

Species mixtures could be deployed as clones. Here it could be argued that sufficient genetic diversity is provided by species differences, and thus the use of fewer clones per species should be more acceptable. It may be possible to find clones that are particularly well suited to deployment in particular species mixtures.

Even where artificial regeneration of forests is intensive, “volunteers” often play an important role in future forest production and the maintenance of genetic diversity. Thus, the diversity of clonal plantations may actually be larger than indicated by the number of (planted) clones. From an environmental perspective, such inadvertent plants may be desirable. In fact, one approach to clonal deployment is to use the clones at wider spacings than would be done with seedlings, thereby achieving acceptable production levels while allowing some space to be occupied by other (also noncommercial) species. Another advantage of this approach is lower establishment costs.

4.6 Relatives

It may be argued that relatives in plantations should be avoided because of similar reasons as monoclonal plantations. I suggest we may be much less concerned with possible adverse effects (in the current generation) of relatives growing together than those of clones growing together. In the natural environment, relatives grow together. Plant species just have to adapt to the fact that neighbors are often relatives. Individuals in a stand (or a region, or a species) share a large fraction of their genes. Relatives may interact with each other in analogous ways to genetically identical plants, but probably at a much less intense level. As far as is known, there are no reports claiming experimental evidence that full-sibs perform more poorly in pure blocks as compared with blocks composed of mixtures of several full-sibs. Thus, adverse effects (in the current generation) of relatives growing together in a stand appear to be small, and if clones from a few families are mixed, any negative effects are likely to be negligible.

In advanced-generation breeding, co-ancestry will be a strong constraint on selection. In seed-orchards, related trees must be avoided, or spaced so as to avoid inbred progeny. In a clonal forest, the intention is usually to regenerate the subsequent forest artificially. Thus, in these respects, a less constrained selection of clones can be made than if seedlings were used, resulting in higher genetic gain. This relaxation on the demand for lack of co-ancestry can be considered one of the major advantages of clonal forestry (Matheson and Lindgren 1985). One of the most promising applications of clonal forestry is the extended use of the best tested full-sib families. However, in many programs there are not many tested unrelated full-sib families (Cotterill 1986). Thus, much potential gain of clonal forestry is lost if it is unacceptable to deploy high frequencies of related clones.

5 Regional and Total Deployment

Complexity of deployment decisions increases when consequences beyond the stand level are considered. Additional factors worth consideration are the diversity among stands, geographic and administrative boundaries, the timing for planting different stands and different clones, and the necessity of a more collaborative decision-making process. It is more difficult to exercise control on a regional level than on a stand level. In situations where clonal forestry is not just a marginal activity, regional considerations may, however, be more important. Useful concepts for the discussion of the interlinks between stand and region are MOMS (Mosaics of Monoclonal Stands) and WIMPs (Widespread Intimately Mixed Plantations), which are discussed elsewhere in this Volume.

5.1 Mosaics of Monoclonal Stands or Widespread Intimately Mixed Plantations?

Environmentalists, ecologists, and pathologists usually regard it as desirable that a large forested area has a mosaic structure. That can be obtained by managing the area so there is variation in age classes, regeneration methods, and species. A genetically diverse mosaic pattern is enhanced when there are many clones employed and each clone is used for a limited time only. Given the same number of clones, it will enhance the mosaic pattern if they are deployed differently in different plantations (like MOMS), rather than using the same mix of clones in all plantations (WIMPs).

Effective clonal number may be better preserved during the rotation time if clones are deployed in monoclonal plantations (MOMS). If mixed (WIMPs), some clones are likely to be outcompeted or lost in thinnings.

If several clones are used at the same time in a region because of legal or biological constraints, the evidence now available suggests that these clones in intimate mixtures (WIMPs) will be more productive and more healthy on an average than the same clones deployed in a series of monoclonal plantations (MOMS) (cf. Sect. 2).

5.2 Minimum Number of Clones

Many of the same arguments used for deploying several clones in a stand also apply at the regional level. Obviously, the minimum number of clones deployed to a region must be larger than that used in a single plantation. A single event could eradicate a single clone. If a sufficient number of clones are used, that would not lead to a regional disaster. If few clones are used, it will be difficult to utilize the knowledge accumulated on them as a basis for selection, because there will not be a sufficient number of well-known alternatives. Within a region, different clones may be best for different sites and different purposes.

Prescribing a large number of tested clones may have several undesirable effects, among them:

a) Gain is reduced;
b) Fewer copies per clone can be used, thereby possibly increasing cost per plant;
c) It is more difficult to control the deployment prescription.

5.3 Temporal Deployment

The constellation of clones used within any given region will vary with time. This is motivated by genetic reasons as better clones become available. There are also biological reasons to change clones, such as avoiding the build-up of parasites and avoiding problems with aging of clones (see Libby and Ahuja, Chap. 1, Vol. 2; Greenwood and Hutchison, Chap. 3, Vol. 1). Different clones will be in service for different length of time. Some may be dropped early due to propagation or aging problems. The better a clone is, the longer it will take to identify a replacement; the more well known it becomes, the longer it may be used. If clones are seldom replaced, a harmful consequence may be that long-term breeding is neglected. A combination of requirement of many clones and requirement of clonal testing (both in principle reasonable) may have a negative synergistic effect. The likely consequence of investments in testing clones and producing few copies of each clone per year (to fill requirements of using many clones) is a long life of a clone. For many situations this is risky, as clones may change over time; expensive, because clonal archives must be kept; and non-stimulating for progress, as it will take too long a time to get a return for investments in testing.

5.4 Maximum Number of Copies of a Clone

Given that it is desirable to place restrictions on the use of clones, a limit on the total number of copies of a clone may be a useful restriction in some (but far from all) situations. It is possible that restrictions on the total number of copies per clone can replace other restrictions and in practice serve the same purpose. The risk of infection build-up in a clone will be reduced. A limit discourages the use of the same clone for a long time, thus the consequences of clonal aging will be reduced. It stimulates the search for new and potentially better clones. It discourages a situation where the same clone can be found over large, contiguous areas. From an administrative standpoint, counting is perhaps the simplest auditing procedure.

In developing a policy regarding the maximum allowable number of copies per clone, the advantage of having many copies must be considered. Allowing a rather high number of copies per clone will justify the investment in testing, documenting, and preserving the clone, including the costs of discarding less favorable clones. A large number of copies used over long time will allow management experience to accumulate and become utilized. A large total number of plants used in the target area or a large size of the target area may justify a higher number of copies. The optimal limit is certainly widely different for different clones. For well-known and tested superior clones, higher numbers are justified. It may be valid to use the algorithm developed for proportions of clones in stands (Sect. 3.1) on the total number of copies, thus making the total use of a clone rather than its relative contribution on the stand level linearly related to its predicted advantages. If a clone remains on the top-list while becoming better known, the acceptable number of copies could be increased (cf. Sect. 3.2), and the active life of the clone prolonged. Administrative and commercial boundaries may create complications. With restrictive policies in place, clonal replacement will sometimes be made by virtue of policy rather than the superiority of newer clones. Conservatism and commercial thinking are likely to oppose replacement of old clones.

For different species there are differences in how easily they can be propagated and how quickly reliable test results can be obtained. For many broad-leaved species, such as many eucalyps, mass multiplication is fast and easy, rotation times are short, test results can be obtained in a short time, and there are no significant problems with clonal aging. For many coniferous species the propagation rate is rather low, rotation times are long and clonal aging is a problem. As there are more biological constraints and it takes a longer time to get reliable answers to questions in the latter case, it seems likely that administrative constraints like maximal numbers of copies per clone will be regarded as necessary to a greater extent for the coming decades.

Acknowledgements. I am grateful to Deborah L. Rogers and W.J. Libby for improving the linguistic style of this manuscript. Helpful comments by Bengt Bentzer, Dusan Gömöry, and Bo Karlsson are also appreciated.

References

Agestam E (1985) A growth simulator for mixed stands of pine, spruce and birch in Sweden. Swed Univ Agric Sci, Dep For Yield Res, Rep 15 (in Swedish with English summary)

Bondesson L (1989) Mathematical principles of optimal utilization of clones under diversity constraints. Swed Univ Agric Sci, Sec For Biometry, Rep 27

Burdon JJ (1987) Diseases and plant population biology. Univ Press, Cambridge

Cotterill P (1986) Breeding strategy: don't underestimate simplicity. In: Conf Proc Joint Meet Work parties on breeding theory, progeny testing and seed orchards, Williamsburg, Virg, pp 8–23

Hall RB (1982) Breeding trees for intensive culture. In: Kleinschmit J (1982) (ed) Proc IUFRO Joint Meet Working parties on genetics about breeding strategies including multiclonal varieties, Sensenstein, Sept 6–10, pp 182–193

Harper JL (1977) Population biology of plants. Academic Press, New York, London.

Hedström BS, Krutzsch P (1982) Regulation on clonal forestry with *Picea abies*. In: Kleinschmit J (ed) Proc IUFRO Joint Meet Working parties on genetics about breeding strategies including multiclonal varieties, Sensenstein, Sept 6–10, pp 109–112

Heybroek HM (1982) Monocultures versus mixtures: interaction between susceptible and resistant trees in a mixed stand. In: Heybroek HM, Stephan BR, von Weissenberg K Resistance to diseases and pests in forest trees. Proc 3rd Int Worksh Genetics of host-parasite interaction in forestry, Wageningen, Netherlands, 14–21 Sept pp 326–341

Högberg KA, Danell Ö (1989) Estimation of genetic parameters and selection gain in a Norway spruce clone trial. In: Stener L-G, Werner M (eds) Norway spruce: provenances, breeding and genetic conservation. Norw Inst For Improv Rep 11, pp 232–244

Huehn M (1985) Theoretical studies on the necessary number of components in mixtures. 1. Number of components and yield stability. Theor Appl Genet 70:383–389

Huehn M (1986a) Theoretical studies on the necessary number of components in mixtures. 3. Number of components and yielding ability. Theor Appl Genet 71:622–630

Huehn M (1986b) Theoretical studies on the necessary number of components in mixtures. 3. Number of components and risk considerations. Theor Appl Genet 72:211–218

Huehn M (1986c) Theoretical studies on the necessary number of components in mixtures. 4. Number of components and juvenile-mature correlations. Theor Appl Genet 73:53–60

Huehn M, Kleinschmit J, Svolba J (1987) Some experimental results concerning age dependency of different components of variance in testing Norway spruce clones. Silv Genet 36(2):68–71

Hühn M 1988. Multiclonal mixtures and number of clones. 1. Number of clones and yield stability (deterministic approach without competition). Silvae Genetica 37:67–73

Kleinschmit J (ed) (1982) Proc IUFRO Joint Meet Working parties on genetics about breeding strategies including multiclonal varieties. Sensenstein, Sept 6–10

Kleinschmit J (1985) Concepts and experiences in clonal propagation of conifers. In: eds. Zsuffa L, Rauter RM, Yeatman CW (eds) Proc Can Tree Improv Assoc 19th Meet, Toronto, p 26–57

Kleinschmit J, Svolba J (1989) The clonal option; the current status and future development. In: Stener L-G, Werner M (eds) Norway spruce: provenances, breeding and genetic conservation. Inst For Improv Rep 11, pp 209–231

Libby WJ (1982) What is a safe number of clones per plantation? In: Heybroek HM, Stephan BR, von Weissenberg K Resistance to diseases and pests in forest trees. Proc 3rd Int Worksh Genetics of host-parasite interaction in forestry, Wageningen, Neth, 14–21 Sept 1980, pp 342–360

Libby WJ (1983) The clonal option. Norsk Inst Skogsforsk Ås, 32 pp

Libby WJ (1987) Testing and deployment of genetically engineered trees. In: Bonga JM, Durzan DJ (eds) Cell and tissue culture in forestry. Nijhoff/Junk, The Hague pp 167–197

Lindgren D, Libby WJ, Bondesson FL (1989) Deployment to plantation of numbers and proportions of clones with special emphasis on maximizing gain at a constant diversity. Theor Appl Genet 77 (6):825–831

Matheson AC, Lindgren D (1985) Gains from the clonal and the clonal seed-orchard options compared for tree breeding programs. Theor Appl Genet 71:242–249

Ohba K (1978) Presence percentage of plus-traits clones and their safety numbers for mixing. For Soc Kyushu Branch Res Pap 31:77–78 (in Japanese with English summary)

Roulund H (1983) Experimentelle undersøgelser med sticklinger og betraktninger over klonskovbrug i piceaarter. Danish PhD dissertation, ISBN 87 7432 2257 (In Danish).

Roulund H, Wellendorf H, Werner M (1985) A clonal experiment in Norway spruce – 15 years' results. Forest tree improvement, vol 17. Arboretet, Hørsholm. Akademisk Forlag, København

Simmonds NW (1979) Principles of crop improvement. Longman, London New York, 408 pp

Scmit RA (1978) Diseases in forest ecosystems: the importance of functional diversity. In: Horsfall JG, Cowling EB (eds) Plant disease, an advanced treatise, vol 2. Academic Press, New York London, pp 287–315

Trenbath BR (1974) Biomass productivity of mixtures. Adv Agron 26:177–210

Umeaerus V (1982) New problems, trends and visions in agriculture resistance breeding. In: Heybrock HM, Stephan BR, von Weissenberg K (eds) Resistance to diseases and pests in forest trees. Proc 3rd Int Worksh genetics of host-parasite interaction in forestry, Wageningen, Neth, 14–21 Sept 1980, pp 381–386

Zobel B (1982) The world's need for pest-resistant forest trees. In: Heybroek HM, Stephan BR, von Weissenberg K (eds) Resistance to diseases and pests in forest trees. Proc 3rd Int Worksh Genetics of host-parasite interaction in forestry, Wageningen, Neth, 14–21 Sept 1980, pp 1–8

5 Selection and Breeding for Extreme Genotypes

G.S. Foster[1]

1 Introduction

In general, tree breeders have not aggressively utilized all of the genetic variation available to them in forest tree populations. This reluctance has been due, in part, to technical limitations such as long generation intervals, strong inbreeding depression, lack of precise estimates of nonadditive variance, and difficulty in vegetative propagation. Research results over the last few decades present tree breeders with knowledge and techniques which together enable them to use more complex improvement plans to exploit the total genetic variation which is available.

One of the products of these techniques is extreme genotypes which can be utilized, via vegetative propagation, for reforestation. Extreme genotypes are those individuals, within a population, that deviate the most from the mean (e.g., greater than 2.5 standard deviations). If the trait of interest is conditioned by polygenes and the frequency distribution approximates the normal distribution, only 1.2% of the population exceeds $\pm$ 2.5 standard deviations (i.e., 0.6% in each of the two ends of the symmetrical distribution). Genotypes may be extreme for more than a single trait; and if the genotype was extreme for more than one trait, all in the positive direction (e.g., superior genotype as opposed to an inferior genotype), it would be even more useful for reforestation. Through complementary breeding, extreme genotypes for multiple traits can be concentrated in a few individuals (Allaire 1980; Cotterill et al. 1988).

Extreme genotypes, especially for multiple traits, will be especially useful for a forest manager for several reasons. Their use in reforestation could enhance various traits including stand productivity, early site capture, resistance to diseases and insects, bole straightness, wood properties, crown morphology, site adaptation, etc. Genetic gain for the traits will far exceed that which is possible for a seed orchard-based program in the same amount of time. Two or more clones could be mixed with contrasting end products in mind: (1) one clone or group of clones which are superior for pulpwood traits (e.g., specific gravity, cellulose content) but have relatively slow growth to serve as trainer trees to be removed in a thinning and (2) one clone or a group of clones which are extreme

---

[1] USDA Forest Service, Southern Forest Experiment Station, Alabama A&M University, Plant and Soil Science Department, P.O. Box 1208, Normal, AL 35762 USA

Clonal Forestry I, Genetics and Biotechnology
Ed. by M.R. Ahuja and W.J. Libby

genotypes for a combination of solid wood product traits (e.g., bole volume growth, straightness, resistance to disease, cold or drought tolerance). Alternatively, clones with contrasting extreme genotypes could be mixed to enhance genetic diversity and better buffer the population against climatic or biotic extremes (e.g., cold tolerance, late flushing, various disease resistance or tolerance mechanisms).

By combining newly available knowledge and techniques for certain species, tree breeders now possess the tools to create extreme genotypes (for one or more traits combined) or ideotypes (Cannell 1978) as well as the techniques to vegetatively propagate them for reforestation in clonal forestry (Burdon 1986; Foster 1986). In contrast, classical tree improvement programs which rely on seed orchards, hence genetic recombination among parents, are not currently capable of faithfully reproducing extreme genotypes for reforestation.

2 Breeding for Extreme Genotypes

Breeding of select trees is used to create new genetic variants in the next generation through genetic recombination. Parent trees may be allowed to breed more or less at random by wind or insect pollination, or may be purposefully crossed in some other type of mating design. Genetic gain or advancement of the population for one or more traits proceeds more rapidly if both parents are from the select population(s) rather than the case where the female parent trees are select and the male parent trees are from the unselected population (e.g., from wind-borne pollen). In general, mating designs useful for stressing the creation of extreme genotypes require that both parents be identified and that several crosses per parent be made to increase the chance of finding good specific combinations. In addition, accurate information regarding the breeding value and pedigree of each parent involved will accelerate the genetic gain achieved.

2.1 Multiple Populations and Sublines

Tree breeding strategies have been developed which subdivide a large population of a species into either multiple populations or sublines. The major difference between the two concepts is the goal of selection (Burdon and Namkoong 1983).

With multiple populations, the goal of selection is either to amplify genetic differences or at least to maintain genetic differences among populations. This may be accomplished by selecting for different traits in the various populations. Multiple populations have been developed to some extent for a species in which several organizations have concurrent tree improvement programs within the same breeding zone but have emphasized slightly different selection criteria (e.g., height growth is almost always chosen but some programs also apply selection pressure to specific gravity and/or disease resistance). Multiple populations are ideal for developing clones which display contrasting arrays of extreme genotypes for one or more traits. Each population could fill different niches in the breeding zone. For example, Namkoong (1976) developed a theoretical basis for a multiple-index selection strategy which utilizes different sets of economic weights to construct a different selection index for each population.

Sublines represent replicate populations of a species within a breeding zone, and the selection criteria are the same within each subline. Often select individuals from among the sublines are combined within a seed orchard (van Buijtenen and Lowe 1979). By definition, selection favors the same set of traits in each subline; therefore, the only possible advantage of sublining toward development of extreme genotypes might arise from the development of unique genotypes through random drift or founder effect (Burdon and Namkoong 1983).

Nucleus breeding is a variant of the former two concepts in which a small subset of individuals (e.g., 40) are copied from the main breeding population into a nucleus population (Cotterill et al. 1988). Breeding efforts continue in the main breeding population although at a lower intensity (e.g., open-pollinated or polymix) while the intense breeding effort is concentrated on the nucleus population (e.g., diallel or factorial). Each generation, selections are transferred from the nucleus population into the main population and vice versa. The likelihood of creating extreme genotypes within the nucleus population is high. More than one nucleus population, each with specific emphasis on a different trait or set of traits, can be developed. Once the different nucleus populations have begun to diverge significantly, inter-population crosses may be used much in the same way that animal breeders concentrate useful genes into the production population through inter-breed crosses. These crosses may involve complementary traits (population or "breed complementation") or heterosis (e.g., two-breed crossing, specific three-breed terminal cross, or rotational crossing) (Dickerson 1973; Hohenboken 1985).

Development of multiple populations or nucleus populations with selection centering on different sets of traits can be a very powerful tool to create contrasting or complementary extreme genotypes.

2.2 Mating Designs

The specific mating design determines the pattern in which parent trees are mated to produce the next generation of genotypes. In the absence of perfect knowledge of the parental genotypes (which is almost always the case for forest trees), creation of extreme genotypes will be favored by using a mating design in which both parents are identified and several male parents are paired with each female parent.

Vegetative propagation of the extreme genotypes for reforestation facilitates the use of the total genetic variation. Total genetic variation (V_G) is composed of additive genetic variation (V_A) and nonadditive genetic variation (V_{NA}). Most progeny test designs will provide estimates of V_A, but only a few will provide accurate estimates of V_{NA}. Either small disconnected half- or partial-diallel designs or small disconnected factorial designs are useful for estimating both V_A and V_{NA} (van Buijtenen and Namkoong 1983).

Since vegetative propagation provides a genetically identical copy of the original tree, the source of the genetic superiority of the extreme genotype (V_A and V_{NA}) is unimportant. (It is important, however, in selecting parents for the next mating cycle; but this will be discussed in Sect. 3.3.) Nonadditive genetic variation (often termed specific combining ability) is still more or less unpredictable in forestry and can only be uncovered by making specific crosses. Unlike many agronomic crop and animal species, pedigrees and knowledge of modes of inheritance, especially nonadditive genetic variation, for various traits are only beginning to be developed for forest tree species. Optimum mating designs to predict both additive and nonadditive genetic variation appear to indicate that there is little additional value in crossing more than six male parents per female parent and vice verse (van Buijtenen and Namkoong 1983). If further crosses are economically feasible, then it is more efficient to include more parents rather than make additional crosses on the existing set of parents.

The same theoretical reasoning for efficiently estimating nonadditive genetic variation applies to efficiently uncovering good specific full-sib families and extreme genotypes within these families. Once detailed pedigrees are developed for forest trees and modes of inheritance are known for certain traits on specific individual trees, then different types of mating designs may become more appropriate (e.g., breed complementation, specific three-breed terminal crosses, etc.; Hohenboken 1985).

2.3 Positive Assortative Mating

In contrast to the traditional practice of random mating, parent trees may be paired for crossing based on their genotypes. This practice is termed assortative mating (Falconer 1981). It is termed positive assortative mating if the mated parent trees are of the same genotype more often than would occur at random, while it is termed disassortative mating if the mated trees have the same genotype less often than would occur at random. Positive assortative mating has no effect on allele frequency. However, homozygosity and genetic variance are quickly increased (within a few generations), leading to a limit of complete homozygosity (Breese 1956; Crow and Felsenstein 1968). Therefore, positive assortative mating mainly affects genotype frequency and genetic variance.

The results of positive assortative mating become more difficult to predict as the genetic models increase in complexity. Nonadditive genetic effects and environmental effects cause the genotypes to become less distinct and therefore reduce the correlation between the genotypes of mates. In a purely additive model with multiple loci and multiple alleles, similar alleles at different loci may be indistinguishable. These factors inhibit the movement of a population toward complete homozygosity following positive assortative mating; yet this mating technique is still quite effective in increasing genetic variance (Crow and

Felsenstein 1968). Fortunately, much of the change in genotype frequency and genetic variance occurs in the first few generations of positive assortative mating.

With polygenic inheritance and a large number of genes, positive assortative mating mainly affects additive genetic variance, with little or no change in nonadditive variance (Reeve 1961; Crow and Felsenstein 1968). Formulas are available to predict changes in additive genetic variance, phenotypic genetic variance, and heritability (Crow and Felsenstein 1968; Falconer 1981; Foster 1986). These formulas show that changes in parameters depend primarily upon the size of the correlation between mates (r) and heritability (h^2). Therefore, to be effective, positive assortative mating must be efficient in pairing "like" mates, and the process is most effective with highly heritable traits.

In a practical sense, positive assortative mating combined with directional selection will increase the frequency of genotypes at the high end of the distribution (extreme genotypes). In a purely additive genetic model, the expected value of the mean of the offspring from a mating is the midparent value (Falconer 1981). With random mating of parents, the probability of mating any two parents is a function of the frequency of their genotypes in the population. If there are few extreme genotypes, then the probability of mating two extreme genotypes is very small with a random mating plan. Even if all the genotypes are extreme, there will be differences among them (e.g., top 50 or 100 individuals). Given the low heritabilities for most traits in forest trees, the correlation between the genotypes of mates will be intermediate at best. By following a positive assortative mating plan, the extreme genotypes will be mated, with a frequency at least a little higher than by chance, thereby increasing the likelihood of producing a higher frequency of extreme genotypes in the offspring generation.

In a simulation study, Mahalovich and Bridgwater (1989) compared genetic gains for positive assortative mating versus random mating in a small (48 genotypes) elite population which was a subset of genotypes from a larger mainline population. The positive assortative mating yielded an increase in gain from 0–6%, as compared with random mating, through 14 generations of mating. They noted a small increase in the number of extreme genotypes due to positive assortative mating.

Positive assortative mating may lead to a higher frequency of matings among related selections than random mating. Typically, a disproportionate number of the extreme genotypes arise from a few families; and if limitations are not placed upon matings between related selections, inbreeding would occur faster than under random mating. To relieve this problem, related matings (especially closely related, e.g., full-sibs) should be avoided. Introduction of some new selections into the nucleus population each generation, as suggested in the nucleus breeding plan (Cotterill et al. 1988), will also help forestall the buildup of inbreeding in the population of extreme genotypes.

Foster (1986) proposed a general tree improvement program that combined assortative mating, separate breeding and production populations (see Sect. 3.3 below), and vegetative propagation (see Sect. 4 below). Foster and Shaw (1987) later presented a specific program for loblolly pine (*Pinus taeda* L.) that incorporated all the features from Foster's (1986) proposal.

2.4 Breeding for Nonadditive and Additive Genetic Variation

Traditionally, tree breeders have selected exclusively for general combining ability (V_A) and have totally ignored specific combining ability (V_{NA}). Reasons for this decision include: (1) little information regarding the relative sizes of general and specific combining abilities, (2) long generation intervals which reduce the practicality of multiple population breeding plans (e.g., reciprocal recurrent selection), (3) sufficient additive genetic variance to enable reasonable gains from selection, and (4) the general lack of an economically feasible method to utilize nonadditive genetic variance in reforestation. With the advancement of knowledge as well as breeding and propagation technology, tree breeders of some species are now able to capitalize on both general and specific combining abilities.

Reciprocal recurrent selection (RRS) was proposed by Comstock et al. (1949) as a technique to utilize both general and specific combining abilities. The RRS procedure requires two populations (e.g., A and B) which are improved simultaneously. It works best if the populations are substantially different, such as the island and mainland populations of radiata pine (*Pinus radiata* D. Don), or even two different species, such as knobcone (*Pinus attenuata* Lemm.) and radiata pines (pers. commun., Dr. W.J. Libby, University of California, Berkeley, CA). Each population is used as a tester for the other, e.g., plants from A are crossed as testers with plants from B and vice versa. Families are identified by maternal parent only (polycross families). In the original scheme for selfable agronomic species, the same plants are also selfed to maintain the genotypes for an additional year. This step is needed for annual species but not for perennial species. In addition, most tree species, when selfed, produce low quantities of viable seed; and the growth of the resultant seedlings demonstrates strong inbreeding depression (e.g., Gansel 1971; Griffin et al. 1986). Superior parents are selected on the basis of results from the interpopulation progeny tests. The superior parents (previously selfed – for agronomic species) are then mated within each population to produce the next generation (e.g., of population A), and the entire process is repeated (Briggs and Knowles 1967). For tree species, either a polycross or some other type of factorial or diallel could be used to reconstitute each population for the next generation instead of selfing. Commercial seed is produced by crossing superior trees between the two improved populations.

A number of variations on the basic RRS scheme exist. Reciprocal full-sib selection (Hallauer and Eberhart 1970) is one of the more interesting for tree improvement. This procedure differs from RRS in that inter-population crosses yield full-sib progenies rather than polycross progenies. In addition, only one set of full-sib progenies is required for progeny testing instead of two sets of polycross progenies (e.g., for RRS, A testers crossed with B parents; and B testers crossed with A parents).

With a few modifications, some form of reciprocal recurrent selection could be used for tree improvement. Given the perennial nature of trees, selfing of the parents is unnecessary. Furthermore, the commercial propagules could be created (1) by crossing select parents from the two populations to produce seed, for example using polycross pollen from the early flowering select males in one population to pollinate later flowering select females in the other population; or (2) by vegetatively propagating superior individuals within the test cross (polycross or full-sib) families or families created in (1) above; or (3) by propagating entire full-sib families (bulking-up) which are shown to be superior in the test crosses or from inter-population crosses between proven superior parents.

A system was proposed by McKeand et al. (1986) that utilized sublines of loblolly pine in a reciprocal recurrent selection system. In their system, a bulked pollen lot from each subline is used to pollinate all trees (polymix) within its paired subline. Within-subline crosses are made using a design such as a small disconnected half diallel. As one of the alternatives listed for development of the production population, superior trees can be chosen from the polymix crosses and vegetatively propagated for inclusion in a clonal testing and selection scheme.

2.5 Inter-Specific Hybridization

Inter-specific hybridization is a popular technique among breeders to create new variation in plant populations. Crosses may be limited by a variety of phenological, mechanical, and chemical barriers (Grant 1963). Breeders are able to utilize techniques to overcome some of these barriers (e.g., Teissier du Cros 1984); however, genetic aberrations may cause the offspring to be sterile and therefore of little value in a breeding program. Through the use of vegetative propagation, however, sterility presents no barrier to large-scale propagation; so the plants may still be commercially useable.

Species may be crossed in random combinations of related species or with an objective to create a hybrid that fits a specific ecological niche better than either of the parent species (Stebbins 1950; Kuser and Knezick 1987). Heterosis may or may not occur, depending upon the specific gene action governing the trait within the species involved. The variance of the F_1 progeny is expected to approximate that of the parental species; however, the variance of the F_2 and later generations is expected to increase dramatically (Stebbins 1950). This added variance should yield at least a few extreme genotypes which are worthy of further propagation or breeding.

Teissier du Cros (1984) proposed a tree improvement program with *Populus* species that utilizes inter-specific hybridization and vegetative propagation. Initially, specific provenances of appropriate species are selected based on field tests. Individual selections are then located within the provenance tests and are vegetatively propagated into clonal field trials. Utilizing clonal test results, the selections serve as parents in intra- and inter-specific crossings. The hybrid progeny enter a new cycle of field testing, and new selections originate from within this population. These selections are vegetatively propagated and are placed in field tests. A second cycle of selection serves to eliminate the poorest clones. The remaining select clones enter a new round of mating to produce another generation of hybrids. The pattern of crossing, testing seedlings, and further testing of the best clones repeats itself. New variation and potentially new plant types are created during each program cycle. Based on the results of clonal field tests each generation, superior clones are vegetatively propagated for reforestation.

3 Selection for Extreme Genotypes

Identification of forest tree genotypes, especially for growth, morphological, or disease resistance traits, is very difficult in natural stands or plantations since most, if not all, of these traits are controlled by polygenes; and all of these traits are strongly affected by the environment. The phenotype, or observable value of a character when an individual is measured, is a function of the genotype and its environment. Statistically designed genetic experiments, often termed progeny tests, are established to differentiate between the genotypic value of an individual and the environmental deviation. Many empirical as well as theoretical studies have led to the establishment of general principles for the design, implementation, and data analysis of progeny tests of forest trees (Progeny Testing of Forest Trees 1983).

Depending upon the model used, analytical results from genetic tests can be used to estimate the genetic value of an individual tree or to estimate population parameters such as genetic and environmental variances (see Chap. 9, Vol. 1). Extreme genotypes within the population can be selected through appropriate genetic testing and selection methodology.

3.1 Genetic Testing

Once the breeding work has been completed (Sect. 2 above), the new genotypes must be tested in an appropriately designed genetic test. The specific design of the genetic test is a function of the trait(s) of interest and the environment(s) in which the trees will be grown.

The major trait for improvement in forestry is yield per unit area of a forest stand (Nance et al. 1987). Unfortunately, evaluation of this trait for a genetic entity (e.g., families or clones) requires large areas (e.g., 0.10 ha or more) and long time periods to reach rotation age (e.g., generally 25 or more years). Therefore, tree breeders have sought to develop indirect selection methods which will lead to genetic improvement in yield per unit area. Tree height and stem diameter at breast height (dbh) are two commonly measured traits, and ones which can in turn be used to estimate a third trait of interest, individual tree volume. The mean of a plot of one or more trees from a genetic entity is then generally assumed to represent the per unit area yield potential of that genetic entity. However, traits measured in small plots (e.g., < 50 trees) do not serve as very good simulators of stand level traits, due to their extreme dependency on stand density, with the possible exception of mean height, which is fairly insensitive to plot size (e.g., Foster 1989). Due to its relative insensitivity to stand density, tree height has been generally accepted by tree breeders as their major trait for improvement; however, many research questions remain unanswered as to its ultimate usefulness as an indicator of yield per unit area (Nance et al. 1987).

On the assumption that the height of an individual tree or mean height of a family or clone plot is a useful trait for selection, much research has been devoted to determining early selection traits which can be used for indirect selection (e.g., see Lambeth 1983 for a review for growth traits). In addition, early screening and indirect selection for resistance to diseases has proven to be successful in some cases (e.g., Sluder and Powers 1986; Foster and Anderson 1989). However, breeders must utilize these techniques with caution since few if any of them yield perfect correlations with mature performance. The mature growth traits that are often used for determining the accuracy of the early testing procedure are usually the means of small plots rather than estimates of yield per unit area (Lambeth 1983). Much research remains to be accomplished in the area of indirect selection in forestry.

A workable testing and selection scheme for use in tree improvement for yield per unit area will require a series of genetic tests rather than a single genetic test prior to operational use of either seedling families or clones. Libby (1987a) has proposed four levels of clonal testing. Level I, *initial screening*, involves large numbers of genotypes (e.g., thousands) with only a single seedling or a very few ramets per genotype (Shaw and Hood 1985). Traits such as early survival, growth, form, and resistance to damage agents (biotic and abiotic) are evaluated at this level. Level II, *candidacy testing*, involves large numbers of genotypes (e.g., hundreds or low thousands) with generally low numbers (e.g., two to six) of ramets per clone (Shaw and Hood 1985; Russell and Libby 1986). Often those genotypes that were shown to be superior in the initial screening will proceed to candidacy testing and will be tested with more ramets which are distributed over more environments. The same traits will be evaluated as in Level I, but in addition the traits will be assessed at older ages. Level III, *clonal performance testing*, involves moderate numbers of genotypes (e.g., less than 200) with large numbers (e.g., tenth-ha plots) of ramets per clone. Extreme genotypes which are strong candidates for large-scale reforestation (i.e., have successfully passed through candidacy testing) will proceed to this level, and evaluated traits will include those from Levels I and II plus genotype × environment interactions and yield per unit area. Tests will generally last for large portions of a rotation. Level IV, *compatibility trials*, includes small numbers of genotypes (e.g., 20 to 50) with very large numbers of ramets per clone. Mixing trials will be installed to attempt to discover mixtures of clones that may be particularly compatible, potentially resulting in overcompensatory yield per unit area (Tauer 1975). Tests will generally last most, if not all, of an entire rotation length.

3.2 Selection Procedures

Once measurement data become available from the genetic tests, selection systems must be developed. Index selection (Baker 1986) can be used to select extreme genotypes for one or more traits. Analysis of variance is first used to derive genetic and phenotypic variances for traits and covariances between traits. These values are combined along with appropriate economic weights to form an optimum selection index. For example, Foster et al. (1985) used index selection in a western hemlock [*Tsuga heterophylla* (Raf.) Sarg.] clonal trial to compare genetic gains when selecting solely for early height growth versus selection for early height growth and rooting ability traits. If multiple populations (Sect. 2.1) are utilized, then the economic weights and possibly traits may vary among populations (Namkoong 1976).

A point must be made regarding the efficiency of selection for multiple traits. Selection for extreme genotypes for a single trait should prove to be no particular problem; however, genotypes which are extreme in several traits will be difficult to develop unless the genetic correlations between pairs of the traits are 1.0. Under index selection, gain in economic terms is greatest compared to other multiple trait selection procedures (e.g., tandem selection or independent culling level) and probably compared to single-trait selection in forestry where several traits have economic value; however, genetic gain in individual traits is only $n^{-1/2}$ as great as selection for a single trait (Turner and Young 1969). Hence, without several generations of breeding to concentrate superior genes for multiple traits in the same individuals, the breeder may have initially to accept extreme genotypes for some traits while the genotype for other traits may have to be superior but less extreme. This can be done by altering the economic weights to maximize (or at least favor) genetic gain for one trait over the other traits (Turner and Young 1969; Baker 1986).

Genetically engineered trees will create some new problems for testing and selection. Genes which originally may have arisen from different species of trees or even nontree species are being engineered into forest trees (e.g., Riemenschneider et al. 1988). These new genotypes must be rigorously tested for long periods of time prior to selection for reforestation (Libby 1987b) both for the effects of the new genes as well as the effects of disrupting the original genome in the tree.

3.3 Separate Breeding and Production Populations

Extreme genotypes may or may not be useful in a breeding population depending upon: (1) the genetic nature of the superiority and (2) the breeding plan. For example, a genotype which is superior mainly due to nonadditive gene action would be of little value in a breeding program that exclusively emphasized general combining ability. However, selection of superior genotypes for reforestation (the production population) need not be constrained by the genetic nature of the superiority or of the breeding plan.

Separation of the breeding and production populations allows the breeder to clearly focus on the objective for each population (Libby 1973; Burdon 1986; Foster 1986). The purpose of the breeding population is to create new genetic variation from which superior selections can be made (see Sect. 2). The production population, however, is the vehicle through which commercial production is enhanced. In an example with eastern cottonwood (*Populus deltoides* Bartr.), Foster (1986) estimated an additional 6.2% genetic gain in height by using a production population (14.1% genetic gain) versus the breeding population (7.9% genetic gain) for reforestation.

The production population can take a variety of forms including wind pollinated seeds from a production seed orchard, controlled cross seed from a good specific cross, vegetative propagules used to bulk up a good full-sib family, or vegetative propagules of superior genotypes. The main obstacle to the use of the various forms becomes production of a sufficient quantity of propagules to meet reforestation needs.

4 Propagation of Extreme Genotypes

Vegetative propagation, by tissue culture or rooted cuttings, facilitates the multiplication of a genotype. Therefore, once an extreme genotype is identified, it can be reproduced for reforestation purposes provided a feasible vegetative propagation technique is available for the species (see Chaps. 11, 13, Vol. 1, and 4–12, Vol. 2). As discussed in Section 1, reforestation with clones of extreme genotypes will greatly enhance genetic gain in forest productivity. Foster (1986) and Foster and Shaw (1987) promote the idea of developing a production population in each generation which contains the extreme genotypes. The extreme genotypes would be vegetatively propagated via rooted cuttings or tissue culture plantlets for reforestation.

The vegetative propagation process does not include meiosis; hence, ramets are genetically identical to the ortet. Phenotypic variation among ramets within a clone still exists. The cause of this variation is environmental and is due to a number of factors related to the propagule [e.g., size of cutting or season in which cutting was collected (Burdon and Shelbourne 1974)] or the nursery environment [e.g., vigor of the propagule or root system quality (Wisniewski et al. 1983)]. Burdon and Shelbourne (1974) indicated that this effect is a type of maternal effect, and they use the symbol "m" to represent it.

Another type of maternal effect manifests itself as persistent differences among clones and leads to an inflated estimate of clonal variance. Lerner (1958) used the term "C effects" while Burdon and Shelbourne (1974) used the symbol "M" to refer to this effect. C effects seem to be a real phenomenon, but appear to be important mainly for traits measured relatively soon after propagation (Libby and Jund 1962; Wilcox and Farmer 1968; Cannell et al. 1988). The stage of maturation (ontogeny) has a great effect on ease of propagation and subsequent growth of rooted cuttings or tissue culture plantlets (Sweet and Wells 1974; Greenwood 1984; McKeand 1985) (also see Talbert et al., Chap. 11, Vol. 1). Cuttings from mature trees may root with extreme difficulty and, once rooted, may express low vigor and poor growth rate. Techniques to either maintain or reduce juvenility are a key to any successful vegetative propagation program. Differential stages of maturation among clones may appear as C effects and would result in artificially increased levels of clonal variance hence inflated estimates of genetic gain from selection. Further discussion of handling C effects in clonal testing is given in Chapter 9, Volume 1.

5 Deployment of Extreme Genotypes

During the last decade, large amounts of genetically improved planting stock have finally become available for reforestation in some parts of the world (e.g., southeastern USA). Forest managers have therefore quickly moved from utilizing a scarce supply of genetically improved seed to having an abundance and even an oversupply in some cases. This large supply of genetically improved planting stock has stimulated questions from forest managers regarding deployment strategies for the genetic entities (e.g., open-pollinated families).

These questions are not as pressing for vegetative propagules since the supply of planting stock has not been as abundant as with seed for most species; however, most of the deployment issues are similar to and equally as relevant as those with seedlings.

5.1 Number of Clones

One of the basic questions relates to the number of clones to deploy in a single forest stand. Long-term health of the forest stand and the adaptation to the local environment (e.g., climatic, soils, and pest attack) are important issues as well as inter-genotypic competitive effects which influence stand structure development and yield per unit area.

Theoretical studies have been conducted which assess either the risk of plantation failure or stability of stand performance. Libby (1987b) sums up these studies with the following generalizations: "(a) mixtures of large numbers of clones are about as safe as similar mixtures of genetically diverse seedlings; (b) mixtures of very small numbers of clones are not safe, and mixtures of two to four clones are often worse than monoclonal deployment; and (c) mixtures of 7–30 clones are about as safe as similar mixtures of larger numbers of clones."

Few empirical results are available regarding the optimum number of clones per stand. In a 5-year-old Norway spruce [*Picea abies* (L.) Karst.] study in southern Sweden, Bentzer et al. (1988) found that mean height of mixtures of about 50 clones was actually much less interactive with test locations than mean height of provenance-level seedling checklots. Due to the young age of the test, these results must be viewed as being preliminary. However, in a subsequent analysis of the same study at age 7 years, Bentzer et al. (1990) found, as a general result (over four different mixtures of clones), that the mixture × test location interaction remained nonsignificant for height differences when the number of clones per mixture was reduced at random from 56 clones to 20 clones. If the number of clones was reduced by eliminating either slow-growing clones or clones which were highly interactive with locations, then the number of clones in the mixtures could be further reduced to five or ten without any evidence of significant mixture × test location interaction. Again, the tests are still at only a fraction of total rotation length and height tends to be less interactive with locations than other traits such as diameter or volume; still, the results seem to indicate that mixtures of clones, at least for Norway spruce, are quite stable in growth performance – much more so than individual clones taken separately (Bentzer et al. 1988).

Within a single breeding zone, theoretical as well as empirical results seem to indicate that from 7–30 clones will provide a reasonable level of risk avoidance of plantation failure, hence of success in stand yield. This area remains in need of research results.

5.2 Stable Versus Nonstable Clones

The relative performance of genetic entities (e.g., provenances, families, or clones) across locations is very important to tree breeders since they want to use planting stock for reforestation which will perform reliably well over large areas. The three general approaches that have been used to assess stability of clones across locations in tree improvement trials are: (1) assessing the significance and level of clone × location interaction (e.g., Burdon 1971; St. Clair and Kleinschmit 1986; Bentzer et al. 1988), (2) utilizing a regression approach which regresses the average performance of each clone against some type of site quality index at each test location (St. Clair and Kleinschmit 1986; Bentzer et al. 1988); a regression slope of 1.0 signifies that the particular clone grows 1 unit better for each unit improvement in site quality; a slope of 1.0 is viewed as average stability; and (3) partitioning the clone × location interaction sums of squares into components due to each clone (St. Clair and Kleinschmit 1986). Two opposing approaches may be taken regarding the stability of performance of clones (Libby 1987b). First, clones may be chosen which are generally stable over the entire breeding zone, thereby constituting a generally useful production population. Conversely, clones may be chosen which are not generally stable but which perform exceptionally well on certain site types. To complicate matters, there is evidence for both clones (Bentzer et al. 1988) and seedling families (Skroppa 1984) of Norway spruce that the faster growing clones or families tend to be less stable.

The tree breeder, therefore, must decide whether to develop a generally useful mixture(s) of clones or to subdivide the sites into areas that are noninteractive with the specific clones and then develop several mixtures, each tailored to a specific site type. Fortunately, as a general rule, the proportion of highly interactive clones is small compared to the majority of clones (Libby 1987b). One reasonable plan is to divide the forest land within a breeding zone into site types which are noninteractive and determine how much of the total area falls within each group. If the vast majority of the land falls within one type, then clones should be selected which match that specific type. The other minority types could be served by a few clones chosen specifically for them. If on the other hand the total land holding is more or less equally divided between two or more contrasting site types (e.g., wet and dry or deficient for a soil element and nondeficient), then it may be more advantageous to choose a mixture(s) of clones that are generally stable across all major site types. These strategies can be compared in a financial analysis once potential genetic gains and land area sizes are known.

5.3 Mixed Versus Pure Clone Deployment

Given that the number of clones and the type of clones have been chosen, the exact configuration of the clones within a stand still must be determined. Two of the extreme choices would be to use "widespread intimately mixed plantations" or to use "mosaics of monoclonal stands", which Libby (1987a, b) has termed WIMPS and MOMS, respectively. Much theoretical and empirical research exists with agronomic species which compares these contrasting deployment strategies (Harper 1977); yet virtually no research results have been published for clonal stands of forest trees.

One of the few research studies which tests the difference between field growth performance of pure clone plots and a mixed clone plot was conducted with eastern cottonwood in Yugoslavia. Markovic and Herpka (1986) reported 4th-year results from a study of five clones which were each planted in pure plots and also a mixed plot. Large plot sizes (100 trees) enabled valid estimates of per ha yield to be calculated. Their results showed an intermediate performance of the mixed plot versus the pure plot performance for volume per ha as well as average height, dbh, and basal area. However, they did detect slightly higher volume per ha (12.8%), mean height, mean dbh, and mean basal area for the mixed plot versus the arithmetic mean of the pure plots. One interpretation of this result is that during natural stand development the faster-growing, dominant clones eventually control the stand; therefore, the volume per ha and mean values for the other stand level traits will be raised closer to that expected of the better clones.

In a simulation study with loblolly pine (*Pinus taeda* L.), Nance (1983) demonstrated that although yield per unit area in binary mixtures of families could be predicted accurately with a purely additive model (i.e., known ratios of each component) the mean size of each component in the mixture could vary largely. In other words, even if total yield is unchanged by mixing, the fastest-growing genetic entities may produce much larger trees; while the slowest-growing ones may produce much smaller trees as compared to the average sizes when they are grown in pure plots. This principle may have some useful applications when thinning is part of the stand management regime and may result in larger average log size in the final crop trees.

At this stage with little empirical research results in forestry, there are few concrete reasons to deploy clones either as WIMPS or MOMS. Mixed clone stands (WIMPS) have an advantage if very low levels (⟨20%) of mortality are expected, due to pest attack or climatic maladaptation, since initial planting density can be increased appropriately. This is especially true if clonal susceptibility to damaging agents is unknown but anticipated to be low, in that a forester would prefer to lose a small percentage of the trees spread throughout a stand rather than all trees in patches (e.g., 10–20% of entire area). Competition induced mortality is normal in forest stands over the life of the plantation (Clutter et al. 1983); and up to a certain limit, trees within stands can grow additionally to compensate for extra growing space. Conversely, if potential mortality, or "Risk to a Random Genotype (RRG)" (Libby 1982), is higher than the "Maximum Acceptable Loss (MAL)" (Libby 1982) that a forest manager is willing to accept, then mosaics of monoclonal stands (MOMS) appear to be the best strategy (Libby 1982). With this strategy, damaged or dead clones can be cut and quickly replaced by resistant clones; however, if the same clones were randomly scattered throughout the stands, removal and replacement would be impractical.

A number of stand management factors argue for MOMS. With this strategy, clonal location can be mapped within the stands. The forest manager can then observe the growth, stem form, resistance to pests, adaptation to climate and environment, etc. of each clone over the rotation length. This knowledge will enable the forest manager to more effectively deploy clones to sites in the future as well as enhancing the ease of logging and greatly improving the efficiency of industrial use of the wood due to uniformity within a clone. The pure clone patches are usually small (e.g., 2–20 ha; Libby 1987b), reducing the risk of a large monoculture. Furthermore, several clones (7–30, see Sect. 5.1) are deployed to a single stand; hence the advantages of genetic diversity within a stand are utilized.

Many management questions regarding clonal deployment remain unanswered. As gauged by the vast amount of research in agronomic crops (e.g., for reviews, Trenbath 1974; Harper 1977), the issue of clonal deployment will become paramount in forestry as tree breeders and forest managers become more familiar with their clones and seek even higher yields. Unfortunately, long-term studies are required; therefore, the issue should receive immediate attention from researchers.

6 Envoi

A number of breeding and selection techniques exist that would enable tree breeders to achieve larger genetic gains in forest productivity. Many technical problems that once served to inhibit tree breeders have been alleviated. Approaches should be followed to utilize the total genetic variance to create positive extreme genotypes. Subsequently, techniques such as vegetative propagation can be used to produce large numbers of these genotypes for reforestation.

Domestication of tree species through tree improvement programs is proceeding rapidly now. Development of the field of tree improvement during the first 20–30 years was concerned with exploring natural genetic variation patterns and genetic control mechanisms for various traits, making first-generation selections from natural stands or plantations, and learning how to establish and manage seed orchards. Research results have shortened the generation cycle dramatically as well as enhanced our ability to propagate (sexually or asexually) genetically superior trees.

The future of tree improvement will challenge forest geneticists and tree breeders to utilize existing procedures as well as newly developed technology to create extreme genotypes, especially for multiple traits, and propagate them for reforestation. Many breeding procedures which are currently used by agronomic crop breeders and animal breeders are now accessible to tree breeders to create new genetic variation and more efficiently utilize it for yield improvement. For some species, tree breeders already are working with third-generation material; and as pedigrees are developed for populations, the ability to breed effectively for particular traits is greatly enhanced. Complementary breeding, as is practised with animal species, will become commonplace as a tool to create specific genotypes which are tailored to particular environments, management styles, and products. The field of molecular genetics is developing rapidly, and it will enhance selection and breeding efficiency as well as the development of novel genotypes, often utilizing genes from different species and even different organisms.

Vegetative propagation will become common for most commercially important species within the next 15 years. Rooted cutting technology will continue to be the most economical procedure for large scale propagation (especially for conifers) for at least the next 10 years; however, for some species, especially hardwoods, economically efficient tissue culture procedures (e.g., embryogenesis) will become available within the next 10 years.

Forest managers will learn to effectively deploy extreme genotypes to capitalize on yield improvement through silvicultural manipulation. Mixing strategies as well as stand density management will be tailored to specific clones to optimize yield per unit area.

References

Allaire FR (1980) Mate selection by selection index theory. Theor Appl Genet 57:267–272

Baker RJ (1986) Selection indices in plant breeding. CRC, Boca Raton

Bentzer BG, Foster GS, Hellberg AR, Podzorski AC (1988) Genotype × environment interaction in Norway spruce involving three levels of genetic control: seed source, clone mixture, and clone. Can J For Res 18:1172–1181

Bentzer BG, Foster GS, Hellberg AR (1990) Impact of clone mixture composition on stability of seventh-year mean height in a series of Norway spruce clone tests. Can J For Res 20:757–763

Breese EL (1956) The genetical consequences of assortative mating. Heredity 10:323–343

Briggs FN, Knowles PF (1967) Introduction to plant breeding. Reinhold, New York Amsterdam London

Burdon RD (1971) Clonal repeatabilities and clone-site interactions in *Pinus radiata*. Silv Genet 20:33–39

Burdon RD (1986) Clonal forestry and breeding strategies – a perspective. In: Proc IUFRO Conf, Joint Meet Working parties on breeding theory, progeny testing, seed orchards, 13–17 Oct. North Carolina State Univ, Raleigh, pp 645–657

Burdon RD, Namkoong G (1983) Multiple populations and sublines. Silv Genet 32:221–222

Burdon RD, Shelbourne CJA (1974) The use of vegetative propagules for obtaining genetic information. N Z J For Sci 4:418–425

Cannell MGR (1978) Improving per hectare forest productivity. In: Hollis CA, Squillace AE (eds) Proc 5th North American forest biology Worksh, 13–15 March. Univ Florida, Gainesville, pp 120–148

Cannell MGR, Sheppard LJ, Cahalan CM (1988) C effects and second generation clone performance in *Picea sitchensis* and *Pinus contorta*. Silv Genet 37:15–19

Clutter JL, Fortson JC, Pienaar LV, Brister GH, Bailey RL (1983) Timber management: a quantitative approach. John Wiley & Sons, New York

Comstock RE, Robinson HF, Harvey PH (1949) A breeding procedure designed to make maximum use of both general and specific combining ability. Agron J 41:360–367

Cotterill P, Dean C, Cameron J, Brindbergs M (1988) Nucleus breeding: a new strategy for rapid improvement under clonal forestry. In: Proc IUFRO Meeting Breeding tropical trees, 28 Nov–2 Dec, Pattaya, Thailand, pp 39–51

Crow JF, Felsenstein J (1968) The effect of assortative mating on the genetic composition of a population. Eugenics 15:85–97

Dickerson, GE (1973) Inbreeding and heterosis in animals. In: Proc Animal breeding and genetics symp Hon Dr. Jay L. Lush, 29 July 1972, Blacksburg. Am Soc Animal Sci L; and Am Dairy Sci Assoc, pp 54–77

Falconer DS (1981) Quantitative genetics. Longman, New York

Foster GS (1986) Making clonal forestry pay: breeding and selection for extreme genotypes. In: Proc IUFRO Conf, Joint Meet working parties on breeding theory, progeny testing, seed orchards, 13–17 Oct 1986. North Carolina State Univ, Raleigh, pp 582–590

Foster GS (1989) Inter-genotypic competition in forest trees and its impact on realized gain from family selection. In: Proc 20th Southern forest tree improvement Conf, 27–29 June, Charleston. NTIS, Springfield, pp 21–35

Foster GS, Anderson RL (1989) Indirect selection and clonal propagation of loblolly pine seedlings enhance resistance to fusiform rust. Can J For Res 19:534–537

Foster GS, Shaw DV (1987) A tree improvement program to develop clones of loblolly pine for reforestation. In: Proc 19th Southern forest tree improvement Conf, 16–18 June, College Stn, Texas. NTIS, Springfield, pp 17–21

Foster GS, Campbell RK, Adams WT (1985) Clonal selection prospects in western hemlock combining rooting traits with juvenile height growth. Can J For Res 15:488–493

Gansel CR (1971) Effects of several levels of inbreeding on growth and oleoresin yield in slash pine. In: Proc 11th Southern forest tree improvement Conf, 15–16 June, Atlanta. NTIS, Springfield, pp 173–177

Grant V (1963) The origin of adaptations. Columbia Univ Press, New York London

Greenwood MS (1984) Phase change in loblolly pine: shoot development as a function of age. Physiol Plant 61:518–522

Griffin AR, Raymond CA, Lindgren D (1986) Effects of inbreeding on seed yield and height growth of *Pinus radiata* D. Don. In: Proc IUFRO Conf, Joint Meet working parties on breeding theory, progeny testing, seed orchards, 13–17 Oct. North Carolina State Univ, Raleigh, p 603

Hallauer AR, Eberhart SA (1970) Reciprocal full-sib selection. Crop Sci 10:315–316

Harper JL (1977) Population biology of plants. Academic Press, New York London

Hohenboken WD (1985) Genetic structure of populations 2. Matings among distantly related individuals. In: Chapman AB (ed) World animal science, A4. General and quantitative genetics. Elsevier, Amsterdam, pp 251–274

Kuser JE, Knezick DR (1987) Pitch × loblolly hybrids after 10 years in New Jersey. N J Appl For 4:207–209

Lambeth CC (1983) Early testing – an overview with emphasis on loblolly pine. In: Proc 17th Southern forest tree improvement Conf, 6–9 June, Athens. NTIS, Springfield, pp 297–311

Lerner IM (1958) The genetic basis of selection. John Wiley & Sons, New York

Libby WJ (1973) Domestication strategies for forest trees. Can J For Res 3:265–276

Libby WJ (1982) What is a safe number of clones per plantation? In: Heybroek HM, Stephan BR, von Weissenberg K (eds) Resistance to diseases and pests in forest trees. PUDOC Wageningen, pp 342–360

Libby WJ (1987a) Testing for clonal forestry. Ann For 13:69–75

Libby WJ (1987b) Testing and deployment of genetically engineered trees. In: Bonga JM, Durzan DJ (eds) Cell and tissue culture in forestry, 2nd edn. Nijhoff/Junk, The Hague, pp 167–197

Libby WJ, Jund E (1962) Variance associated with cloning. Heredity 17:533–540

Mahalovich MF, Bridgwater FE (1989) Modeling elite populations and positive assortative mating in recurrent selection programs for general combining ability. In: Proc 20th Southern forest tree improvement Conf, 27–29 June, Charleston. NTIS, Springfield

Markovic J, Herpka I (1986) Plantations in short rotations. In: Poplars and willows in Yugoslavia. Poplar Res Inst, Novi Sad, pp 182–198

McKeand SE (1985) Expression of mature characteristics by tissue culture plantlets derived from embryos of loblolly pine. J Am Soc Horic Sci 110:619–623

McKeand SE, Foster GS, Bridgwater FE (1986) Breeding systems for pedigree-controlled production populations of loblolly pine. In: Proc Worksh Advanced generation breeding: current status and research needs, 6–7 June 1984, Baton Rouge. S Coop Ser Bull. 309. Louisiana State Univ, Baton Rouge, pp 53–61

Namkoong G (1976) A multiple-index selection strategy. Silv Genet 25:199–201

Nance WL (1983) Simulated growth and yield of single family versus multi-family loblolly pine plantations. In: Proc 2nd Biennial Southern silviculture research Conf, 4–5 Nov 1982, Atlanta. US For Serv, SE For Exp Stn Gen Tech Rep SE-24, pp 446–453

Nance WL, McCutchan BG, Talbert CB, Buford MA, Foster GS, Sprinz P (1987) Experimental approaches for evaluating genetic effects on stand growth and yield. In: Proc Worksh Statistical considerations in genetic testing, 25–26 June 1986, Gainesville. S Coop Ser Bull 324. Univ Florida, Gainesville, pp 22–39

Progeny Testing of Forest Trees (1983) Proc Worksh Progeny testing, 15–16 June 1982, Auburn. S Coop Ser Bull 275. Univ Press, Auburn

Reeve ECR (1961) A note on non-random mating in progeny tests. Genet Res 2:195–203

Riemenschneider DE, Haissig BE, Sellmer J, Fillatti JJ (1988) Expression of an herbicide tolerance gene in young plants of a transgenic hybrid poplar clone. In: Ahuja MR (ed) Somatic cell genetics of woody plants. Kluwer, Dordrecht Boston London, pp 73–80

Russell JH, Libby WJ (1986) Clonal testing efficiency: the trade-offs between clones tested and ramets per clone. Can J For Res 16:925–930

Shaw DV, Hood JV (1985) Maximizing gain per effort by using clonal replicates in genetic tests. Theor Appl Genet 71:392–399

Skroppa T (1984) A critical evaluation of methods available to estimate the genotype × environment interaction. Stud For Suec 166:3–14

Sluder ER, Powers HR (1986) Further comparisons between infection of loblolly and slash pines by fusiform rust after artificial inoculation or planting. US For Serv, SE For Exp Stn Res Note SE-342, 3 pp

St. Clair JB, Kleinschmit J (1986) Genotype-environment interaction and stability in ten-year height growth of Norway spruce clones (*Picea abies* Karst.). Silv Genet 35:177–186

Stebbins GL (1950) Variation and evolution in plants. Columbia Univ Press, New York London

Sweet GB, Wells LG (1974) Comparison of the growth of vegetative propagules and seedlings of *Pinus radiata*. N Z J For Sci 4:399–409

Tauer CG (1975) Competition between selected black cottonwood genotypes. Silv Genet 24:44–49

Teissier du Cros E (1984) Breeding strategies with poplars in Europe. For Ecol Manag 8:23–39

Trenbath BR (1974) Biomass productivity of mixtures. Adv Agron 26:177–210

Turner HN, Young SSY (1969) Quantitative genetics in sheep breeding. MacMillan Aust, South Melbourne

van Buijtenen JP, Lowe WJ (1979) The use of breeding groups in advanced generation breeding. In: Proc 15th southern forest tree improvement Conf, 19–21 June. Mississippi State Univ. NTIS, Springfield, pp 59–65

van Buijtenen JP, Namkoong G (1983) Mating designs. In: Progeny testing of forest trees. Proc Worksh Progeny testing, 15–16 June 1982, Auburn. S Coop Ser Bull 275. Univ Press, Auburn

Wilcox JR, Farmer RE (1968) Heritability and C effects in early root growth of eastern cottonwood cuttings. Heredity 23:239–245

Wisniewski LA, McKeand SE, Brooks RE (1983) Growth of tissue culture plantlets of loblolly pine in a nursery and greenhouse. In: Proc 17th Southern forest tree improvement Conf, 6–9 June, Athens. NTIS, Springfield, pp 186–193

6 Clonal Material as a Focus for Genetic and Physiological Research in Forest Trees

R.F. Stettler[1] and R.J. Ceulemans[2]

1 Introduction

Most people who need to know about forest trees would agree that our understanding of the biology of these fascinating but recalcitrant organisms is still utterly sketchy. This is especially true for the physiological and biochemical mechanisms that underlie such common phenomena as dormancy, spring phenology, shoot elongation, root foraging, to name a few; or responses to stress induced by excess or shortage of water and nutrients, extreme temperatures, and pests. Our ignorance is even more profound when it comes to the genetic control of these mechanisms.

It is the basic premise of this chapter that such information will be increasingly needed as we move from extensive to intensive forestry; as we make assumptions about adaptability of forest trees to rapidly changing conditions; and as we make decisions on how to manage genetic diversity in our forest ecosystems. The two tenets of this chapter are that major insights into the developmental biology of trees will require a closer-than-traditional collaboration of physiologists and geneticists, and that clonal material will provide an attractive object for the many lines of collaborative research that develop.

In a book on clonal forestry it seems obvious to emphasize the virtues of clones for physiological/genetic study. What we suggest in addition is that clonal studies will be attractively useful regardless of whether a particular tree species is managed via vegetative or sexual propagules; and that data from clonal studies in species A may also be useful for species B, perhaps even more useful than had they been generated from nonclonal material of the latter. We do not suggest, however, that cloning makes up for sexual recombination. Clearly, the two have to be combined if we strive for both repeatability and generality in research.

We have chosen three aspects from which to examine the merits of clonal material for physiological/genetic studies. The first is *productivity*, a subject of central importance in forestry and increasingly so in a world where needs for fuel, fiber, and shelter of an expanding human population have to be met from a shrinking land base (Spears and Ayensu 1985). In their concluding statements at a 1975 conference on tree physiology and yield improvement, Cannell and Last (1976) aptly pointed to the paucity of physiological studies with a genetic

[1] College of Forest Resources, University of Washington, Seattle, WA 98195, USA
[2] University of Antwerp, Department of Biology, Universiteitplein 1, 2610 Wilrijk, Belgium

Clonal Forestry I, Genetics and Biotechnology
Ed. by M.R. Ahuja and W.J. Libby

dimension. "If [physiological] samples were taken from known ecotypes, or, better still, from clones or full-sib families, there would automatically be more information of value to breeders, even if it were impossible to determine the patterns of inheritance" (Cannell and Last 1976, p. 531). Fourteen years later, this plea is still valid.

The second aspect we singled out is how cloning may be useful in generating a better understanding of *hybridization*. With rapidly developing technologies of genetic transfer between genomes, more opportunities present themselves to produce novel combinations. What are the physiological "ingredients" of successful combinations? How early can we recognize them, and how can we be more effective "match makers"? To the extent that hybridization is often used to achieve higher productivity, there is some overlap of this topic with the first section. However, we see the physiological study of hybridization in its broadest sense as a test of the prediction of hybrid intermediacy, at the level of its components, and therefore warranting separate treatment.

In the third section we draw attention to the unique opportunities for genetic/physiological study that are offered by *natural clones*, that is, by clones in nature, especially old ones. Much information is contained in such genotypes, some of it extractable from their study in situ, even more from their incorporation in experiments. This information will shed light on the relative advantages and disadvantages of the clonal habit itself and on the mechanisms by which long-living plants achieve adaptability vis-a-vis physical and biotic changes.

Given the diversity of the three subjects, we have limited ourselves to some selected examples and have referenced pertinent reviews for additional details.

2 Advantages and Limitations of Clonal Material for Physiological Research

High signal-to-noise ratio and high repeatability are desiderata in all experimentation. In physiological research they are often hard to attain because of the variation introduced by instrument error, by the instrument user, by sampling errors at the tissue or organ level, and finally by the variation encountered among phenotypes and underlying genotypes in the material under study. Access to clonal material thus offers the experimenter a welcome reduction in the overall sum of errors. This is especially true in the case of forest tree species, which tend to have high levels of genetic polymorphism and high average heterozygosity (Hamrick 1979). The within-population polymorphism leads to great variability among seedlings of the same seedlot, and high heterozygosity leads to great variability even among those related through one or both parents. The unavailability in most forest tree species of inbred lines for study purposes, and their likely aberrant growth and poor health even if available, adds to the argument for clonal material from trees of normal or desired genetic constitution.

The increased repeatability associated with the use of clonal material affords important opportunities. For example, it allows the parallel study of the same genotype (s) in different environments. Such comparison is essential for the study of phenotypic plasticity, or stability – phenomena of great interest in long-living plants. Experiments with several parallel treatments also gain greater precision from clonal material. For example, a commonly adopted protocol in physiological research calls for parallel measurements being taken on plants raised in the growth chamber, the greenhouse, and the field (Nelson and Ehlers 1984; Jarvis 1985). Given the constraints associated with each of these experimental environments, a combination of the three seems essential for inferences to be made about genotype/environment interaction. There is a similar need for sequential studies, since science progresses in a step-wise fashion. For example, it is often useful and appropriate to conduct follow-up studies on the same material, either as a repeat or as extension or elaboration of a promising experimental result. Cloning lends unequalled precision to this process.

Much insight can be gained from the study of extreme or abnormal genotypes and artificial cloning may be a prerequisite for it (Foster, Chap. 5; Vol.1). Mutant phenotypes have played a crucial role in the decyphering of countless human disorders and, equally importantly, have helped identify many basic biochemical pathways in plants and animals. Until recently, such material has received little attention in forestry research. Both the new methodologies to generate such material and those to analyze its constitution at the molecular and physiological level, will give new impetus to this approach. The vast array of existing horticultural varieties and cultivars in many of our major timber species may become a useful resource to be tapped in this effort. Beyond unusual forms we may well strive to better understand the physiology of *small* trees. Forestry, and especially tree improvement, has a distinct predilection for the upper end of the normal distribution curve. What we tend to discard may contain important information on carbon allocation, water relations, nutrition, defense chemistry, etc., and the underlying genetics. Clonal propagation may often be the only way to save short trees from early elimination through competition by taller genotypes. Short trees may be propagated and deployed either in monoclonal blocks or in mixture with other clones of their kind in plantations specifically established for study purposes. Even in practice, and for other reasons such as increased harvest value, additional arguments can be made against the over-emphasis on tall trees (Libby 1987a).

Cloning gives greater experimental flexibility, especially in numbers and also in timing. Under permissive conditions, material can be vegetatively propagated in virtually unlimited number year round. By contrast, even under permissive conditions, there is a limit to the numbers of seed in a given seedlot, especially one of an individual parent or parental combination. The need for sexual maturity and the vagaries of fruiting and seed production impose additional constraints.

However, there are also some inherent limits to the use of clonal materials for physiological/genetic research. The most serious one concerns representation. The number of forest tree species in which almost any genotype can be asexually propagated is small and currently excludes many commercially important conifers. Commonly clonal propagation is selectively successful in some genotypes of a species but not in others, or only during the juvenile phase, or only in heterografts on alien rootstock or, finally, only obtainable via elaborate in vitro regimes (Talbert et al., Chap. 11, Vol. 1). There are also species for which successful cloning has yet to be reported. Thus, for the immediate future, the choice of material is limited and so is the genetic representation. Physiological experiments that require clonal propagules are, therefore, likely to generate information that will apply to a small subset of genotypes, and possibly to an atypical subset whose physiology allows it to be cloned. Associated with this selective principle is the natural tendency for researchers to "run up the score" on familiar material and thus to learn more and more about less and less, rather than to complicate matters with a broader spectrum of genotypes.

Nor does clonal material satisfy all postulates of uniformity. Indeed, it would be surprising if the spatially and temporally heterogeneous environment of a tree crown were not in some way reflected in the growth and development of propagules recruited from different positions or ontogenetic phases of this crown. Attempts to identify the source of this phenotypic "memory" and to quantify the effects have been made (Libby and Jund 1962; Libby 1974; Burdon and Shelbourne 1974; Hood and Libby 1978; Farmer et al. 1986) and are also reviewed by Greenwood and Hutchison (Chap. 3, Vol. 1). Of potentially greater significance than this often transitory physiological variability of clonal material is any genetic variation that may have arisen in it from somatic mutations. As discussed in a subsequent section, there is increasing evidence for the existence of somatic variation among tissues or ramets of the same clone (Halperin 1986; Paige et al. 1989). Quite likely, the broader application of high-resolution tools will reveal more of this somatic variation within contiguous phenotypes of forest trees. The implications of such variants in terms of clonal fidelity, and the enhancement and capturing of them through novel in vitro procedures (see Ahuja, Chap. 10, Vol. 1) are of great interest to both geneticists and physiologists.

In spite of new developments in live-tissue preservation of woody plants (Aitken-Christie and Singh 1987; Chen and Kartha 1987), long-term storage of clonal material in a dormant state is more difficult than that of seed, and the post-thaw regrowth less certain (see also Millar, Chap. 3, Vol. 2). Thus, problems associated with clonal senescence as well as with the dissemination of disease vectors through clonal propagation, may keep setting limits to the long-term use of clonal material in physiological research.

3 Genetics/Physiology of Productivity

The analysis of yield components and their genetic control has played a prominent part in agricultural crop improvement (e.g., Watson 1952; Loomis and Williams 1961; Loomis et al. 1971). Yield is a composite trait and can be selected and bred for either directly or via the individual traits that contribute to it. The latter approach has been especially successful in annual seed crops in which "ideotypes" – hypothetical plant models (Donald 1968) – have served as targets for the breeding and selection process (reviewed by Donald and Hamblin 1983; Rasmusson 1987). Ideotype breeding differs from traditional breeding in that goals are specified for each trait of interest, resulting in an idealized model plant for a set of environmental and cultural conditions. It presupposes detailed knowledge of the plant and its yield components, the range of variation that is available in each of them, and their heritability. The same information is useful to, but not mandatory for, traditional breeding for yield. Among tree crops, fruit-tree improvement has followed a similar pathway, if still at a lower genetic resolution. Much advance has also been made in augmenting the productivity of rubber (*Hevea brasiliensis*) through clonal selection, based on detailed studies of latex chemistry, its biosynthesis and response to various management and harvest procedures (Templeton 1969). While much of the information on these trees crops has yet to be interpreted, most of it will be specific to the individual species and the unique features of their products.

In contrast to grain and fruit crops, the direct relationship between forest products and vegetative yield makes forest crops particularly well suited for productivity improvement, especially when combined with shorter harvest cycles. The system that has pushed this concept the furthest is short-rotation-intensive culture, or SRIC for short (McAlpine et al. 1966; Steinbeck et al. 1972). It refers to wood production on productive sites, using fast-growing hardwoods and intensive management regimes such as site preparation, weed control, fertilization, and mechanized harvesting at rotations of 3–10 years. The SRIC concept received a major impetus during the oil shortage of the 1970s, when new research was initiated to explore alternative sources of energy. Since 1978, the US Department of Energy (DOE) has screened woody species in well over 100 locations in the US for their growth, coppicing ability, site adaptability, disease resistance, and response to intensive cultural regimes (Ranney et al. 1987). Promising results have been obtained with several poplar species (*Populus*), *Acer saccharinum*, *Liquidambar styraciflua*, *Platanus occidentalis*, and *Robinia pseudoacacia* (Ranney et al. 1987). Along with *Alnus*, *Salix*, and *Eucalyptus*, several of these genera have proved to be well suited for SRIC outside the United States (Campinhos 1980; Perttu 1984; Hummel et al. 1988).

Although the SRIC concept has been implemented with both clonal material and seedlings, clones confer to it a measure of predictability and uniformity that is almost demanded by the high investment in this system. In fact, it is safe to predict that SRIC with hardwoods will increasingly move towards the clonal approach, and this because of several benefits: (1) the superior early growth of cutttings as compared to seedlings; (2) the gains from early ideotype formulation and the application of early selection criteria; (3) the more effective exploitation of root variation; (4) the option to coppice; (5) the greater uniformity of performance in monoclonal stands, and (6) the greater precision with which simulation models can make predictions for such systems.

Production physiology within the SRIC context has been most thoroughly studied in *Populus*. The long list of structural and functional components into which the overall process has been resolved can be grouped as follows:

Leaf growth physiology:	phyllotaxy, leaf initiation, leaf growth, cell wall extensibility, leaf senescence.
Stomatal physiology:	morphology, anatomy, distribution, regulation of stomatal function.

Carbon physiology: photosynthesis, partitioning, crown geometry, leaf orientation, leaf duration.
Root physiology: morphology, anatomy, growth, branching.

The interaction of many of these components has been studied under normal (e.g., Isebrands et al. 1988) as well as stress conditions, such as moisture stress (Ceulemans et al. 1978; Scarascia-Mugnozza et al. 1986; Schulte et al. 1987), or oxygen deprivation during flooding or water-logging of soil (Smit 1988; Smit and Stachowiak 1988), and has led to several predictive models at the individual plant level, such as ECOPASS (Ceulemans and Impens 1982) and ECOPHYS (Rauscher et al. 1988, 1989). Since clonal material makes for the simplest possible aggregate stand, it will be of great theoretical interest to extend these individual tree models to stand models. It will provide for a welcome comparison of "bottom up" vs. "top down" modeling, that is, the estimation of growth from a calculation of all actions and interactions of the relevant physiological process (bottom up) vs. the quantification of response of the system as a whole to changes in its main driving variables (top down; Landsberg 1986, pp. 166–170). It will also provide for empirical tests that bear on competition, commensalism, and spatial homeostasis (Bunting 1976; Ford 1976). Furthermore, this information will aid in formulating appropriate selection criteria and in deciding on optimal clonal deployment (see also Lindgren, Chap. 4, Vol. 1). However, what remains as one of the least understood domains in tree productivity, and severely limits any stand modeling, is the physiology of the rhizosphere. Here, too, clonal analysis, as possible in SRIC poplar plantations, is beginning to reveal remarkable clonal differences in carbon allocation to roots and in root "foraging" which greatly affect aboveground yield and stress response (Mazzolini and Dickmann 1988; Smit 1988). Especially promising are new tools such as remote sensing with the aid of fiber optics, that permit monitoring the dynamics of the intractable, yet so important fine roots (Taylor 1987).

Resolving poplar productivity into its components has not yet progressed to the point where the effects of these components have been isolated from those of associated traits. That is, most individual components such as leaf size, or leaf duration, have formed part of an ensemble of traits rather than having been tested in various genetic backgrounds. Nor has the material in which they were studied offered much insight on the underlying genetics: physiologists have shown a predilection to study hybrid poplar clones, often of unknown pedigree or origin. By contrast, geneticists have tended to focus on open-pollinated progenies from natural populations. A coordination of materials and protocols will be necessary to make real progress in elucidating the role of the various productivity components.

Broad-sense heritability estimates (H^2) in several North American poplars (*P. balsamifera*, *P. deltoides*; and *P. trichocarpa*) have been moderately high ($H^2 = 0.4$–0.5) for juvenile height and diameter growth, and even higher ($H^2 = 0.4$–0.7) for growth phenology and leaf-size traits (Wilcox and Farmer 1968; Mohn and Randall 1971; Randall and Cooper 1973; Weber et al 1985; Farmer et al. 1986; Nelson and Tauer 1987). Needed are now narrow-sense heritability estimates for individual growth components as a measure of the additive genetic variance associated with individual growth components. These estimates will help focus selection on those traits from which a positive genetic response can be expected at a reasonable selection intensity.

Finally, one area in SRIC that deserves critical attention by both physiologists and geneticists – in fact, it really requires a broader range of expertise than these two groups can offer – is the interaction between plant material and cultural regime. Implicit in the SRIC concept is the selection of plant material that responds well to intensive culture. Intensive culture means initial site preparation, weed control until crown closure, and eventually mechanized harvesting; it also may include control of rodents, insects, and pathogens, soil amendments, and irrigation. Costs and benefits are closely tied to the material employed. For example, clones may be selected (Shirley 1988) or engineered (Haissig 1987; Haissig and Riemenschneider 1987) to withstand effective doses of specific herbicides, but with what impact on net gain in biomass? Similarly, certain clones may require fertilization whereas others may be more nutrient efficient (Heilman and Stettler 1986). Management and ecological concerns will influence the choice of regimes, but will need background information on the trade-offs. While such refinements as "fine tuning" clones to local site heterogeneities based on detailed soil mapping (Libby and Rauter 1984), or programmed planting sequences of complementary clones to minimize site exhaustion (i.e., "Level IV trails", Libby 1987b) are conceptually attractive, we suspect they will be too costly for most programs. Operational reality will typically demand some flexibility about material deployment and, thus, will give preference to clones with high adaptability (see Sect. 4 below). Continued breeding will also provide for more gains to be made through regular clonal replacement, thereby truncating investment returns on excessive clonal information.

Are the insights gained from SRIC research on productivity useful for systems with longer rotations? In a limited way, yes. For a variety of reasons, there has been a general shift in SRIC from the mini-rotation concept (1–2 years) of the early seventies (e.g., Schreiner 1974) towards 8–10 year cycles (Ranney et al. 1987), thus shrinking the gap to the 20–40-year rotations now practised with certain conifers (e.g., *Pinus radiata, P. taeda, Cryptomeria japonica*). However, major discrepancies remain between these management regimes, especially the greater role that reproductive allocation and stress avoidance or tolerance mechanisms play in longer rotations. These systems will thus have to be studied on their own terms for quantitative models of productivity to be developed.

4 Genetics/Physiology in Relation to Hybridization

To date, artificial hybridization has played only a modest role in applied forest tree improvement. The reasons are many and have both biological and operational origins (see reviews in Wright 1976; Zobel and Talbert 1984). Simply stated, it generally requires a greater effort to produce hybrid seed, with no guarantee that the added cost is offset by increased performance of the hybrid material. Nor is there greater uniformity in the F_1 generation, one of the major incentives in agronomic hybridization between pure lines. The generally high levels of heterozygosity harbored in forest tree species (Hamrick 1979; Loveless and Hamrick 1984; Mitton and Grant 1984) tend to produce highly variable seedling populations in the F_1 and, of course, even more so in the F_2 generation (Wright 1976).

It has been generally recognized that vegetative propagation will play a decisive role in the future application of hybridization in forestry (Brown 1967; Zobel and Talbert 1984). Cloning not only permits the capture of both additive and nonadditive variance but, if combined with hybridization, also permits the exploitation of new, favorable genome combinations while at the same time maintaining uniformity and repeatability in the planting stock. Variability in an F_1 generation is thus no deterrent, provided there are a few outstanding individuals that can be mass propagated. The prominence in the annals of hybridization success of genera such as *Salix*, *Populus*, and *Chamaecyparis*, that are easy to propagate vegetatively, is therefore no coincidence.

We anticipate a far more significant effort to be made in this combined hybridization/cloning approach to forest tree improvement in the future, and this for several reasons: (1) a broader array of techniques that are now available to combine genomes, or parts thereof; (2) greater success in vegetatively propagating the more recalcitrant tree species; (3) increased availability of disturbed and new habitats, as a function of anthropogenic impacts, to which existing species are poorly adapted, (4) climatic changes; and (5) increased specialization of production forestry and a concomitant de-emphasis of seedling material (see Sect. 3 above).

From a conceptual point of view it may be constructive to consider hybridization in the broad sense as an exchange of gene material among unlike but related individuals, where relatedness may vary on a continuum from siblings to more distant members of the same genus. In other words, rather than considering hybridization as qualitatively different from a "normal" mating, we may view it as a mere extension of the latter. This is not to ignore the barriers that may intercede with increasing genetic distance, but rather to consider them violable and to find means to circumvent them. This is consistent with the view that species are dynamic "ad-hoc" entities and seems to be especially appropriate for outcrossing, wind-pollinated and wind-dispersed forest trees in many of which the bulk of genetic variation is contained within populations – at the allozyme level often over 90% (Ledig 1986) – and in which interpopulation gene flow seems to be highly effective (Guries and Ledig 1982; Loveless and Hamrick 1984; see also Libby and Critchfield 1987). A more integrated approach that combines intraspecific selective breeding with racial and interspecific hybridization will also help incorporate the gains made through population and family selection in existing improvement programs into the hybridization effort. It has long been argued that hybridization is no "miracle solution" but, while providing for novel, perhaps unprecedented, gene combinations, its outcome ultimately depends on the quality of the individual parents used (e.g. Wright 1976; Zobel and Talbert 1984).

The two most common goals in hybridization are (1) to combine desirable traits from two biological entities (races, species, etc.) into one; and (2) to obtain hybrid vigor or "heterosis," i.e., hybrid superiority relative to the mid-parent value (average between the two parents) or to the better parent (Sinha and Khanna 1975). A third objective may be to obtain increased developmental homeostasis, i.e., greater phenotypic stability in fluctuating environments, which has often been associated with heterozygosity or heterosis (Lerner 1954; Mitton and Grant 1984). These goals will have to be realized in essentially new, or different, environments from those in which parental species evolved. Heterotic hybrids may thus be viewed as novel genotypes that excel in a set of new physical and biotic conditions.

If hybridization with cloning is worth pursuing, in which genera and at which taxonomic level is it likely to be of promise? It is this broad question that can be significantly elucidated through close collaboration of geneticist and physiologists. What is needed goes beyond the customary description of growth and morphology of hybrids; it focuses on the processes underlying growth and differentiation both under field and controlled conditions, stressed and unstressed; and it requires pedigreed hybrids, their parents, and their control progenies within the respective species. The iterative nature of these studies and the synchronization needed in the study material makes cloning an essential prerequisite.

An example of this approach is the comprehensive research program that has been conducted over the past ten years at the University of Washington and Washington State University, which has focused on the study of *Populus trichocarpa* × *P. deltoides* hybrids and the two parental species. Alerted to the promise of this hybrid combination from earlier exploratory crosses (Stettler 1968; Heilman and Stettler 1985), researchers were able to verify the superior growth performance of these hybrids relative to the better parent species in an expanded crossing and field testing program (Stettler et al. 1988). Parallel physiological studies, conducted in the field, greenhouse and growth chamber, (summarized by Hinckley et al. 1989), identified a sufficient but not exhaustive list of components to explain hybrid superiority. These components include (1) the more efficient ability of stomata to respond to decreasing soil moisture and/or increasing evaporative demand; (2) the coupling between leaf morphological/anatomical features and the response of stomata to drought; (3) leaf size as affected by cell number and cell size; (4) diurnal growth pattern of leaves; (5) osmotic maintenance of foliar cell turgor and cell wall extensibility; (6) the interaction between crown architecture, leaf phenology and distribution, photosynthesis and carbon allocation; (7) the rate and nature of root development as a function of carbon allocation; and (8) root structure and function in response to either lack of oxygen or drought stress. A ninth component that may mediate one or several of the eight above, concerns the role of gibberellic acid in hybrid vigor (Rood et al. 1984, 1988; Bate et al. 1988), and is being investigated by Rood and his coworkers in this hybrid material.

As to the genetic transmission patterns in these studies, "simple" components in the hybrids, such as cell size, cell number, or leaf orientation, displayed either dominance or intermediacy between parental species, and overdominance – values clearly exceeding the better of the two parents – was found only in "complex" traits such as leaf size, total leaf area, and leaf area index. Hybrid performance in this case seems to be a result of parental complementation in a number of components, having a cumulative effect on total biomass production.

Furthermore, this complementation seems to provide for increased performance stability or developmental homeostasis over time and space (Hinckley et al. 1989).

How could such "instant" complementation come about and why did a similar "package" of traits not evolve in one of the parental species, for example in *P. trichocarpa* in whose native range the hybrids do so well? The first question may be easier to answer than the second. In spite of the assignment of *P. trichocarpa* to the *Tacamahaca* section of poplars, and of *P. deltoides* to the *Aigeiros* section, the latter seems to have evolved in North America as late as the middle Miocene from a common *Tacamahaca* ancestor (Eckenwalder 1977). Thus, the two species may have preserved many gene sequences that now find harmonious expression in the hybrid. Further evidence for this commonality is the excellent crossability between the two species (Guries and Stettler 1976), the abundance of natural hybrids in contact zones between them (Eckenwalder 1977) and the vigor reported about hybrids of the two elsewhere (Teissier du Cros 1984; Ceulemans et al. 1987). Many other species-pairs in *Populus* have given rise to vigorous hybrids although the underlying mechanisms are less well understood (see Zsuffa et al., Chap. 5, Vol. 2).

The dissection of *P. trichocarpa* × *P. deltoides* hybrid superiority conforms to data from many agronomic crops, as compiled by Sinha and Khanna (1975). Heterotic hybrids (i.e., hybrids exceeding the performance of the better of the two parents) commonly show additive or dominance inheritance in their individual traits rather than overdominance, and it is the multiplicative effects of these individual traits that generate the heterotic ensemble. Ledig (1986), in a recent review of heterozygosity, heterosis, and fitness in outbreeding plants considers the dominance model the norm for heterosis in forest trees, overdominance the exception (see also Jinks 1983). He attributes the benefits of heterozygosity to the masking of deleterious recessives which may be unusually frequent in trees as a consequence of their long lifespan and the concominant accumulation of mutations (Ledig 1986). By contrast, Mitton and Grant (1984) interpret the literature quite differently and suggest a major role for heterozygosity per se, i.e., overdominance, in the control of growth rate and developmental homeostasis. Clearly, the subject is far from resolved and deserves more careful study. What is beyond doubt, however, is that forest trees, because of their cumulative growth pattern and unusually heterogeneous environment in time and space, are uniquely predisposed to benefit from developmental homeostasis, whatever the underlying mechanisms (Mitton and Grant 1984; Ledig 1986).

If hybridization between species has proved to be rewarding in certain angiosperm genera, such as *Populus*, *Eucalyptus*, *Platanus*, and *Ulmus* (Heybroek 1968; Townsend 1975; Campinhos 1980), it has been less predictable and generally less successful in the conifers (Ledig 1986; Critchfield 1988). But even in that more ancient and evolutionarily conservative group there are enough reports of heterotic hybrids in *Picea* (Hoffman and Kleinschmit 1979), *Larix* (Keiding 1962), *Abies* (Mergen and Gregoire 1988; but also Critchfield 1988), and *Pinus*; (Hyun and Hong 1969; LaFarge and Kraus 1980) to demand attention. What would be especially helpful is a better understanding – in physiological terms – of genomic congruity/incongruity (Hogenboom 1984). The concept of incongruity was coined by Hogenboom (1973) in relation to reproductive isolation and was aimed at distinguishing the inhibition of the pistil-pollen relationship (incompatibility) from the more encompassing mismatch or "incompleteness of the partner relationship" (incongruity). In contrast to incompatibility, which is an outbreeding mechanism and under control of one or two loci, incongruity is an expression of evolutionary divergence and genetically more complex (Hogenboom 1984). The concept has proved useful in describing pre-fertilization reproductive failures in *Solanum* (Hermsen et al 1974; Hermsen and Sawicka 1979), *Cucumis* (Kho et al. 1980), and *Populus* (Guries and Stettler 1976; Stettler et al. 1980), but may also be extended to later malfunctions during embryogeny or early seedling growth of hybrid progeny (Hogenboom 1984).

It would be unrealistic to expect significant research efforts to be devoted to hybrid clones of poor performance, although the reasons for that poor performance might be interesting and provide valuable insights. It would be equally unrealistic to expect significant insights from the exclusive study of heterotic hybrids. Much will be gained from a more balanced documentation of physiological "packages" that span the range of the continuum between perfect complementation and development disharmony. This will be increasingly possible as advanced-generation material (F_2-hybrids, F_3-hybrids, backcrosses to parental species) becomes available as a consequence of earlier hybridization work. Developmental studies of advanced-generation hybrids may also shed light on the recurring question of the genetic basis of species differences in plants. Far from being resolved, this question is a challenge to both physiological and quantitative geneticists. The hypothesis that plant speciation may come about by changes in a few "major genes", with subsequent "elaboration" of multiple "modifier" genes, has recently received renewed attention (see review by Gottlieb 1984; but also Coyne and Lande 1985; Gottlieb 1986). As methods of molecular analysis become more routine, the hypothesis will be increasingly accessible to testing. Apart from its intrinsic interest, the question has important implications for biotechnology (see Chaps. 10–15, Vol. 1).

5 Physiology/Genetics of Natural Clones

Plants are modular organisms, or metamers, (Harper and Bell 1979; White 1979), that is, their growth pattern is iterative, with dispersed meristems adding new modules to the existing body. No other plant form displays this principle more vividly than the repetitive architecture of a tree. Clonal plants extend modularity one step further and thus are capable of stretching both the biomechanical size constraints and life span of aclonal plants (Watkinson and White 1986; Watkinson 1988). The mechanisms of natural cloning are manifold and widely distributed in the plant kingdom; they have enlisted structures as diverse as tubers, tuberous roots, bulbs, stolons, runners, root sprouts, stump sprouts, layered branches, shoot segments, inflorescence bulbils, and apomictic seed. Clearly, asexual propagation – whether we view it as an expression of growth or reproduction – has played a major role in the evolution of plant diversity. The fact that it is also well represented among tree genera bodes well for its potential as a tool in the more controlled evolution of trees in the future. What can we learn from natural clones?

For obvious reasons, the form of cloning in trees that has received the greatest attention is the spatially continuous cloning which is displayed for example by aspen (*Populus tremuloides*). Individual clones of this species, comprising as many as 47 000 ramets, have been found to cover up to 80 ha, and may have originated as far back as the Pleistocene (Kemperman and Barnes 1976). Based on root suckering, that is the generation of adventitious shoots from roots, this mode of clonal spread is common in members of the Salicaceae; less prominent but evident in such genera as *Alnus*, *Fagus*, *Ulmus*, *Sassafras*, *Prunus*, *Liquidambar*, *Robinia*, and *Rhus*; and it is occasionally found also in *Araucaria*, *Sequoia*, and *Chamaecyparis*. Far more widespread in trees is the ability to re-sprout from the stem base or root crown, as a replacement strategy. It is common in most genera of temperate hardwoods, both early and late successional, and includes also *Sequoia*, young *Sequoiadendron*, and some pines (Harlow et al. 1979). Multiple stems with basal fusion are the most obvious indicator of this mode of cloning, although basal fusion may also result from the early joining of twin seedlings and may occasionally even involve different species. If we add branch layering as a third form of cloning in trees, which also comprises several conifer genera (*Abies*, *Picea*, *Pinus*, *Chamaecyparis*, *Thuja*), we can say that trees are among the most extreme metamerous organisms in existence, and that they embody cloning both in space (as additional plants) and in time (prolonging the existence of a genotype).

One question this raises is what is the true age of tree genets? We suspect that for many trees it is considerably older than indicated by the ring count of an extant specimen. While age of current ramets may be a sufficient measure for certain demographic and ecological studies, it is inadequate for population genetic and evolutionary studies, which focus on changes in gene frequencies per generation. Since it is of more than academic interest to understand evolutionary patterns in one of our major life forms, we encourage paleobotanists to provide us with more reliable data on genet age in trees, especially those with cloning in time.

If, as we suspect, many tree genotypes are substantially older than their current genome carriers, this raises two further questions, namely (1) how is clonal fidelity maintained, and (2) how can such long-living clones adapt to changing conditions?

Clonal fidelity and its violation, i.e., the nature and source of somatic variation in plants, have been the subject of much recent research, with exciting implications (see reviews by Larkin and Scowcroft 1983; Orton 1984; Halperin 1986). There is now sufficient evidence to discard the long-held view that the plant body is an aggregate of genetically uniform cells. For example, in most angiosperms studied, endopolyploidy seems to be the rule for those mature tissues that are not destined to undergo further cell division (Nagl 1976; Halperin 1986). Differential DNA synthesis is common in cell and organ differentiation, often involves "discrimination" of heterochromatin, and may be part of the "programmed senescence in determinate organs" (Halperin 1986). Additional genetic variation may be generated by point mutations, structural rearrangements of chromosomes, transpositional events, somatic crossing-over, and gene amplification. Needless to say that these phenomena are still poorly understood in trees (Ledig 1986). Clones of sufficient age and with sufficiently heterochronic ramet populations will be ideal study material for this purpose (Cheliak and Dancik 1982; Whitham et al. 1984; Paige et al. 1989).

Genetic constancy, if not the normal state in differentiated cells of the plant body, seems to be more likely maintained in the plant's "germ line", that is, its population of meristems (e.g., Hardwick 1986). Mechanisms have been suggested that eliminate mutant cells from these tissues though diplontic selection (Klekowski and Kazarinova-Fukshansky 1984). Again, old tree clones may be especially effective practitioners of this principle and, therefore, worth studying. The assumption of genetic constancy is of great consequence for artificial cloning. Old clones such as the pyramidal cultivars of *Populus nigra*, including var. '*thevestina*,' *italica*; '*Afghanica*,' some of which originated centuries ago in Asia minor (Bugala 1967) and have since been spread around the world, will provide challenging test cases. Molecular scrutiny may reveal variation that escaped morphological detection and which, in fact, may have contributed to local adaptation of the clone. The nature of such variation or the lack thereof, and the physiological consequences, will be of much interest to geneticists.

Phenotypic variation within a clone, with a genetic or epigenetic basis, or both, may be an effective strategy to adapt to changing biotic conditions. As cogently argued by Whitham (1981) and Whitham et al. (1984) long-living plants form "monocultures in time" and thus risk predation by rapidly evolving pests and pathogens. Any mechanism that would generate variation within a plant might, therefore, have selective value. In a survey conducted to verify this proposition, Whitham and his coworkers found at least three types of mechanisms by which individual plants present themselves as temporal or spatial mosaics of resistance, all with a genetic basis. They include (1) genetic chimerism or mosaicism; (2) patterns of differential gene expression as a function of development (e.g., juvenile vs. adult tissues, vegetative vs. flowering branches, etc.); and (3) induced plant defenses that are tiggered by the pest or pathogens (or by wounding) and generate the production of defense chemicals in localized tissues or on a systemic scale (Whitham et al. 1984). Relevant forest-tree examples reported include *Abies balsamea* resistance to spruce budworm (Blais 1958, Kimmins 1971); *Larix laricina* response to larch bud moth attack (Niemela et al. 1980); *Betula pubescens* (Haukioja and Niemela 1976) and *Populus trichocarpa* × *P. deltoides* response to wounding (Parsons et al. 1989); *P. angustifolia* response to gall aphids (Williams and Whitham 1986); and the phytochemical deterrence mechanisms to the Alaska snowshoe hare by four early succcessional hardwoods (Bryant 1981).

An understanding of defense chemistry, its triggering and deployment, and its physiological cost, is essential for our basic understanding of forest trees and even more so for our efforts at manipulating them towards specific purposes. While this information is relevant to both clonal and aclonal trees, it is likely to be more easliy generated by clonal species, and old "successful" clones may be especially deserving attention. In a similar way, we may learn from such material more about the physiology of adaptability to changing physical conditions. Clonal stability across sites ranks high on the list of desiderata in any tree-improvement program. It may well be outranked in the future by the criterion of climatic tolerance, as stresses due to extreme weather conditions and climatic change assume greater significance (Schneider 1989). Comparative experiments involving "old" and "young" clones in a wide network of sites, including extreme conditions, and going well beyond the establishment phase, would be highly informative, especially if complemented by in-depth studies of physiological variables.

Finally, natural clones of the spatially expansive type such as in *Populus tremuloides*, *P. balsamifera*, and *Liquidambar styraciflua*, afford opportunities to test specific hypothesis of foraging behavior and resource exploitation, with special emphasis on intragenet competition (or cooperation). This is an area of active investigation by plant and animal ecologists (see reviews by Watkinson 1988; Sutherland and Stillman 1988; also Hardwick 1986) and warrants study with suitable tree material. Again, in situ studies, combined with experimentally constructed stands of the same clones in mono- and polyclonal mixtures, in resource-abundant and resource-limited conditions, will yield important insights, if perhaps more in relation to the physiology of fitness than of productivity.

6 Concluding Remarks

The coordination of genetic and physiological research requires communication and planning across disciplinary boundaries. It also calls for compromises in approach. Genetic protocols typically call for large numbers, broad sampling, replication in space, and long time spans; physiological protocols demand precision, repeatability in time, verification under controlled conditions, and prefer to have a shorter time scale. A frequently used compromise is to begin with the study of few, sharply contrasting phenotypes; to then refine measurement techniques on them; and to subsequently extend the refined measurements to a broader array of interpolating phenotypes for validation.

Geneticists can make a significant contribution to the initiation of joint studies; they are at the source of the material and are familiar with its origin, its genetic architecture and its distribution in plantations. Thus, they can alert physiologists to research opportunities. If they go one step further, by deliberately generating clonal research material, rather than only accepting it as a byproduct of their tree improvement effort, they can help set the stage for a more profound understanding of physiological mechanisms in the species at hand.

Acknowledgments. We thank D. Sprugel, W. Libby, and P. Mathias for their thoughtful comments on an earlier version of this manuscript. Many ideas expressed in this chapter had their origin in discussions with P. Heilman, T. Hinckley, and J. Isebrands. Support from the US. Department of Energy (Contract DE-AC05-840R21400, Subcontract 19X-43382C, ORNL with Martin Marietta Energy Systems, Inc.) is gratefully acknowledged.

References

Aitken-Christie J, Singh AP (1987) Cold storage of tissue cultures. In: Bonga JM, Durzan DJ (eds) Cell and tissue culture in forestry, vol 2. Nijhoff, Dordrecht Boston, pp 285–304

Bate NJ, Rood SB, Blake TJ (1988) Giberrellins and heterosis in poplar. Can J Bot 66:1148–1152

Blais JR (1958) The vulnerability of balsam fir to spruce budworm attack in northwestern Ontario with special reference to the physiological age of the tree. For Chron 34:405–422

Brown CL (1967) The application of tissue culture techniques to various problems in forest tree improvement. Proc S Conf Forest Tree Improv 9:108–110

Bryant JP (1981) Phytochemical deterrence of snowshoe hare browsing by adventitious shoots of four Alaskan trees. Science 213:889–890

Bugala W (1967) Taxonomy of Euroasiatic poplars related to *Populus nigra* L. Arbor Kornickie (Poznan) 12:45–219

Bunting AH (1976) Maximizing the product or how to have it both ways. In: Cannell MGR, Last FT (eds) Tree physiology and yield improvement. Academic Press, London New York, pp 1–20

Burdon RD, Shelbourne CJA (1974) The use of vegetative propagation for genetic and physiological information, pt 6. The use of vegetative propagules for obtaining genetic information. NZ J For Sci 4:418–425

Campinhos E (1980) More wood of better quality through intensive silviculture with rapid growth improved Brazilian *Eucalyptus*. TAPPI 63:145–147

Cannell MGR, Last FT (eds) (1976) Tree physiology and yield improvement. Academic Press, London New York

Ceulemans R, Impens I (1982) ECOPASS: a multivariate model used as an index of growth performance of poplar clones. For Sci 28:862–867

Ceulemans R, Impens I, Lemeur R, Moermans R, Samsuddin Z. (1978) Water movement in the soil-poplar-atmosphere system. II. Comparative study of the transpiration regulation during water stress situations in four different poplar clones. Oecol Plant 13:139–146

Ceulemans R, Impens I, Steenackers V (1987) Variations in photosynthetic, anatomical, and enzymatic leaf traits and correlations with growth in recently selected *Populus* hybrids. Can J For Res 17:273–283

Cheliak WM, Dancik BP (1982) Genetic diversity of natural populations of a clone-forming tree, *Populus tremuloides* "trembling aspen." Can J Genet cytol 24:611–616

Chen THH, Kartha KK (1987) Cryopreservation of Woody species. In: Bonga JM, Durzan DJ (eds) Cell and tissue culture in forestry, vol 2. Nijhoff, Dordrecht Boston, pp 305–319

Coyne JA, Lande R (1985) The genetic basis of species diferences in plants. Am Nat 126:141–145

Critchfield WB (1988) Hybridization of the California firs. For Sci 34:139–151

Dickmann DI, Stuart KW (1983) The culture of poplars in eastern North America. Michigan State Univ, East Lansing, 168 pp.

Donald CM (1968) The breeding of crop ideotypes. Euphytica 17:385–403

Donald CM, Hamblin J (1983) The convergent evolution of annual seed crops in agriculture. Adv Agric 36:97–143

Eckenwalder JE (1977) Systematics of *Populus* L. (Salicaceae) in Southwestern North America with special reference to Sect. Aigeiros Duby. PhD Diss, Univ Cal, Berkeley

FAO (ed) (1958) Poplars in forestry and land use. FAO forestry and forest products studies 12. FAO, Rome, 511 pp

Farmer RE, Jr Reinholt RW, Schneckenberger F (1986) Environmental preconditioning and variance in early growth of balsam poplar. Silv Genet 35:129–131

Ford ED (1976) Competition, genetic systems and improvement of forest yield. In: Cannell MGR, Last FT (eds) Tree physiology and yield improvement. Academic Press, London, New York, pp 463–472

Gottlieb LD (1984) Genetics and morphological evolution in plants. Am Nat 123:681–709

Gottlieb LD (1986) The genetic basis of plant form. Philos Trans R Soc London Ser B 313:197–208

Guries RP, Ledig FT (1982) Genetic diversity and population structure in pitch pine (*Pinus rigida* Mill.). Evolution 36:387–402

Guries RP, Stettler RF (1976) Pre-fertilization barriers to hybridization in the poplars. Silv Genet 25:37–44

Haissig BE (1987) Tissue culture-based biotechnology for *Populus* clones. In: Klass DL (ed) Energy from biomass and wastes X. Inst Gas Technol 11th Annu Meet, Chicago, IL, pp 155–175

Haissig BE, Riemenschneider DE (1987) Genetic engineering of hybrid poplars for herbicide tolerance. In: Proc Forest products research Conf, Madison WI, 21–23 Oct 1986. US Dep Agric For Serv For Prod Lab, Madison, pp 37–46

Halperin W (1986) Attainment and retention of morphogenetic capacity in vitro. In: Vasil IK (ed) Cell culture and somatic cell genetics of plants, vol 3. Academic Press, London New York, pp 3–47

Hamrick JL (1979) Genetic variation and longevity. In: Solbrig OT, Jain S, Johnson GB, Raven PH (eds) Topics in plant population biology. Columbia Univ Press, New York, pp 85–113

Hardwick RC (1986) Physiological consequences of modular growth in plants. Philos Trans R Soc London Ser B 313:161–173

Harlow WM, Harrar ES, White FM (1979) Textbook of dendrology, 6th edn. McGraw Hill, New York

Harper JL, Bell AD (1979) The population dynamics of growth form in organisms with modular construction. In: Anderson RM, Turner BD, Taylor LR (eds), Population dynamics. Blackwell, Oxford, pp 29–52

Haukioja E, Niemelä P (1976) Does birch defend itself actively against herbivores? Rep Kevo Subarct Res Stn 13:44–47

Heilman PE, Stettler RF (1985) Genetic variation and productivity of *Populus trichocarpa* and its hybrids. II. Biomass production in a 4-year plantation. Can J for Res 15:384–388

Heilman PE, Stettler RF (1986) Nutritional concerns in selection of black cottonwood and hybrid clones for short rotation. Can J For Res 16:860–863

Hermsen JGT, Sawicka E (1979) Incompatibility and incongruity in tuber-bearing *Solanum* species. In: Hawkes JGR, Lester N, Skelding AD (eds) The biology and taxonomy of the Solanaceae. Academic Press, New York London, pp 445–453

Hermsen JGT, Olsder J, Jansen P, Hoving E (1974) Acceptance of self-compatible pollen from *Solanum verrucosum* in dihaploids from *S. tuberosum.* In: Linskens HF (ed) Fertilization in higher plants. North Holland Biomedical Press, Amsterdam, pp 37–40

Heybroek HM (1968) Taxonomy, crossability and breeding of elm. In: Int Symp Dutch elm disease, Ames IA, 26–28 Feb 1968

Hinckley TM, Ceulemans R, Dunlap JM, Figliola A, Heilman PE, Isebrands JG, Scarascia-Mugnozza G, Schulte PJ, Smit B, Stettler RF, van Volkenburgh E, Wiard BM (1989) Physiological, morphological and anatomical components of hybrid vigor in *Populus.* In: Kreeb HH, Richter H, Hinckley TM (eds) Structural and functional responses to environmental stresses. Academic Publ, The Hague, Neth, pp 119–217

Hoffman D, Kleinschmit J (1979) A utilization program for spurce provenance and species hybrids. In: IUFRO Norw Spruce Meet, Bucharest pp 216–236

Hogenboom NG (1973) A model for incongruity in intimate partner relationships. Euphytica 22:219–233

Hogenboom NG (1984) Incongruity: non-functioning of intercellular and intracellular partner relationships through non-matching information. In: Linskens HF, Heslop; Harrison J (eds) Encyclopedia of plant physiology, vol 17. Springer, Berlin Heidelberg New York, pp 640–654

Hood JV, Libby WJ (1978) Continuing effects of maturation state in radiata pine and a general maturation model. In: Hughes KW, Henke R, Constatin M (eds) Propagation of higher plants through tissue culture. Proc Int Conf Univ Tenn, Knoxville, April 16–19, US Dep Energ pp 220–232

Hummel FC, Palz W, Grassi G (eds) (1988) Biomass forestry in Europe: a strategy for the future. Elsevier, Essex, UK, 600 pp

Hyun SK, Hong SH (1969) The growth performance of pitch-loblolly hybrid pine produced by different races of loblolly pine in their early age. Inst For Genet Suwon, Korea, Res Rep No. 7, pp 35–44

Isebrands JG, Ceulemans R, Wiard BM (1988) Genetic variation in photosynthetic traits among *Populus* clones in relation to yield. Plant Physiol Biochem 26:427–437

Jarvis PG (1985) Transpiration and assimilation of tree and agricultural crops: the "Omega" factor. In: Cannell MGR, Jackson JE (eds) Attributes of trees as crop plants. Nat Environ Res Counc, Huntingdon, UK, pp 460–480

Jarvis PG, McNaughton KG (1986) Stomatal control of transpiration: Scaling up from leaf to region. Advances in ecological research, vol 15. Academic Press, New York London, 49 pp

Jinks JL (1983) Biometrical genetics of heterosis. In: Frankel R (ed) Heterosis – a reappraisal of theory and practice. Monogr Theor Appl genet 6: 1–46

Keiding H (1962) Krydsningsfrodighed Hos Laerk. Dans Skovforen Tidsskr 47: 139–157

Kemperman JA, Barnes BV (1976) Clone size in American aspen. Can J Bot 54: 2603–2607

Kho YO, den Nijs APM, Franken J (1980) Interspecific hybridization in *Cucumis* L. II The crossability of species: an investigation of in vivo pollen tube growth and seed set. Euphytica 29: 661–672

Kimmins JP (1971) Variation in the foliar amino acid composition of flowering and non-flowering balsam fir [*Abies balsamea* L. (Mill)] and white spruce [*Picea glauca* Moench (Voss)] in relation to outbreaks of the spruce budworm (*Choristoneura fumiferana* Cleus). Can J Zool 49: 1005–1011

Klekowski EJ, Kazarinova-Fukshansky N (1984) Shoot apical meristems and mutation: fixation of selectively neutral cell genotypes. Am J Bot 71: 22–27

LaFarge T, Kraus JF (1980) A progeny test of (short leaf × loblolly) × loblolly hybrids to produce rapid-growing hybrids resistant to fusiform rust. Silv Genet 29: 1979–200

Landsberg JJ (1986) Physiological ecology of forest production. Academic Press, New York London

Larkin PJ, Scowcroft WR (1983) Somaclonal variation and crop improvement. In: Kosuge T, Meredith CP, Hollender A (eds) Genetic engineering of plants: an agricultural perspective. Plenum, New York, pp 289–314

Ledig FT (1976) Physiological genetics, photosynthesis and growth models. In: Cannell MGR, Last FT (eds) Tree physiology and yield improvement. Academic Press, New York London, pp 21–54

Ledig FT (1986) Heterozygosity, heterosis, and fitness in outbreeding plants. In: Soulé MT (ed) Conservation biology. Sinauer, Sunderland, Mass, pp 77–104

Lerner IM (1954) Genetic homeostasis. Oliver & Boyd, London

Libby WJ (1974) The use of vegetative propagules in forest genetics and tree improvement. NZ J For Sci 4: 440–447

Libby WJ (1987a) Do we really want taller trees? Adaptation and allocation as tree-improvement strategies HR MacMillan Lectureship in forestry, Jan 22, 1987. Univ BC, Vancouver, 16 pp

Libby WJ (1987b) Testing for clonal forestry. Ann For (Zagreb) 13 (1–2): 69–75

Libby WJ, Critchfield WB (1987) Patterns of genetic architecture. Ann Fore (Zagreb) 13 (1–2): 77–92

Libby WJ, Jund E (1962) Variance associated with cloning. Heredity 17: 533–540

Libby WJ, Rauter RM (1984) Advantages of clonal forestry: breeding, production efficiency, silviculture. For Chron 60: 145–149

Loomis RS, Williams WA (1961) Maximum crop productivity: an estimate. Crop Sci 3: 67–72

Loomis RS, Williams WA, Hall AE (1971) Agricultural productivity. Annu Rev Plant Physiol 22: 431–468

Loveless MD, Hamrick JD (1984) Ecological determinants of genetic structure in plant populations. Annu Rev Ecol Syst 15: 65–95

Mazzolini S, Dickman DI (1988) Differential physiological and morphological responses of two hybrid *Populus* clones to water stress. Tree Physiol 4: 61–70

McAlpine RG, Brown CL, Ruark HE (1966) "Silage" Sycamore. For Farm 26: 607

Mergen F, Gregoire TG (1988) Growth of hybrid fir trees in Connecticut. Silv Genet 37: 118–124

Mitton JB, Grant MC (1984) Associations among protein heterozygosity, growth rate, and developmental homeostasis. Annu Rev Ecol Syst 15: 479–499

Mohn CA, Randall WK (1971) Inheritance and correlation of growth characters in *Populus deltoides*. Silv Genet 20: 182–184

Nagl W (1976) Early embryogenesis in *Tropaeolum majus* L.: evolution of DNA content and polyteny in the suspensor. Plant Sci Lett 7: 1–8

Nelson CD, Tauer CG (1987) Genetic variation in juvenile characteristics of *Populus deltoides* Bartr. from the Southern Great Plains. Silv Genet 36: 216–221

Nelson ND, Ehlers P (1984) Comparative carbon dioxide exchange for two *Populus* clones grown in growth room, greenhouse and field environments. Can J For Res 14: 924–932

Niemelä P, Tuomi J, Haukioja E (1980) Age-specific resistance in trees: defoliation of tamaracks (*Larix laricina*) by larch bud moth [*Zeiraphera improbana* (Lep. Tortricidae)] Rep Kevo subarc Res Stn 16: 49–57

Orton TJ (1984) Genetic variation in somatic tissue – method or madness? Adv Plant Pathol 2: 153–189

Paige K, Whitham TG, Keim P, Lark KG (1989) The use of restriction fragment length polymorphism to study the ecology and evolutionary biology of aphid-plant interactions. In: Campbell RK, Eikenbary RD (eds) Aphid-plant genotype interactions. Elsevier, Amsterdam

Parsons TJ, Bradshaw HD, Gordon MP (1989) Systematic accumulation of specific mRNAs in response to wounding in poplar trees. Proc Natl Acad Sci USA 86: 7895–7899

Perttu KL (ed) (1984) Ecology and management of forest biomass production systems. Swed Univ Agric Sci Rapp 15, Uppsala

Piccarolo G (1949) Experiences faites dans la culture du peuplier en Italie. CR Congr Int Un For Res Org, Zurich, 1948, pp 97–140

Randall WK, Cooper DT (1973) Predicted genotypic gain from cottonwood clonal tests. Silv Genet 22: 165–167

Ranney JW, Wright LL, Layton PA (1987) Hardwood energy crops: the technology of intensive culture. J For 85: 17–28

Rasmusson DC (1987) An evaluation of ideotype breeding. Crop Sci 27: 1140–1146

Rauscher HM, Isebrands JG, Crow TR, Dickson RE, Dickmann DI, Michael DA (1988) Simulating the influence of temperature and light on the growth of juvenile poplars in their establishment year. In: Proc IUFRO Forest growth modeling Conf, Minneapolis MN, 24–18 Aug 1987, pp 331–339

Rauscher HM, Isebrands JG, Dickson RE, Dickmann DI, Crow TR, Michael DA (1989) ECOPHYS: an ecophysiological growth process model of poplar. Tree physiol 7: 255–281

Rood SB, Daicos G, Blake TJ (1984) Giberellic acid induced growth acceleration in *Populus* hybrids. Can J For Res 14: 850–854

Rood SB, Buzzell RI, Mander LN, Pearce D, Pharis RP (1988) Giberrellins: a phytohormonal basis for heterosis in maize. Science 241: 1216–1218

Scarascia-Mugnozza G, Hinckley TM, Stettler RF (1986) Evidence for non-stomatal inhibition of net photosynthesis in rapidly dehydrated shoots of *Populus*. Can J For Res 16: 1371–1375

Schneider SH (1989) The greenhouse effect: science and policy. Science 243: 771–780

Schreiner EJ (1974) Poplars can be bred to order for mini-rotation fiber, timber, and veneer production and for amenity plantings. In: Proc 21st NE Forest tree improvement conf 21, pp 85–96

Schulte PJ, Hinckley TM, Stettler RF (1987) Stomatal responses of *Populus* to leaf water potential. Can J Bot 65: 255–260

Shirley DM (1988) Herbicide studies with *Populus trichocarpa* (T&G) and *P. trichocarpa* × *P. deltoides* hybrids in the Pacific Northwest. MS Thesis, Univ Washington, Seattle

Sinha SK, Khanna R (1975) Physiological, biochemical and genetic basis of heterosis. Adv Agron 27: 123–174

Smit B (1988) Selection of flood-resistant and susceptible seedlings of *Populus trichocarpa* (T&G). Can J For Res 18: 271–275

Smit B, Stachowiak M (1988) Effects of hypoxia and elevated carbon dioxide concentration on water flux through *Populus* roots. Tree Physiol 4: 153–165

Spears J, Ayensu ES (1985) Resources, development and the new century: Foresty. In: Repetto R (ed) The global possible. Yale Univ Press, New Haven London, pp 299–335

Steinbeck K, McAlpine RG, May JT (1972) Short rotation culture of sycamore: a status report. J For 70: 210–214

Stettler RF (1968) Irradiated mentor pollen: its use in remote hybridization of black cottonwood. Nature (London) 219: 746–747

Stettler RF, Koster R, Steenackers V (1980) Interspecific crossability studies in poplars. Theor Appl Genet 58: 273–282

Stettler RF, Fenn RC, Heilman PE, Stanton BJ (1988) *Populus trichocarpa* × *P. deltoides* hybrids for short rotation culture: variation patterns and four-year field performance. Can J For Res 18: 745–753

Sutherland WJ, Stillman RA (1988) The foraging tactics of plants. Oikos 52: (239–244).

Taylor HM (ed) (1987) Minirhizotron observation tubes: methods and applications for measuring rhizosphere dynamics. Am Soc Agron Spec Publ 50: 143 pp

Teissier du Cros E (1984) Breeding strategies with poplars in Europe. For Ecol Manag 8: 23–39

Templeton JK (1969) Where lies the yield summit for *Hevea*? Planters Bull 104: 220–225

Townsend AM (1975) Crossability patterns and morphological variation among elm species and hybrids. Silv Genet 24: 18–23

Watkinson AR (1988) On the growth and reproductive schedules of plants: a modular view-point. Oecol Plant 9:67–81

Watkinson AR, White J (1986) Some life-history consequences of modular construction in plants. Philos Trans R Soc London Ser B. 313:31–51

Watson DJ (1952) The physiological basis of variation in yield. Adv Agron 4:101–145

Weber JC, Stettler RF, Heilman PE (1985) Genetic variation and productivity of *Populus trichocarpa* and its hybrids. I. Morphology and phenology of 50 native clones. Can J For Res 15:376–383

White J (1979) The Plant as metapopulation. Annu Rev Ecol syst 10;109–145

Whitham TG (1981) Individual trees as heterogeneous environments: adaptation to herbivory or epigenetic noise? In: Denno RF, Dingle H (eds) Insect and life history patterns: habitat and geographic variation. Springer, Berlin Heidelberg New York, pp 9–27

Whitham TG, Williams AG, Robinson AM (1984) The variation principle: individual plants as temporal and spatial mosaics of resistance to rapidly evolving pests. In: Price PW, Slobodchikoff CN, Gaud WS (eds) A new ecology: novel approaches to interactive systems. John Wiley & Sons, New York, pp 15–51

Wilcox JR, Farmer RD (1968) Heritability and C-effects in early shoot growth of eastern cottonwood cuttings. Heredity 23:239–245

Williams AG, Whitham TG (1986) Premature leaf abscission: an induced plant defense against gall aphids. Ecology 67: 1619–1627

Wright JW (1976) Introduction to forest genetics. Academic Press, New York London

Zobel B, Talbert J (1984) Applied forest tree improvement. John Wiley & Sons, New York

7 Regeneration and Genetic Variation in Plant Tissue Cultures

Z. Chen[1] and M.R. Ahuja[2]

1 Introduction

Problems related to regeneration and genetic variation in tissue culture are commonly matters of interest and concern to plant geneticists and breeders. Successful application of in vitro technology to clonal forestry depends on the inducibility of growth and differentiation in tissues from woody plants and the regeneration of true-to-type viable plants in selected genotypes. Great differences exist in organogenesis, embryogenesis, and regeneration of plantlets among plant species, varieties, and even individuals of the same variety. In woody plants some materials are more easy to regenerate in vitro than others. A strong genetic component has been implicated in the regeneration capacity of tree species (Ahuja 1983; Ahuja et al. 1988). In addition to genetic differences in the morphogenetic potential between tree species, or within a species, maturation seems to play a crucial role in the regeneration process in vitro. In general, juvenile tissues are more responsive to in vitro conditions; whereas tissues from mature trees of most forest tree species are still very difficult to propagate by in vitro technology. The aging-related micropropagation problems are more pronounced in conifers than in hardwoods. It would appear that in addition to genetic control of regeneration, the stage of maturation also affects the regeneration potential of tree species in vitro. The mechanisms of maturation and rejuvenation in woody plants are discussed by Bonga and von Aderkas (Chap. 12, Vol. 1), and Hutchison and Greenwood (Chap. 3, Vol. 1).

Plants regenerated from somatic cell cultures may exhibit variation, which may have a genetic or a nongenetic basis. Genetic variation that occurs in plant tissue cultures has been termed "somaclonal variation" (Larkin and Scowcroft 1981). For obtaining true-to-type propagules from a selected genotype, somaclonal variation is undesirable. On the other hand, somaclonal variation offers prospects for the recovery of useful mutants in tissue cultures. The variation rate and spectrum are dependent on the genotype, type of culture (for example, organ or callus culture), and the environmental conditions. Variation may occur in chromosomes (number or structure), nuclear genes, or mitochondrial genes. Understanding the causes of variation will enable us to control it, to obtain true-

[1]Institute of Genetics, Chinese Academy of Sciences, Beijing, 100012, China
[2]Institute of Forest Genetics, Federal Research Centre for Forestry and Forest Products, Sieker Landstrasse 2, W-2070 Grosshansdorf, FRG

Clonal Forestry I, Genetics and Biotechnology
Ed. by M.R. Ahuja and W.J. Libby

to-type propagules, or to induce a specific type of variation for plant improvement. The current status of regeneration and genetic variation in woody plant cultures is reviewed in this chapter.

2 Inheritance of Morphogenetic Potentiality Under in Vitro Culture

2.1 Differences in Morphogenetic Potentiality of Cultured Materials

Differences in the degree of morphogenetic potential of cultured materials have been described in woody plants. For instance, among 100 species of *Eucalyptus* with economic importance, more than 40 species are extensively cultivated. However, plants can be regenerated in somatic cell cultures from only about ten species (Ouyang and Pong 1986). In *Hevea* more than 80 varieties and hybrids were selected for anther culture, but embryoids and plantlets could be obtained from only 15 of these after over 10 years of hard work. In anther culture of *Malus*, under the same culture conditions, the embryoid induction frequency of the variety Delicious reached 1.9–2.5%, while that of the variety Rainier was only 0.3–1.2%, and the variety Ralls gave no embryoids (Xua and Niu 1986). A large body of data from different woody plants indicates that the differences in morphogenetic potentiality of cultured materials are determined largely by their genotype. Micropropagation studies involving more than 100 trees of European aspen (*Populus tremula*) and hybrid aspen (*Populus tremula* × *P. tremuloides*) have shown that bud tissues from about 50% of the trees were prone to regeneration (Ahuja 1983, 1984, 1986, 1987a; Ahuja et al. 1988). Some varieties expressed different requirements in culture conditions, and their morphogenesis could be achieved by improving culture techniques through repeated experiments, while others failed to respond even after many attempts. Regeneration can be achieved easily in some families, such as the Solanaceae and Salicaceae. The genotypic difference of morphogenetic potentiality exists not only among families, genera, and species, but also among individuals of the same variety. Because of this difference, biologists spend much time and effort in seeking the optimal culture conditions. Often the materials selected by breeders for obtaining a new variety are not easily regenerable by somatic embryogenesis. On the other hand, the materials which are easily inducible may be least valuable to the breeder. Therefore, the genotypic difference in morphogenetic potentiality has aroused much attention. (Raquin 1982; Wu and Chen 1987). Further understanding of and the solution to this problem will greatly accelerate the application of tissue culture in crop improvement.

2.2 Genetics of Morphogenetic Potentiality

Few studies have been done on the inheritance of morphogenetic potentiality in woody plants, but those on herbaceous plants have led to different conclusions. Firstly, studies on rice and wheat revealed that androgenetic potentiality was determined by nuclear inheritance, but not by a single gene. When a parental plant with good androgenetic potentiality was crossed with a poor one, the androgenetic potentiality of the F_1 plants was mostly between that of the two parents (Bullock et al. 1982; Wu and Chen 1987). Secondly, work on maize indicated that maize somatic embryogenetic ability was controlled to a great extent by a cytoplasmic gene. The embryogenetic ability of F_1 plants from every cross was almost the same as that of the mother plant, and no Mendelian segregation was observed in the F_2 generation (Herman 1987a). A third argument was put forward by Raquin et al. (1982), working on anther culture of *Petunia*. They considered that although pollen embryogenetic ability was determined by a single nuclear gene, and the yield of pollen embryoids was always related to heterozygosity of donor plant, the embryoid yield of F_1 hybrids, as a result of hybrid vigor, was often remarkably higher than that of their parents, and no intermediate case was observed. In rubber, pollen embryoids could be induced from the two varieties PR107 and PB86, but the frequency was very low, only 0.3–5% (number of embryoids produced/number of anthers inoculated). However, the new variety Haikeng 2, selected from their F_1 hybrids, produced 25–140 embryoids from 100 anthers inoculated, which is significantly higher than their parents (Chen et al. 1982). Using some interspecific hybrids of *Populus* for anther culture, the frequency of both callus induction and plantlet regeneration was also considerably high (Lu and Lin 1986). But not all the hybrids showed superiority in embryogenesis. For example, in shoot tip culture of 19 grape materials, one hybrid failed to develop buds (Cao 1986).

In summary, the inheritance of morphogenetic potentiality may be very complicated. Different crops have different morphogenetic patterns. Understanding the causes of differences in the morphogenetic potentiality would be valuable for achieving improvement of this character in the woody plant species.

2.3 Biochemical Basis of Morphogenesis

A series of biochemical changes takes place in a cell during morphogenesis. When cultured carrot cells were transferred from a 2,4-D medium to one without it, after 3 days the following changes were monitored by radioactive labeling: (1) Chromatin and nuclear material replicated rapidly; (2) the content of RNA and protein increased quickly; and (3) the synthesis of poly mRNA speeded up, which is related to the biochemical reaction in embryogenesis (Namura and Komamine 1985). In coniferous trees, it was found that embryogenic callus (EC) of *Picea* and *Pinus* were similar, but distinctly different from nonembryogenic callus (NEC) in terms of ethylene evolution rates (EC low and NEC high), level of total reductants, including glutathione (EC low and NEC high), and protein synthesis rates (EC high and NEC low) (Becwar et al. 1988).

Changes in polyamine metabolism are important for embryogenesis. There is a great difference in polyamine metabolism between embryogenetic and nonembryogenetic cells. For instance, when competent cells were transferred into the medium without 2.4-D, the concentration of putrescine almost doubled within 24 h. Additionally, the concentration of spermidine and spermine and the activity of their corresponding synthetases in embryogenetic cells were also much higher than in nonembryogenetic cells, while at the same time they contained more microtubulins (Sung 1985).

Studies of an embryogenic cell line from citrus callus showed that oxidase activity increased markedly during embryogenesis, and new isozyme bands appeared in its electrophoretogram.

Using radioautography and polyprophlene-dimensional electrophoresis. Sung and Okimoto (1981) found the existence of two specific proteins during embryogenetic process of carrot suspension cells, which they called embryo proteins E_1 and E_2. Embryoids emerged only on the 6th day after the production of these proteins (Sung and Okimoto 1981). Two-dimensional polyacrylamide gel electrophoresis analysis of proteins extracted from embryogenic and nonembryogenic carrot cultures showed that newly synthesized polypeptides labeled with ^{35}S methionine are represented in embryo cultures, but not in nonembryonic cultures. Of approximately 800 polypeptides resolved, less than 2%, or fewer than 20, are either embryo-specific or embryo-modulated (Thomas and Wilde 1986). RNA expression patterns in EC and NEC tissues were examined by hybridizing randomly selected cDNA recombinants with cDNA probes representing different mRNA populations (Thomas and Wilde 1986). These facts show that many investigators are exploring the molecular genetic basis of morphogenesis, and are trying to find out if there is a set of genes controlling morphogenesis. If this mystery is resolved, control of plant morphogenesis is not far off.

2.4 Improvement of Morphogenetic Potentiality

Much work has been done with certain crops on the improvement of morphogenetic potential. Impressive achievements have been made by using breeding techniques and developing and improving in vitro culture techniques.

2.4.1 Sexual Hybridization in Material with Readily Inducible Morphogenesis

Good morphogenetic potential can be transmitted from a parental plant to its F_1 hybrids. Thus, easily inducible materials can be achieved from recalcitrant ones through hybridization, a method that has been widely used in herbaceous plants. However, in view of the longer life cycle in woody plants, and the difficulties in the identification of hybrids in early stages, it is better to use the available superior hybrids obtained from a cross between readily inducible varieties, or at least with one readily inducible parent (Chen et al. 1986).

2.4.2 Selection of Explants for Regeneration

Song et al. (1987) germinated rape seeds and selected vigorously growing seedlings for callus induction from hypocotyl, and then selected the cell which could produce shoots for further culture. A somaclonal variant with high regenerative ability was obtained through repeated selection during subculture.

Similar examples can also be cited for woody plants. A somatic clone on which it is relatively easy to initiate embryogenesis was selected from *Eucalyptus* (Ouyang and Pong 1986). An embryogenetic cell clone with differentiation potential was also isolated by anther culture in litchi (Fu 1986), and in Chen's lab, somatic clones with high-frequency regenerating potential have been selected from pollen plants of the rubber tree. Differences in inducibility can be very great between trees of different ages, and different organs and locations on one and the same tree. In most cases, juvenile materials are more easily regenerated than mature explants. Floral organs are also better for induction than other organs.

2.4.3 Improvement of Culture Conditions

Improvement in culture conditions to meet the requirements of different genotypes might be an effective method of increasing inducible genotypes. For example, different hormone levels are required in the anther culture of different genotypes of *Malus pumila*. When anthers of the variety Delicious are inoculated in MS medium supplemented with 0.4 $mg\,l^{-1}$ 2, 4-D, 0.2 $mg\,l^{-1}$ kinetin and 4 $mg\,l^{-1}$ IAA, embryoids can be regenerated every time, but it is not adaptable to other varieties. If these hormones are removed, and instead an appropriate amount of BA and GA is added, successful regeneration occurs in the varieties Ralls, Jonathan, American Summer, and Rainier (Xua and Niu. 1986).

2.5 Pretreatment of Explants with Physical and Chemical Factors

It is known that low temperature treatment of anthers of cereal crops can greatly increase pollen embryo frequency, and this was also reported in woody plants. For example, in anther culture of *Lycium chinense*, the anthers were pretreated with low temperature for 120 h, and the induction frequency of pollen embryos was increased threefold. In anther culture and somatic cell suspension culture of grape, a period of dark treatment remarkably increased the frequency of embryoid regeneration. It was observed that anthers from seven grape varieties could not produce embryoids under light culture conditions, whereas an embryoid frequency of 0.6–9% was obtained when a period of dark culture was provided (Gu et al. 1984). In recent years, some papers reported that increased osmotic pressure in the medium was very effective in promoting the initiation of embryogenesis. When suspension cultures of *Carica papaya* were pretreated with 0.18 M NaCl or 0.3–0.45 M mannitol, and then transferred onto a medium with low osmotic pressure for subculture, the induction frequency of somatic embryoids could be doubled (Herman 1987b). It was considered that the addition of NaCl or mannitol can synthesize certain proteins, and the changes of protein synthesis in cultures under high osmotic pressure may help embryo formation. At the same time, high osmotic pressure can raise the polyamine concentration, which may also favor the occurrence of embryogenesis.

It was reported recently that electroporation stimulates embryogenesis or shoot differentiation. For example, in wheat, 1 or 2 μA electric current passing through the medium to the callus caused an increase in shoot differentiation up to four times higher than in the control (Williams and Maheswaran 1986). It was suggested that electric current may help to regulate the physiological polarity of cells (Herman 1987c).

3 Variation Under in Vitro Culture

The fact that in vitro culture can cause genetic variation in regenerated plants aroused much interest and discussion, and this topic has been frequently reviewed (Evans and Bravo 1986, Evans and Sharp 1986, Scowcroft et al. 1986). Some new varieties were selected from variants in annual crops (Evans and Sharp 1983). However, since the reproductive cycle of woody plants is very long and the influencing factors are complicated, it will take a long time to obtain results from studies on genetic variations in trees, so that there are very few reports on genetic variations in woody plants (Ahuja 1987b). Nevertheless, the achievements made in herbaceous plants can be used as reference for woody plants.

3.1 Concepts

Larkin and Scowcroft (1981) detailed various sources of variation in regenerated plants. They considered that plants derived from any form of cell and tissue culture should be termed somaclones, and genetic variations produced in the clones should be called somaclonal variations. This term was accepted by many investigators. However, the situation in the culture of male and female gametes or other haploid cells is different from that in diploids. Segregation occurs in the gametes because of genetic recombination through synapsis and crossing over during meiotic division, which is one of the variation sources. Furthermore, variation will also occur during in vitro culture. Therefore, haploid materials of culture have two sources of variation, and it is hard to distinguish between them. Some investigators termed the variation in haploid cultures gametoclonal variation in order to distinguish it from somaclonal variations, which originated from sporophytic materials during in vitro culture (Evans and Sharp 1986). Both somaclonal and gametoclonal variations refer to genetic variations, such as variations in chromosome number, chromosome aberrations, nuclear or cytoplasmic gene mutations. During culture various environmental factors may lead not only to genetic variation, but can also influence gene expression, or cause variation in certain interactions between genetic materials. For instance, changes in gene function (switch on or off), synthesis of certain proteins, cell differentiation, or protein distribution can cause a phenotypic variation which is called epigenetic phenomena. Although this kind of variation cannot always be transmitted to offspring via sexual reproduction, it can be maintained by asexual propagation.

3.2 Causes of Variations

3.2.1 Cytogenetic Variation

Cytological analyses of cells from tissue culture have shown frequent chromosomal instability. Apart from chromosome number changes, such as aneuploidy, polyploidy, monosomy, structural alterations such as translocation, inversion, and centric fusions have been observed at the morphological level. Cytogenetic variation routinely occurs at frequencies above 10% in many plants derived from culture. Cells with aberrant karyology often occur in a major fraction of callus and suspension cultures. (Gould 1986). Since the growth cycle of woody plants is very long, many variations can be accumulated in their somatic cells. For instance, poplar trees are mostly diploid ($2n = 2x = 38$) in nature, while natural triploid or tetraploid trees still exist. But much more variation can be induced by abnormal culture conditions, especially in plantlets developed via callus formation. In the last 10 years, much work has been done on in vitro induction of haploids in trees, and it was observed that the pollen plants derived in the anther culture of several tree species were mixoploid (Chen et al. 1986). In the different developmental stages the chromosome number varies; for instance, between the development of callus or embryoid to the formation of the plantlet. For example, the observation of a root-tip cell of a pollen-derived grape plant indicated that although the cells were mostly haploid (i.e., 19 chromosomes), there were also aneuploid, diploid, as well as tetraploid cells (Zou and Li 1981). Pollen-derived plants of *Proncirus trifoliata* were also mostly mixoploids. In the root-tip cells, besides cells with chromosome number $x = 9$, others had chromosome numbers of $x + 1$, $x + 2$ and $2x$ (Hidaka et al. 1979). Cytological observation of embryoids and plantlets of the rubber tree showed that although most of the cells were haploid, there were also mixoploids, and the chromosome number was often a multiple of basic $x = 9$, i.e., $2x = 18$, (gametic chromosome number), $3x = 27$, $4x = 36$ and $5x = 45$ (Cin et al. 1981). These studies show that a mixoploid chromosome number commonly exists in woody pollen plants before transplantation.

In woody plants, when pollen-derived plantlets survived after transplanting, the chromosome number of the somatic cells continued to change. For instance, on average, haploid cells of pollen-derived poplar plants at the time of transplanting accounted for 58 %, diploid cells 20.1 %, and the rest were aneuploids. After 1 year, diploid cells reached 58 %, haploid cells were reduced to 30 %, tetraploid cells accounted for 8 %, and the aneuploid cells became fewer. The plant thus developed into a mixoploid, containing mainly diploid cells. In addition, an aneuploid tree containing mostly aneuploid cells was observed (Lu et al. 1977). Chromosome counts in leaves of a transplanted pollen-derived rubber tree plant showed that the chromosome numbers of its cells were mostly between 18–27 right after transplantation, but more than 60 % cells contained 28–36 chromosomes when the plant had reached 160 cm in height. These results showed that the gradual diploidization of somatic cells was observed in more than one tree species. In pollen-derived plants of the rubber tree, there was one plant which contained mainly aneuploid cells (Chen et al. 1982). Its shoot tip grew slowly, and although the lateral shoots developed, their tip buds grew slowly too, so that secondary lateral shoots also appeared later, and its multi-branched growth was not vigorous.

In different pollen plants of the same species, the speed of diploidization also differs. Some trees develop into mixoploids chiefly with diploid cells in the 2nd year, while others become diploidized from mixoploids after 3–4 years. The complete natural chromosome doubling of poplar needs 8 years (Lu and Liu 1986).

Chromosomal mixoploidy was also exhibited in endosperm cultures. It was reported that chromosomal diversity existed even in endosperm cells of trees growing in the field. Mixoploid plants regenerated from endosperm cultures of *Malus pumila*, *Prunus*, and *Actinidia chinensis* (Lu and Liu 1980; Gui et al. 1982; Mu 1984) consisted of mainly polyploid and some triploid cells. This might be the result of an interaction between original chromosome number variations before culture and the variations during culture. Up to now, plants have been regenerated from endosperm cultures of many tree species, such as *Citrus grandis* and *Eriobotrya japonica* (Wang and Zhang 1978; Chen et al. 1983). These authors reported that the plantlets were triploid. However, no report dealt in detail with the characteristics of chromosome variation.

Many investigators studied chromosome variations which occurred during in vitro culture. Some investigators proposed that the chromosome number of haploid cultures changes more readily than that of diploids (Sacristan 1971). The cause of variation in chromosome number in embryoids calli, and plantlets derived from pollen grains by anther culture was also analyzed, and it was considered that the chromosome number variation might be related to endomitosis, nuclear fusion, and multipolar mitosis (McCorb 1978). The chromosome number of pollen-derived embryoids and plants of the rubber tree usually increased at a basic rate of 9. Only a few cells had a chromosome number which was not a multiple of 9. The rubber tree is an amphidiploid plant, with $x = 9$ and $n = 18$. It was observed that the chromosome number in some cells was less than 18, and it was suggested that the chromosome number decreased during the development of the embryoids and plantlets, such cells accounting for 4–5 % of the total observed phases. Poplar is a diploid tree with 38 chromosomes and $n = 19$. Although the chromosome number of cells in pollen embryoids and plantlets is obviously not related to the multiple of any basic number, cells with chromosome number less than 19 were also often observed. Sometimes, somatic cells undergoing multipolar mitosis were observed during cytological observation of embryoids and plantlets of the rubber tree. For example, when a cell with 18 chromosomes divided, there were two poles on one side of the equatorial plate, one with 18 chromosomes and the other with 9, while on the other side, there was another pole with 9 chromosomes. As a result of such division, one cell with 27 chromosomes and another cell with 9 chromosomes were produced. The cells with 27 chromosomes could be observed more often than those with 9 chromosomes, possibly because they multiplied more quickly. If cells with 27 chromosomes underwent multipolar mitosis again, two daughter cells were produced with 36 and 18 chromosomes, or with 45 and 9 chromosomes, respectively. In fact, cells with 36 or 45 chromosomes were observed (Chen et al. 1986). Multipolar mitosis was also observed in poplar cultures (Lu, pers. commun.). At the same time, Lu proposed that the chromosome doubling of haploid cells might be the result of endomitosis. However, Lu also found cells with chromosome number less than n in pollen-derived plants of poplar. Among 17 poplar plants of 4–5 years, cells containing 6–14 chromosomes accounted for 9.7% (Lu and Lin. 1986). This phenomenon can only be interpreted by multipolar mitosis.

From these facts, we consider that multipolar mitosis may be a major mechanism in the occurrence of cells with a chromosome number more or less than n in pollen-derived embryoids and plantlets of trees.

As already stated, the rubber tree is an amphidiploid species. If a haploid cell ($2x = 18$) contains two sets of chromosomes, i.e., $x = A$ and $x = B$, multipolar mitosis will probably lead to the formation of three types of normal diploid daughter cells, i.e., AAAB, AABB, and ABBB. Only the cells with AABB chromosome are really diploid pure line.

Two conclusions can be drawn from the foregoing analysis: firstly, that not all the plants obtained via multipolar mitosis are pure lines. Therefore at the multicellular or embryoid stages, artificial chromosome doubling should be initiated with colchicine, without waiting for the long process of natural doubling. Secondly, the pollen-derived embryoids or plantlets of trees are all mixoploids, while under natural conditions, diploid cells often multiply quickly and gradually replace the cells with different chromosome number, so that buds or shoots usually cannot develop from such cells. However, when shoot tip, bud, tender leaf, meristem, etc. from pollen-derived embryoids or young plants are used for isolated cell culture, the single cells with chromosome number variation will then form callus or embryoid, from which, after chromosome doubling, homozygous plants with specific chromosome numbers can be obtained. This is an area worth investigating in the chromosome engineering of woody plants.

3.2.2 Somaclonal Mutants and Other Variations

Single gene mutations have been recovered in several species of herbaceous plants such as maize ADH somaclonal mutant (Scowcroft et al. 1986), tomato somaclonal mutant including dominant, semi-dominant, and recessive nuclear mutations (Evans and Sharp. 1986). There have been some reports on somaclonal variation in woody plants, such as blackberry somaclonal variant, with altered characteristics of leaf size and pattern, fruit size, and chlorophyll pigment (Swartz et al. 1983), but it is difficult to obtain results of genetic analyses in trees. There have also been some reports on gametoclonal variation in woody plants.

It was shown that there is great diversity in the morphological features of leaves, growth vigor, circumference, etc. among pollen-derived trees originating from the same donor tree. For example, the highest 1-year-old pollen tree of poplar was 80 cm, the smallest only 20 cm i.e., one fourth that of the highest one (Chen et al. 1981). The height of 1-year-old pollen-derived rubber trees ranged from 19.5 cm to 131 cm.

The investigation also showed that poplar trees derived from pollen from the identical donor were morphologically diversified. A total number of 101 pollen trees of *Populus simonii* Carr. × *P. nigra* L. could be divided morphologically into five types: donor type (49 trees), *P. nigra* L. type (2 trees), *P. simonii* Carr.

type (2 trees). donor-like type with fine modification (46 trees) (Wu et al. 1981). These observations indicated that the characters of the pollen tree *P. simonii* Carr. × *P. nigra* L. were indeed segregating. Moreover, the phenomenon of character segregation was also observed among pollen trees H_1 of grape and Chinese crabapple, and in addition, some wild characters also appeared (Wu et al., pers. commun.).

It is worth noting that there may be some after-effects from hormones and other culture factors on pollen plant H_1. Wu et al. (1981) observed that among pollen plants of poplar, part of the morphological leaf-shape variations disappeared gradually, while some variations were still retained in successive generations (Wu et al. 1981). A similar phenomenon was also observed among pollen plants in the rubber tree.

The difference in ploidy level among pollen trees could lead to a difference in morphology and growth rate in the pollen trees. In pollen plants of poplar and rubber, the majority of cells had a haploid chromosome number, and these cells grew more slowly than those with diploid chromosome number. Therefore it is necessary to take the ploidy of the plant into account when comparing growth vigor and morphological characters, especially the quantitative characters of pollen trees.

Nevertheless, the morphological differences may be due to the techniques of in vitro manipulation, such as stage of pollen plantlet at transplantation, soil conditions, microatmosphere, duration of potting, etc.

In comparison to seedlings, the pollen plantlets usually lack strong roots. Therefore, the recovery time needed for transplanted pollen plants is much longer and their growth rate is slower (10–12 months for rubber) than that of a seedling derived from the same inflorescence, and it is difficult to choose a suitable control for comparison with pollen trees (Chen et al. 1986). There is thus an urgent need to design an accurate methodology, including a suitable control, for the study of transplanted pollen trees.

3.2.3 Investigation and Analysis of Genetic Variation

A comparison of clones of S_1 (plantlets obtained from somatic cells transplanted in the field) and H_1 (plantlets derived from pollen and transplanted in the field) with those of donor trees appears to be an ideal method for evaluating regenerated trees, because this method can minimize the lack of uniformity in cultivation techniques and eliminate the after-effects of culture conditions on regenerated plants, and the difference in recovery time between the regenerated plantlets and seedlings. This approach was used in investigations on poplar and rubber pollen trees (Chen et al. 1981; Wu et al. 1981; Xe et al., pers. commun.). Four vigorously growing pollen trees H_1 of rubber were selected on the basis of periodical measurements of their height in 1980 (Xu et al. pers. commun.; Table 1), and later cloned by budding simultaneously with their donor trees. The mean girth measured 4 years after budding showed that the growth vigor of four clones of the pollen tree had greatly surpassed that of the donor clone ($P > 0.01$). Among clones of pollen tree, the mean girth of clone 9 was significantly greater than those of others ($P > 0.05$, Table 1).

Table 1. The girth of pollen tree clones and their donor clone. (Zhenghua Chen et al., unpubl.)

Clone no.	No. of plants investigated	Girth (mm mean ± SE) (4-year-old)
3	24	238.6 ± 4.02
6	30	223.1 ± 3.42
9	39	254.4 ± 3.92
15	18	213.5 ± 1.48
Donor	27	203.4 ± 4.23

Table 2. Latex and dry rubber yield of pollen tree No. 1[a] and its donor clone. (Xe Xu, pers. commun.)

Month		4	5	6	7	8	9	10	11	12	Total	%
Latex yield	No. 1	4.2	23.8	34.3	86.1	30.6	88.0	118.5	59.0	96.2	540.7	122
(ml/tipping)	Donor	2.4	33.3	26.2	83.3	33.3	59.3	78.6	56.3	71.9	444.6	100
Dry rubber	No. 1	1.7	9.7	15.3	32.5	11.8	35.2	39.9	22.4	38.0	206.5	129
yield (g/tipping)	Donor	1.0	13.7	11.1	28.4	12.9	22.8	25.5	19.9	24.3	159.6	100

[a]The number of random samples was 36 trees per treatment (7-year-old trees).

The latex and dry rubber yield of a selected clone (pollen tree No. 1) and the donor clone were compared (Table 2). The results showed that the latex and rubber yield of pollen tree No. 1 were significantly higher than those of the donor clone (22 and 29 % respectively). On the basis of growth and height, elite clones from pollen trees of poplar were also selected (Chen et al. 1981).

4 Obtaining True-to-Type Propagules and Control of Variation

The word true-to-type refers to the phenotypic characters of propagules/offspring which are very similar to their parents. If such a population comes from a superior donor, a high-yielding and good quality cultivar will be achieved, and additionally, field management can be easily handled. True-to-type propagules can be obtained by rapid clonal propagation employing tissue culture technology. In vitro clonal production systems in conifers (see Talbert et al., Chap. 11, Vol. 1), and hardwoods, for example *Populus* (Ahuja 1983, 1986, 1987a; Barocka et al. 1985; Ba and Guo 1986), *Eucalyptus* (Quyang and Pong 1986; McComb and Bennet 1986), and *Vitis* (Cao 1986).

4.1 Methods for Obtaining True-to-Type Plants

4.1.1 Use of Mature Material as Explant

It is desirable to vegetatively propagate superior trees by a rapid in vitro multiplication method, because the economic traits of these mature trees have been tested under field conditions. However, propagation of such material is much more difficult than of juvenile explants. Reliable techniques need to be developed for micropropagation of mature trees (Bonga 1982).

4.1.2 Use of the Right Culture System

If callus develops first and then buds differentiate from callus, variation occurs more frequently. It is better to induce bud proliferation on bud explants in order to achieve rapid propagation and thus avoid variation. In order to reduce variation, two aspects can be considered. Firstly, the culture conditions should be made as similar to the natural endogenous environment as possible. For instance, in order to promote cell proliferation and differentiation, the hormone levels should be reduced to a minimum limit within the effective range. Secondly, the number of cell divisions should be controlled. It has been suggested that in vitro culture in fact introduces a large number of extra cell divisions between the fertilized egg and the next gametogenetic cycle, thus increasing the probability of errors in mitosis. If the frequency of cell division can be controlled to a minimum required number, and at the same time, the plantlets are regenerated rapidly, then it might be possible to minimize variation in the in vitro-produced regenerants (Gould 1986). Production of genetically stable regenerants is a prerequisite for genetic engineering in trees and their subsequent clonal propagation.

References

Ahuja MR (1983) Somatic cell genetics and rapid clonal propagation of aspen. Silv Genet 32:131–135

Ahuja MR (1984) A commercially feasible micropropagation method for aspen. Silv Genet 33:174–176

Ahuja MR (1986) Aspen. In: Evans DE, Sharp WR, Ammirato PJ (eds) Handbook of plant cell culture, vol 4. Macmillan, New York, pp 626–651

Ahuja MR (1987a) In vitro propagation of poplar and aspen. In: Bonga JM, Durzan DJ (eds) Cell and tissue culture in forestry, vol 3. Nijhoff, Dordrecht, pp 207–223

Ahuja MR (1987b) Somaclonal variation. In: Bonga JM, Durzan DJ (eds) Cell and tissue culture in forestry, vol 1. Nijhoff, Dordrecht, pp 272–285

Ahuja MR, Krusche D, Melchior GH (1988) Determination of plantlet regeneration capacity of selected aspen clones in vitro. In: Ahuja MR (ed) Somatic cell genetics of woody plants. Kluwer, Dordrecht, pp 127–135

Ba YL, Guo JZ (1986) Propagation of *Populus* in vitro. In: Chen ZH (ed) Tissue culture of woody plants and its application. High Education Press of China, Beijing, pp 187–195

Barocka KH, Baus M, Lontke E, Sievert F (1985) Tissue culture as a tool for in vitro mass propagation of aspen. Z Pflanzenzücht 94:340–343

Becwar MR, Wann SR, Johnson MA, Verhagen SA, Feirer RP, Nagmani R (1988) Development and characterization of in vitro embryogenic systems in conifers. In: Ahuja MR (ed) Somatic cell genetics of woody plants. Kluwer, Dordrecht, pp 1–18

Bonga JM (1982) Vegetative propagation in relation to juvenility, maturation and rejuvenation. In: Bonga JM, Durzan DJ (eds) Tissue culture in forestry. Nijhoff, Dordrecht, pp 387–412

Bullock WP, Baenziger PS, Schaeffer GW, Bottino PJ (1982) Anther culture of wheat (*Triticum aestivum* L.) F_1's and their reciprocal crosses. Theor Appl Genet 62:155–159

Cao ZY (1986) In vitro propagation of grape. In: Chen ZH (ed) Tissue culture of woody plants and its application. High Education Press of China, Beijing, pp 328–344

Chen HD, Dong Y, Guo YF, Dong HH (1981) A study on the inheritance, selection and utilization of poplar pollen tree H_1. Yangshu Keji Tongxun 5:7–15 (in Chinese)

Chen ZG, Lin S, Lin Q (1983) A preliminary report on formation of plantlets from endosperm culture of loquat. J Fujian Agric Coll 12(4):343–346 (in Chinese with English Abstr)

Chen ZH (ed) (1986) The achievements and distinguishing features of haploid induction in woody plants. In: Tissue culture of woody plants and its application. High Education Press of China, Beijing, pp 75–91

Chen ZH, Qian CF, Cin M, Xu XE, Xiao YL (1982) Recent advances in anther culture of *Hevea brasiliensis* (Muell.-Arg). Theor Appl Genet 62:103–108

Chen ZH, Xu XE, Pang RS (1986) Anthet culture of *Hevea* and study on pollen tree. In: Chen ZH (ed) Tissue culture of woody plants and its application. High Education Press of China, Beijing, pp 481–500

Cin M, Chen ZH, Qian CF, Wang CH, He YT, Xiao YL (1981) Investigation of ploidy in the process of anther culture of *Hevea brasiliensis*. Acta Genet Sin 8(2):169–174 (in Chinese with English Abstr)

Evans DA, Bravo JE (1986) Phenotypic and genotypic stability of tissue culture plants. In: Zimmerman RH, Griesbach RJ, Hemmerschlag FA, Lawson RH (eds) Tissue culture as a plant production system for horticulture crops. Nijhoff, Dordrecht, pp 73–96

Evans DA, Sharp WR (1983) Single gene mutations in tomato plants regenerated from tissue culture. Science 221:949–951

Evans DA, Sharp WR (1986) Somaclonal and gametoclonal variation. In: Evans DA, Sharp WR, Ammirato PV (eds) Handbook of plant cell culture, vol 4, Macmillan, New York, pp 97–132

Fu LF (1986) Anther culture of litchi. In: Chen ZH (ed) Tissue culture of woody plants and its application. High Education Press of China, Beijing, pp 396–407

Gould AR (1986) Factors controlling generation of variability in vitro. In: Vasil IK (ed) Cell culture and somatic cell genetics of plants, vol 3, Academic Press, New York London, pp 549–567

Gu SR, Gui YL, Xu TY (1984) Effects of physical and chemical factors on induction frequency of the pollen plantlet and changes in starch formation in anthers. Acta Bot Sin 26(2):156–162 (in Chinese with English Abstr)

Gui YL, Mu XJ, Xu Ty (1982) Studies on morphological differentiation of *Actinidia chinesis* Planch in vitro. Acta Bot Sin 24, 3 (in Chinese with English Abstr)

Herman EB (1987a) Cytoplasmic control of regeneration. Agricell Rep 8(3):20

Herman EB (1987b) Osmotic stress induces embyogenesis. Agricell Rep 8(2):11

Herman EB (1987c) Electrical stimulation of protoplast embryogenesis. Agricell Rep 8(3):20

Hidaka T, Yamada Y, Shichijo T (1979) In vitro differentiation of haploid plants by anther culture in *Poncirus trifoliata* (L.) Raf Jap J Breed 29(3):248–254

Larkin PJ, Scowcroft WR (1981) Somaclonal variation – a novel source of variability from cell cultures for plant improvement. Theor Appl Genet 60:197–214

Lu SQ, Liu JQ (1980) Callus induction and embryoid formation in endosperm culture on *Prunis persica*. Acta Bot Sin 22(2):198–199

Lu ZH, Liu YX (1986) Anther culture of *Populus* and study on ploidy of pollen tree. In: Chen ZH (ed) Tissue culture of woody plants and its application. High Education Press of China, Beijing, pp 158–178

Lu ZH, Zhang XF, Liu YX, Zhang PG (1977) Observation on the chromosomes of somatic cells of poplar pollen plants. J NE For Inst 2:14–22 (in Chinese with English Abstr)

McComb JA Bennet IJ (1986) Eucalypts (*Eucalyptus* spp). In: Bajaj YPS (ed) Biotechnology in agriculture and Forestry, vol 1: Trees I. Springer, Berlin Heidelberg New York, pp 340–362

McCorb JA (1978) Variation in ploidy levels of plants derived from anther culture. In: Proc Symp on Plant tissue culture. Science Press, Beijing, pp 167–180

Mu SK (1984) Advances in endosperm culture of angiosperms. In: Plant cell and tissue culture. Contrib Inst Bot, Acad Sin China, pp 74–79

Namura K, Komamine A (1985) Physiological and biochemical aspect of somatic embryogenesis from single cells. In: Terzi M, Pitto L, Sung ZR (eds) Somatic embryogenesis. Incremento Produttinita Resorse Agricole, Rome, pp 1–5

Ouyang Q, Pong HZ (1986) Tissue culture of *Eucalyptus*. In: Chen ZH (ed) Tissue culture of woody plants and its application. High Education Press of China, Beijing, pp 196–213

Raquin C (1982) Control of embryo production and embryo quality in anther culture of *Petunia*. Theor Appl Genet 63:151–154

Sacristan MD (1971) Karyotype changes in callus culture from haploid and diploid plants of *Crepis capillars* (1.) Wallr. Chromosoma 33:273–283

Scowcroft WR, Brettel RIS, Ryan SA, Davis PA, Pallotta MA (1986) Somaclonal variation and genomic flux. In: Green CE, Somers DA, Hackett WP, Biesboer DD (eds) Plant tissue and cell culture, plant biology, vol 3. Liss, New York, pp 275–288

Song YH, Chen ZH, Li WB, Han BW, Lu F (1987) Plant regeneration from cell suspension culture and establishment of a somaclone with high ability of regeneration. Acta Genet Sin 24:243–248

Sung ZR (1985) Developmental states of embryogenic culture. In: Terzi M, Pitto L, Sung ZR (eds) Somatic embryogenesis. Incremento Porduttinita Resorse Agricole, Rome, pp 117–121

Sung ZR, Okimoto R (1981) Embryoic proteins in somatic embryos of carrot. Proc Natl Acad Sci USA 78:3683–3687

Swartz HJ, Galletta GJ, Zimmerman RH (1983) Field preformance and phenotypic stability of tissue-culture-propagated thorn blackberries. J Am Soc Hortic Sci 108:285–290

Thomas TL, Wilde D (1986) Analysis of carrot embryo gene expression program. In: Green CE, Somers DA, Hackett WP, Biesboer DD (eds) Plant tissue and cell culture, plant biology, vol 3. Liss, New York, pp 83–93

Wang TY, Zhang JR (1978) Triploid citrus plantlets from endosperm culture. Sintia Sin 4:452–455 (in Chinese)

Williams EG, Maheswaran G (1986) Somatic embryogenesis: factors influencing coordinated behaviour of cells as an embryogenic group. Ann Bot (London) 57:443–462

Wu CY, Chen Y (1987) A study on the genotypical differences in anther culture of keng rice (*Oryza sativa* subsp. Keng) Acta Genet Sin 14:168–174 (in Chinese with English Abstr)

Wu KX, Xu MZ, Ma HP, Li W, Zhu WH (1981) Preliminary analysis on characteristic of pollen plants of *Populus simonii* Carr. Linye Keji 2(5):5–6 (in Chinese)

Xua GR, Niu JZ (1986) Anther culture of apple. In: Chen ZH (ed) Tissue culture in woody plants and its application. High Education Press of China, Beijing, pp 241–253

Zou CJ, Li PF (1981) Induction of pollen plants of *Vitis vinifera* L. Acta Bot Sin 23:79–81

8 Clone Identification

W.M. Cheliak[1]

1 Introduction

Conservation, selection and breeding, and propagation constitute the three basic phases of a tree improvement program. To be effective, each of these depends upon accurate pedigree and germplasm documentation. Conceptually, germplasm and their associated pedigrees can be considered genetic resources to be used and managed to improve a domesticated population. These resources typically include various types of open- and control-pollinated families, as well as clones.

Continued progress in improvement programs is directly linked to breeding, testing, and selection. In the past, it was assumed that the identity of germplasm was known without error; but two types of error can occur. First, there can be error in physically identifying the plant. Second, mistakes can be made in the process of recording pedigree, after identifying the plant. Virtually every phase of tree improvement programs critically analyzed so far has carried errors in identity or pedigree (Hunter 1977; Adams 1983; Cheliak and Pitel 1984a; Friedman and Adams 1985; Strauss 1985; Cheliak et al. 1987; Pitel et al. 1987; Adams et al. 1988; Kossuth et al. 1988). Such errors can arise from a variety of sources, ranging from simple labeling and data entry mistakes to pollen contamination during controlled breeding, and vandalism.

It is thus desirable, at least in principle, to periodically certify all germplasm used in crop-improvement programs. Certification, in this context, means confirmation or rejection of the recorded identity of germplasm. In practice, the need for certification will depend upon the type and stage of an improvement program. Bulked seed collections from natural stands need only be certified at the provenance level, or perhaps the species level in areas of natural hybrid swarms (Szmidt et al. 1988). As the degree of genetic management and investments in domesticated populations and germplasm increases, so does the importance and specificity of certification. The objectives of a tree improvement program will also influence the need for certification. A population improvement program based on general combining ability is, in theory, less seriously affected by errors in pedigree of identity than a comparable program based on specific combining ability. Errors in pedigree or identity are especially damaging for clonal programs.

[1]Biorational Control Agents Program, Forest Pest Management Institute, 1219 Queen Street East, P.O. Box 490, Sault Ste. Marie, Ontario P6A 5M7, Canada

Clonal Forestry I, Genetics and Biotechnology
Ed. by M.R. Ahuja and W.J. Libby

Three types of "plants" can be recognized in a clonal forestry program and include the foundation stock, expansion stock, and test plants (Libby 1983). The foundation stock is usually the ortet and some small number of ramets and are the ultimate source of all members of a clone. The expansion stock usually consists of the immediate donors of propagules used in production planting. Test plants are the propagules entered in various levels of clonal trials and, ideally, derive from the foundation stock, not expansion stock. Thus, an error in the identity of a single foundation plant affects a large fraction of its "clone", even all of it if there is only one foundation plant. Wrongly identifying a single expansion plant affects that portion of the production run produced from it. The mistaken identity of a single test plant simply puts some unwanted bias into a well-designed experiment. Identity error in a single production-propagule, even if it is not further propagated, can cause some annoying variation in the production plantation; on the other hand, if the production plantation is a WIMP, there would be no problem. An error made in the identification of a single plant during serial propagation will be spread to all plants in that series, which is one argument against serial propagation and in favor of stable, mapped donor-hedges (Libby 1983). Consequently, clone identity should be safeguarded by certifying clones used for experimentation or in producing the next generation. Furthermore, in some jurisdictions, laws or regulations exist that prescribe the number of clones to be legally deployed in a specified area (Muhs, Chap. 13, Vol. 2). In such situations the ability to verify identity and number in plantations may be important when prosecuting a violation or defending against one. We now turn our attention to ways of certifying germplasm and the concept of including or excluding a stated member, or ramet, from a particular clone.

2 Clonal Inclusion and Exclusion

What problems do we face when trying to certify clones? Firstly we have to ascertain three facts; (1) whether the individual plant has a genotype consistent with its recorded identification, (2) whether it is a member of one of the other characterized clones known to be present, or (3) whether it is somehow a "contaminant" that is not supposed to be there.

2.1 Inclusion

Two individuals can be considered members or ramets of the same clone if, and only if, their genotypes are identical at all loci. In the strict sense, this would require a complete description of the genotypes. But, because of redundancy of genotypes due to segregation, complete correspondence of two individuals at all gene loci does not absolutely guarantee they are members of the same clone. Thus, if two plants are members of the same clone, they must have identical marker alleles, barring any mutation or ascertainent error. However, other clones may also be "included" in the group characterized by those marker alleles. The more loci that are characterized and used, the more alleles per locus, and the more the allele frequencies have central values, the lower the probability that two different clones will be "included" in the same identically characterized group. Statistical methods, combined with appropriate traits, can quantify this probability of error. This allows us to make practical statements such as "there is a 95% chance that individual A and individual B are members of the same clone". This demonstrates the principle of inclusion.

2.2 Exclusion

If two individuals are claimed to be identical but differ at a minimum of one gene locus, an unequivocal statement can be made that they are not members of the same clone. Thus, the case of exclusion is always exact, and does not need reference to a statistical test. Depending upon the types and amounts of genetic variation in the suite of traits considered, there will be one group of individuals who are putative members of a clone, and another group who are not members of the clone. The principles of inclusion and exclusion are, therefore, usually used in combination.

3 Criteria for Identification

Characters used in identification and certification must be quantifiable and be under genetic control. These two key factors must be considered regardless of whether the objective is to certify clones, varieties, or families. It should also be remembered that environmental influences on a character reduce its usefulness for this application (Mast 1975).

3.1 Morphological Traits

Morphological traits are the most obvious and readily available. Depending upon the mode of genetic control, we distinguish between two classes of morphological traits; those controlled by a single gene and those controlled by more than one gene. Single-gene traits may often show dominant/recessive patterns of expression, and usually follow classical Mendelian rules of inheritance. In the case of dominance or near-dominance it is usually not possible to distinguish between genotypes of the homozygous dominant class and the heterozygote without progeny testing. Determination of parental genotypes by controlled crosses and progeny testing is time-consuming and labor-intensive, and may not provide a definitive identity for certification. Quantitative traits, those controlled by multiple genes, show a complex pattern of inheritance. Usually only a part of the observed trait's variation, quantified by the heritability of the trait, can be attributed to genetic causes. The remainder is the result of environmental modification and genotype-by-environment interaction.

Morphological traits, despite their general ease of access, have been of limited use in certification programs. This is because few of these traits contain sufficiently precise genetic variation. Often, single-gene traits are associated with deleterious or lethal mutants, which further limits their application. Because of environmental influences quantitative traits are rarely used for certification.

3.2 Biochemical Traits

Many secondary metabolites (e.g., alkanes, tannins, monoterpenes) have been used to study the genetics of forest trees (Miksche 1976). Within this class of metabolites, monoterpenes have been used extensively as traits. The mode of inheritance (Mendelian single-locus ratios vs. substrate-competitive multiple loci with more complex ratios) of monoterpenes is still uncertain (Hanover 1966; Birks and Kanowski 1988). But there is no doubt that this variation is under genetic control, and their high broad-sense heritabilities are well established. Despite uncertain modes of inheritance, monoterpenes and other secondary metabolites can serve as traits for clonal identification and certification.

Even though the mode of genetic control is uncertain, a biochemical fingerprint or phenotype can be developed for the clones of interest. If the nongenetic (environmental) variation of such traits among ramets within a clone is substantially less than the among-clone component of genetic variation, the use of these traits in identification and certification is straightforward. Within the monoterpenes, there appears to be sufficient variation to warrant their use for certification. Monterpenes for clonal identification have been used by Thorin and Nommik (1974), Radwan and Ellis (1975), Esteban et al. (1976), Rottink and Hanover (1977), and Kossuth et al. (1988). But remember that, among forest trees, terpenes exist only in conifers.

In those studies where a formal certification analysis has been performed using terpenes, a large proportion of the clones could be uniquely differentiated from each other. It is important, as demonstrated by Kossuth and Muse (1986), to carefully synchronize the collections and standardize a sampling strategy to minimize the amount of environmental and/or developmental and/or tissue-specific variation among ramets in a clone.

3.3 Codominant Single Gene Traits

Single gene traits, expressed codominantly, are well suited for certification and identification. Such traits are likely to be employed by most certification programs. The nature of the information gathered from such traits and some of the basic sampling problems are discussed below.

When identifying and certifying clones, a primary concern is the likelihood of observing a unique genotype. As discussed by Cheliak and Pitel (1984b), this is a problem with technical, statistical, and biological components. Three parameters need to be considered: (1) the number of loci analyzed, (2) the number of alleles observed at each locus, and (3) the frequencies of those alleles in the group of clones being studied. It is important to recognize that all statements made with respect to an individual certification relate to a specific reference sample of clones.

The number of diploid genotypes possible when alleles are expressed codominantly is

$$Ng = \prod_{i=1}^{L} [n_i(n_i + 1)/2],$$

where Ng is the number of unique genotypes, ni is the number of alleles at the ith locus, i = 1, . . . , L, and L is the number of loci. Thus, two alleles at one locus, for example, A and a, will generate three genotypes AA, Aa, and aa. Two alleles at each of two loci, A and a and B and b, will generate nine different genotypes. The likelihood of observing the total array of genotypes is dependent upon the frequency distribution of alleles and the size of the sample of clones from the population. Unless the sample is extremely large and/or the frequency distribution of alleles is equal, it is likely that some genotypes will be missing. This likelihood increases if some of the loci are linked, and if the sample of clones is not in linkage equilibrium.

3.3.1 Isozymes

Since their popularization as a means of detecting single gene variation in the mid-1960s, isozymes have been used widely in forest genetics research (El-Kassaby and White 1985). In the majority of studies, abundant variation has been detected for a large number of single-gene loci. Most loci have been shown to be under simple Mendelian control and, except for null alleles or modifiers (Harry 1983), most demonstrate a codominant type of expression. Because of this it is possible to directly observe heterozygotes and both classes of homozygotes at a diallelic locus.

Isozymes can be readily detected in angiosperms and gymnosperms, even from minute amounts of material (Cheliak and Pitel 1984c), allowing essentially nondestructive sampling. A variety of tissues can be used; these include leaves, buds, embryos and, in conifers, megagametophytes. In those species where studies have assessed the stability of expression for various tissues from the same individuals, a consistent inferred or observed genotype has been observed (Adams 1983; Cheliak and Pitel 1984a; Neale et al. 1984). The greatest likely use for isozyme analysis is for certification of foundation and expansion stock. However, the stability of genotype expression, combined with the minute amount of tissue required for isozyme analysis, makes it feasible to consider certification of individual test plants prior to field tests.

Isozymes have been successfully used on numerous occasions to identify mislabeled clones. Most of this work has dealt with problems in seed orchards. Ramets of the clones could be one of two types, those produced by grafting clonal scions onto root stock, or those produced from their own roots by rooting cuttings or tissue culture. Seed orchards specifically and clone banks more generally represent a unique situation in that clones developed for these applications are sometimes grafts, which are genetic mosaics. That is, the root stock is genetically different than the scion. Figure 1 shows two instances of

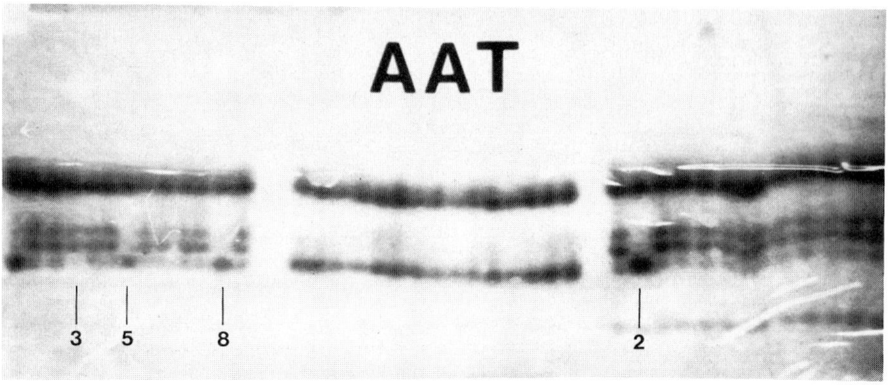

Fig. 1. Gel of the isozymes of Aspartate Amino Transferase (*AAT*) from needle tissue of Scots Pine (*Pinus sylvestris* L.). This example shows three groups that represent "three" clones from a production seed orchard. It is obvious from the banding patterns that the 3rd, 5th, and 8th ramets of the first group are different from the other ramets of the "clone". Similarly, the second ramet of third group is genetically different from the other ramets of the "clone"

clonal misidentification among the ramets of two clones. Errors were detected by an isozyme of aspartate amino transferase (AAT). In the first clone, only one individual was excluded from its stated clonal identity. In the second clone, two individuals were excluded. These two excluded individuals were also different from each other at a number of loci. Furthermore, in both of the cited cases, the individuals also varied at a number of other loci. The ramets which differed from the consensus genotype (Cheliak and Pitel 1984b) originated either from mislabeled scions or, perhaps more likely in the second case, from the root stock overtaking the grafted scion. Unfortunately, it is almost impossible to determine the exact cause of error after the fact. We have observed similar results in clonal material of *Populus* currently used in production. In this case a large number of different genotypes were detected in a sample of 25 cuttings from a single stool bed.

3.3.2 Nucleic Acid Probes

DNA or RNA fragments can be used to detect genetic variation. This technique has become commonly known as Restriction Fragment Length Polymorphisms or RFLPs (Botstein et al. 1980). The use of RFLPs is currently the most powerful method available to detect genetic variation. Although still rare in forest tree applications, RFLPs have been utilized in many other crops (Bernatzky and Tanksley 1986; Helentjaris et al. 1986; Landry et al. 1987). It has been amply demonstrated that RFLPs share many of the same positive attributes as isozymes. However, RFLPs, unlike isozymes, are essentially unlimited in number and variation. They have been found to vary at levels not only useful for large-scale taxonomic work, but also at the individual level. As first demonstrated in humans (Jeffreys et al. 1985a), but later extended to other mammals (Jeffreys and Morton 1987) and birds (Burke and Bruford 1987), probes can be found that are able to genetically fingerprint an individual. These have come to be known as hypervariable minisatellite DNAs or Variable Number Tandem Repeat (VNTR) loci. Such probes, with their unprecedented power of discrimination, are revolutionizing forensic medicine (Gill et al. 1985).

In many instances it is possible to use these techniques to make a positive statement (i.e., quantified inclusion) about the identity of an individual suspected of some event (Jeffreys et al. 1985b). There is no reason to expect these probes to be any less powerful in plant populations. Disadvantages of nucleic acid probes for plants include the amount of tissue currently required for analysis and the cost of extraction and analysis. Many of these problems are, however, associated with the novelty of the technology. For example, new techniques and developments, such as sequence amplification (Li et al. 1988; Fahrlander and Klausner 1988) or the use of allele-specific oligonucleotide probes (Beckmann 1988), should significantly reduce both the amount of tissue required and cost.

4 Model Program

A certification and testing program for forest trees has recently been established by the USDA Forest Service. It is centered in Placerville, California and is mandated to serve the Forest Service as well as other agencies. According to Dr. S. Friedman, the recent director of this service, it has been decided to use isozymes as markers to certify crosses, families, and clones. Eventually, it is expected that the suite of markers will expand to include RFLPs. Presently, this is the only central certification service available for forest trees. Several companies have in-house capacity to do isozyme and certification analyses for their propriety materials.

5 Summary

Several conclusions can be drawn from these experiences. First, assumptions of nil errors in identity or pedigree in the germplasm of a crop-improvement program are not supported by the available data. Such errors are likely to be a part of most, if not all programs, and should not be taken as harsh criticism of the breeding program. Rather, identity screening and other appropriate measures should be taken from now on to rectify any pedigree or identity errors. There are numerous ways that errors can enter a crop improvement program. Once an incorrectly identified plant is established among the donor plants in a production line or among parents in a breeding line, it becomes virtually impossible to detect without access to an independent external screening or testing procedure.

There are several procedures available for germplasm certification. The choice of an appropriate screening procedure depends upon factors such as the nature of the specific problem (certification of provenance, family, clone, etc.), the type of material available for analysis (for example, seeds or vegetative tissues), the amount of material available, and the number of individuals to be screened. There is no single most suitable technique. However, isozymes and nucleic acid probes offer the most genetically definitive, flexible, and powerful means to identify clones. Given the state of technology today, isozymes offer the best cost/effective solution to most germplasm certification problems.

References

Adams WT (1983) Application of isozymes in tree breeding. In: Tanksley DD, Orton TJ (eds) Isozymes in plant genetics and breeding, pt A. Elsevier, Amsterdam, pp 381–400

Adams WT, Neale DB, Loopstra CA (1988) Verifying controlled crosses in conifer tree-improvement programs. Silv Genet 37: 147–152

Beckmann JS (1988) Oligonucleotide polymorphisms: a new tool for genomic genetics. Bio/Technology 6: 1061–1064

Bernatsky R, Tanksley SD (1986) Towards saturated linkage maps in tomato, based on isozymes and random cDNA sequences. Genetics 112: 887–898

Birks JS, Kanowski PJ (1988) Interpretation of the composition of coniferous resin. Silv Genet 37: 29–39

Botstein D, White RE, Skolnick M, Davis RW (1980) Construction of a genetic linkage map in man using restriction fragment length polymorphisms. Am J Human Genet 32: 314–331

Burke T, Bruford MW (1987) DNA fingerprinting in birds. Nature (London) 327: 149–152

Cheliak WM, Pitel JA (1984a) Genetic control of allozyme variants in mature tissues of white spruce trees. J Hered 75: 34–40

Cheliak WM, Pitel JA (1984b) Electrophoretic identification of clones in trembling aspen. Can J For Res 14: 740–743

Cheliak WM, Pitel JA (1984c) Techniques for starch gel electrophoresis of enzymes from forest tree species. Can For Serv Inf Rep PI-X-42

Cheliak WM, Yeh FC, Pitel JA (1987) Use of electrophoresis in tree improvement programs. For Chron 63: 89–96

El-Kassaby Y, White EE (1985) Isozymes of forest trees: an annotated bibliography. Can For Serv Inf Rep BC-X-267

Esteban I, Bergmann F, Gregorius H-R, Hutinen O (1976) Composition and genetics of monoterpenes from cortical oleoresin of Norway spruce and their significance for clone identification. Silv Genet 25: 59–66

Fahrlander PD, Klausner A (1988) Amplifying DNA probe signals: a "Christmas tree" approach. Bio/Technology 6: 1165–1168

Friedman ST, Adams WT (1985) Estimation of gene flow into two seed orchards of loblolly pine (*Pinus taeda* L.). Theor Appl Genet 69: 609–615

Gill P, Jeffreys AJ, Werrett DJ (1985) Forensic application of DNA 'fingerprints'. Nature (London) 318: 577–579

Hanover JW (1966) Inheritance of 3-carene concentration in *Pinus monticola*. For Sci 12: 447–450

Harry D (1987) Identification of a locus modifying the electrophoretic mobility of malate dehydrogenase isozymes in incense-cedar (*Calocedrus decurreus*), and its implications for population studies. Biochem Genet 21: 417–434

Helentjaris T, Slocum M, Wright S, Schaefer A, Nienhuis J (1986) Construction of genetic linkage maps in maize and tomato using restriction fragment length polymorphisms. Theor Appl Genet 72: 1229–1231

Hunter SC (1977) An electrophoretic analysis of isozyme variation in a Piedmont loblolly pine seed orchard. MSc Thesis, NC State Univ, Raleigh

Jeffreys AJ, Morton DB (1987) DNA fingerprints of dogs and cats. Anim Genet 18: 1–15

Jeffreys AJ, Wilson V, Thein SL (1985a) Hypervariable "minisatellite" regions in human DNA. Nature (London) 314: 67–73

Jeffreys AJ, Brookfield JFY, Semeonoff R (1985b) Positive identification of an immigration test-case using human DNA fingerprints. Nature (London) 317:818–819

Kossuth SV, Muse D (1986) Cortical monoterpene variation among slash pine ramets by season, aspect, crown position, and bud vigour. For Sci 32:605–613

Kossuth SV, McCall E, Ledbetter J (1988) Clone certification by use of cortical monoterpenes as biochemical markers. Silv Genet 37:73–76

Landry BS, Desseli RV, Farrara B, Michelmore RW (1987) A genetic map of lettuce (*Lactuca sativa* L) with restriction fragment length polymorphism, isozyme, disease resistance and morphological markers. Genetics 116:331–337

Li H, Gyllensten UB, Xiangfeng C, Saiki RK, Erlich HA, Arnheim N (1988) Amplification and analysis of DNA sequences in single human sperm and diploid cells. Nature (London) 335:414–417

Libby, WJ (1983) The clonal option. Norw For Res Inst 1432 Ås, NLH

Mast H (1975) The organization and work of UPOV. Seed Sci Technol 3:377–386

Miksche JP (ed) (1976) Modern methods in forest genetics. Springer, Berlin Heidelberg New York

Neale DB, Weber JC, Adams WJ (1984) Inheritance of needle tissue isozymes in Douglas-fir. Can J Genet and Cytol 26:459–468

Pitel JA, Cheliak WM, Barrett J (1987) Inheritance of allozymes in a black spruce diallel cross. Silv Genet 37:149–153

Radwan MA, Ellis WD (1975) Clonal variation in monoterpene hydrocarbon of vapours of Douglas-fir foliage. For Sci 21:63–67

Rottink GR, Hanover JW (1975) Identification of blue spruce cultivars by analysis of cortical oleoresin monoterpenes. Phytochemistry 11:3255–3257

Strauss SH (1985) The relation of heterozygosity to growth rate and stability among inbred and crossbred knobcone pine (*Pinus attenuata* Lemm.). PhD Thesis, Univ Cal, Berkeley

Szmidt AE, El-Kassaby YA, Sigurgeirsson A, Alden T, Lindgren D, Hallgren J-E (1988) Classifying seedlots of *Picea sitchensis* and *P. glauca* in zones of introgression using restriction analysis of chloroplast DNA. Theor Appl Genet 76:841–845

Thorin J, Nommik H (1974) Monoterpene composition of cortical oleoresin from different clones of *Pinus sylvestris*. Phytochemistry 13:1879–1881

9 Field Testing Vegetative Propagules

L.J. Frampton, Jr.[1] and G.S. Foster[2]

1 Introduction

1.1 The Need for Field Testing

The potential benefits of producing clonal planting stock for reforestation are many and well documented (e.g., Brix and van den Driesche 1977; Libby 1983; Zobel and Talbert 1984). Worldwide interest in cloning has given rise to many programs researching and/or applying rooted cutting and tissue culture technology for forest regeneration (e.g., Brix and van den Driesche 1977; Kleinschmit and Schmidt 1977; Arnold and Gleed 1985; Zobel 1985; Amerson et al. 1988). While some forest tree species are clonally propagated with little difficulty, experience with other species has shown that cloned propagules do not duplicate faithfully the growth and development of the trees from which they were derived. In such cases, the utilization of vegetative propagules in production plantations may result in various degrees of reduction in the potential benefits of cloning, or may even become a liability to forest productivity. Thus, well-planned field testing of vegetative propagules is necessary in order to knowledgeably assess the real (as opposed to expected) benefits of utilizing clonal planting stock. The purpose of this chapter is to present concepts and demonstrate the importance of testing vegetative propagules under field conditions, in part, by using examples from the forestry literature.

1.2 Problems Associated with Vegetative Propagation

For production forestry, the ultimate purpose of cloning is to produce genetically identical replicates of trees possessing desirable characteristics. Unfortunately, many factors complicate achievement of this goal. A brief discussion of these complications will provide a better appreciation of the necessity and objectives of field testing vegetative propagules.

During their course of development, trees undergo a succession of morphological and physiological changes. This development generally appears as a

[1]Weyerhaeueser Corp., Southern Forestry Research, P.O. Box 1060, Hot Springs, AR 71902, USA
[2]US Forest Service, Southern Forest Experiment Station, Alabama A&M University, Plant and Soil Science Department, P.O. Box 1208, Normal, AL 35762, USA

Clonal Forestry I, Genetics and Biotechnology
Ed. by M.R. Ahuja and W.J. Libby

gradual and continuous accumulation of minor changes, although some characteristics appear to undergo sudden changes at particular times of development (Sussex 1976). Since meristems are the centers of growth and organization in trees, they are intimately involved in these developmental changes (Greenwood and Hutchison, Chap. 3, and Bonga and von Aderkas Chap. 12, Vol. 1). The processes controlling development are complex and not fully understood, but appear to involve the following: (1) reactions of meristems to competition or stimuli from different parts of the tree, (2) ontogenetic aging of meristems as a function of the amount of growth they have produced (i.e., the number of cell divisions they have undergone), and (3) reactions of meristems to factors external to the tree (Hackett 1976; Hood and Libby 1978; Mott 1981).

Although a discussion of developmental control is not within the scope of this chapter, several important points must be recognized by those interested in testing or utilizing vegetative propagules. The first point is that ramets propagated from different parts of the same tree may grow and develop differently from each other and/or from the ortet. Generally, propagules from the lower and central regions of a tree possess more juvenile characteristics than those originating from the upper and peripheral regions (Bonga 1982). Minor variations within this general pattern may also occur (Franclet 1985).

A further complication to propagation is the chronological age of the tree. Propagules from older trees generally grow differently than those derived from younger trees and do not always duplicate expression of characteristics associated with juvenile growth. Furthermore, propagules derived from younger donors are expected to more closely resemble the growth and development of their ortet and to display less variation among ramets than propagules from mature donors (Franclet 1985; Kleinschmit 1985). Finally, the environment in which the donor tree develops can dramatically alter its development, and hence may influence the performance of its ramets (Libby and Jund 1962).

Researchers involved with rooted cuttings of forest species have developed terminology to describe the influence of these various factors on growth and development. These terms have been confused and changed with time but currently accepted definitions follow. Cyclophysis is the process of maturation of apical meristems (Olesen 1978). Topophysis is a state resulting from differentiation in developmental and physiological potential of apical meristems among branch hierarchical orders, independent of the processes of maturation of terminal meristems (Dodd and Power 1988). Periphysis is the effect of the environment in preconditioning tissue (Hallé et al. 1978).

The use of cell, tissue or organ culture technology to clone forest tree species in vitro may produce a similar but potentially more complex and less predictable set of problems regarding genetic expression. Spontaneous variation in plants recovered from in vitro cloning methods has been termed somaclonal variation (Larkin and Scowcroft 1981; Chen and Ahuja, Chap. 7, Vol. 1). Such variation may be associated with the developmental state of the original plant explant similar to the case of rooted cuttings, or induced by the use of growth regulators or other features of in vitro cloning techniques. This variation may be (1) genetic, (2) epigenetic and/or (3) physiological in nature with its magnitude and frequency dependent upon such factors as species, propagation method, source of initial explant, and the length of exposure to, or concentration of, growth regulators (Berlyn et al. 1986). Careful field testing of propagules cloned in vitro is necessary to understand and develop methods of manipulating any desirable or undesirable long-term factors.

1.3 Rejuvenation

Methods of reducing the above effects are readily available for some forest tree species. The use of stump sprouts as a source of cuttings for rooting is common for many species. Some such sprouts arise from suppressed buds laid down by a juvenile meristem while others arise adventitiously (Kramer and Kozlowski 1979). Subsequent to felling, both types of sprouts develop under relatively uniform conditions and are, therefore, less affected by cyclophytic, topophytic, and periphytic influences. Other species possess structures such as lignotubers or produce root suckers which similarly provide a source of juvenile and uniform growing shoots for propagation (Hartney 1980).

Unfortunately, many species do not provide easily propagated sources of cuttings or tissue except at young ages. Efforts to develop techniques to rejuvenate mature tissue from such species are underway and in some cases appear promising (Chap. 10 in Franclet 1985; Fouret et al. 1986; Bonga and von Aderkas, Chap. 12, Vol. 1). Such methods include (1) severe hedging, (2) cytokinin applications, and (3) serial propagation via grafting, rooted cuttings or in vitro micropropagation. Prior to the routine application of these methods, research and development must address not only restoration of propagative ability, but also the restoration of desirable growth and development.

2 Goals of Field Testing

Vegetative propagation has various potential uses in forestry. However, as discussed above, there may be problems associated with utilizing this form of propagation. Field testing may assess those problems and/or address numerous other objectives. The following objectives of field testing are discussed below: comparison of propagule types, clonal evaluation, assessing clone-by-environment interactions, estimating quantitative genetic parameters, estimating C effects, and demonstrating/promoting clonal forestry.

2.1 Comparison of Propagule Types

An important objective to be addressed early in a field testing program is quantification of any differences between vegetative propagules and seedlings. Since the age of the ortet from which vegetative propagules are derived has such a dramatic influence on the outcome of such comparisons, testing vegetative propagules from ortets of different ages will also be considered in this section. Generally, the more difficult a species is to propagate, the more important and complex this objective of field testing becomes.

2.1.1 Genetic Considerations

Ideally, in such comparisons, many replicates of the same genotypes would be tested as each propagule type. Since it is not possible to replicate seedling genotypes as seedlings, a large number of individuals of comparable genetic value for each propagule type must be compared instead.

One strategy to accomplish this objective is to produce vegetative propagules from each seedling that will be tested. Two drawbacks to this procedure are: (1) the seedlings will be larger in size at the time of establishment and (2) the seedling performance may be altered by removing explants to produce the vegetative propagules. Despite these drawbacks, this strategy is desirable, since identical individual genotypes are tested as each propagule type.

In a more commonly used strategy, vegetative propagules and seedlings are produced from a common set of seedlots. Thus, the genetic unit of comparison between propagule types may be open-pollinated or full-sib families, geographic provenances or a genetically less-defined bulk seedlot. Although identical individual genotypes of each propagule type are not tested in this strategy, the means of the genetic units employed are expected to be equal among propagule types tested if there are no effects of the propagation methods and a sufficiently large number of genetic units are tested.

Another strategy which has sometimes been used for species which readily propagate at maturity is comparison of vegetative propagules and seedling progeny (usually open-pollinated) derived from the same tree. This strategy is less desirable, since the units of comparison among propagule types are genetically different. Depending on the type of genetic control involved for the traits of interest and type of progeny seedlings being tested, an extremely large sample of each propagule type may be required for the expectation of their means to be equal. Further, depending on the same criteria, the expectation of the genetic rank correlations among propagule types may be low, regardless of the sample size employed.

Caution must be exercised not to confound the degree of genetic selection with the types of propagules being tested. It may be enticing to compare vegetative propagules from select trees with seedlings from standard planting stock. Any difference in the performance of these sources provides an estimate of gains to be derived from a vegetative versus seedling propagation program. Such information may be extremely useful in decision making; however, interpretation of the results is difficult since differences in the field performance due to the propagule types are confounded with the differences in genetic improvement of the propagules. Trials comparing vegetative propagules originating from ortets genetically less improved than the seedling comparison have been reported (e.g., West 1984; and to a lesser extent Sweet 1973; Sweet and Wells 1974) but establishment of such studies should be avoided.

2.1.2 Size and Physiological Considerations

Another consideration in comparing types of propagules is their initial planting size and physiological condition. Ideally, each pair or set of propagule types should be of a similar size and more importantly, physiological condition, at the time of establishment. Ideally, each propagule type should have apical buds in the same developmental stage and equally developed root systems.

In practice, this is not usually the case. For example, cuttings are often set into nursery beds during winter and early spring while seed are sown later in the spring. This may result in rooted cuttings larger in size than seedlings at the time of lifting and study establishment. When culturing propagules under greenhouse conditions, coordinating the growth of the propagule types is often difficult. This co-ordination is complicated by factors such as differences in the growth rates of propagule types, the large range in transfer dates of tissue culture propagules from the laboratory to the greenhouse, and the variation in time of root formation by cuttings. For these reasons, many field trials have been established with rooted cuttings or tissue culture propagules different in size and vigor than seedlings (e.g., Sweet 1973; Amerson et al. 1987; Foster et al. 1987).

In order to account for initial differences in size, several researchers have made use of a parameter termed relative growth rate in comparing different types of propagules or propagules derived from ortets of differing ages (Sweet 1973; Shelbourne and Thulin 1974; Sweet and Wells 1974; Foster et al. 1987). This parameter is calculated as follows:

$$\text{Relative Growth Rate} = \frac{\ln(X_{t2}) - \ln(X_{t1})}{t2 - t1}, \tag{1}$$

where X_{t2} = size at time t2, and X_{t1} = size at time t1.

By definition, a propagule type with a smaller overall mean size but same absolute growth (e.g., annual increment) as another propagule type will have the larger relative growth rate. In part due to this effect, relative growth rates do not remain constant but generally decrease as the size of the tree increases. For this reason, covariance analysis may be used to adjust for size differences before relative growth rates are compared. Using the average within-group regression of relative growth rate between times t1 and t2 on size at time t2, the mean relative growth rate between t1 and t2 is adjusted on the basis of the overall mean size at time t2 for each type of propagules (Sweet and Wells 1974).

Another, simpler method of comparing the relative growth rates among types of propagules established at different sizes has been employed. This method involves plotting the ratio of the mean size of the propagule types over time (Frampton 1986; Amerson et al. 1988; Frampton and Isik 1987). Despite initial differences in size, this ratio will remain constant over time if the two propagule types are growing at the same relative growth rate. If seedlings are considered the standard of comparison, their mean size is in the denominator of the ratio. An increase or decrease of this ratio over time represents an increase or decrease of the relative growth rate of the propagule type to that of the seedlings, respectively.

Initial size and vigor differences between propagule types create substantial difficulties in the analysis and interpretation of subsequent growth. None of the techniques to adjust for these differences appears to completely equalize comparisons (Burdon and Sweet 1976). Therefore, all possible efforts should be made to standardize the size and vigor of propagule types at test establishment.

2.1.3 Ortet Age Considerations

As previously mentioned, ortet age has a profound influence on the subsequent field performance of vegetative propagules. To fully exploit the genetic benefits of cloning, using previously selected and tested individuals as ortets is desired. The appropriate age for selection in a tree improvement program varies with species, improvement objectives, and numerous other factors. Often, the age of selection is old enough to detrimentally affect the field performance of vegetative propagules.

Unless coppice growth or another juvenile-behaving source from select trees can be utilized, it is recommended that new testing programs initially establish vegetative propagules from seedling material in addition to propagules from selection-age material. This extra effort will provide a critical evaluation of propagation techniques unbiased by age effects.

2.1.4 Design Considerations

When comparing propagule types or vegetative propagules from ortets of different ages, dramatic differences in initial growth rates should be anticipated. One serious problem arising from a row- or single-tree plot design is competition between propagule types or ages. Trees displaying rapid early growth rates will have a decided competitive advantage, particularly during and after crown closure. To prevent this problem, the use of block plots is recommended. Unfortunately, paired or sets of row-plots have more commonly been utilized probably because (1) other objectives are commonly addressed in these types of studies, (2) relatively few plants have been available for testing, and (3) selection before or at crown closure may yield optimum correlated genetic responses (McKeand 1988), and thus lessen the interest in later competition.

Regardless of type of plot configuration used, tests of this type should be replicated several times at each of at least three or four locations. Further design considerations that are relevant to these and other types of genetic tests are discussed later in this chapter and elsewhere (Zobel and Talbert 1984; Foster 1988).

2.1.5 Measurement Age

Due to the likelihood of differences in initial planting sizes and large differences in growth patterns over time between types of propagules, early measurements in tests of this type are not necessarily indicative of later results. Caution should be exercised in making decisions based on early measurements in such tests. It is highly recommended that tests comparing types of propagules be measured at early ages and at frequent intervals thereafter, in order to quantify temporal differences in growth patterns. Such information should prove useful in deciding advantages and disadvantages of each propagule type and in developing new silvicultural practices for vegetative propagules, if necessary.

2.1.6 Examples from the Literature

Monterey Pine Rooted Cuttings. Monterey pine (*Pinus radiata* D. Don) is the most notable example of a species where differences between the field performance of rooted cuttings and seedlings has been documented for many characteristics. Table 1 summarizes reported growth comparisons between seedlings and cuttings of this species. The field heights of rooted cuttings generally are similar to or exceed those reported for seedlings for ortet ages less than or equal to 8 years and likely, for slightly older ortet ages. Cuttings from ortets aged 11 to 17 years display less height growth relative to seedlings (about 8% after five growing seasons). Beginning at 7 or 8 years, greater ortet age results in more dramatic decreases (up to 35% less at age 5) in the diameter growth of rooted cuttings relative to seedlings.

Monterey pine rooted cuttings generally have thinner bark, straighter and less tapered stems, smaller branch diameters, and fewer branches with greater ortet age up to maturity (Fielding 1970; Pawsey 1971; Sweet 1973; Tufor and Libby 1973; Sweet and Wells 1974; Eldridge 1981; West 1984; Arnold and Gleed 1985). Even cuttings from very young ortets show definite form superiority relative to seedlings. For these reasons, silvicultural advantages have been attributed to cuttings over seedlings, such as lower initial stockings and shorter pruning times (Arnold and Gleed 1985; Klomp and Hong 1985). Cuttings from older ortets have been reported to be less susceptible to infection by *Dothistroma pini* Hulbary var. *linearis* than seedlings (Power and Dodd 1984) but more susceptible to root rot caused by *Armillaria* spp. (Klomp and Hong 1985).

The stem-wood properties of Monterey pine rooted cuttings generally mimic those of the upper stem of their seedling ortet (Nicholls and Brown 1971; Nicholls et al. 1974; Nicholls et al. 1976, 1977; Arnold and Gleed 1985). The height at which the stem-wood properties of seedlings matches that of the base of the cuttings' stem increases with ortet age. These differences between cuttings and seedlings are greatest at the lower few meters of the stem and negligible in the middle and upper portions of the stem. Thus, basal or whole-tree wood-specific gravity of cuttings from older ortets is slightly lower than that of seedlings. Similarly, the tracheid length and fiber angle of cuttings increase with ortet age, with the greatest differences at the base of the stem.

Loblolly Pine Tissue Culture Propagules. The field performance of loblolly pine (*Pinus taeda* L.) tissue culture propagules has been studied intensively, perhaps more so than any other forest tree species, and therefore, will be discussed here as an example.

Successful reports of tissue culture propagation of loblolly pine began to appear in the late 1970s and early 1980s (Mehra-Palta et al. 1977; Mott and Amerson 1981; Abo El-Nil 1982). Similar to the results from other conifers, these early methods involved organogenesis of adventitious shoots and roots using embryonic explants, typically cotyledons, as starting material. Currently, research efforts for this and other forest species have shifted to more productive tissue culture methods: micropropagation via axillary and/or fascicular buds using shoot explants as a starting source or somatic embryogenesis using immature zygotic embryos as a starting source (Amerson et al. 1988). However,

Table 1. Summary of reported growth comparisons between Monterey pine rooted cuttings and seedlings[a]

Ortet age (years)	Measurement age (years)	Cutting mean / Seedling mean × 100: Height	Diameter	Genetic quality of cuttings relative to seedlings	Reference
2	7	99.3	107.1	Similar	Eldridge (1981)
4	1	136.9	—	Inferior	Sweet (1973); Sweet and Wells (1974)
	2	128.9	—		
	3	117.6	120.1		
	5	110.3	114.1		
5	5	98.1	85.0	Inferior	West (1984)
5	6	104.5	93.1	Inferior	Fielding (1970)
5	1	101.5	—	Similar	K. Tufuor, W.J. Libby and D.G. Arganbright 1981, unpubl. data
	2	102.1	—		
	5	107.0	109.2		
	8	103.4	105.0		
	14	103.9	104.4		
6–7	4	113.6	109.8	Superior	Shelbourne and Thulin (1974)
7	5	97.7	88.1	Superior	Fielding (1970)
7	1–15	~ 100.0	—	Similar	Klomp and Hong (1985)
	6	—	85.5		
	15	—	93.0		
7–8	4	112.3	102.1	Superior	Fielding (1970)
8	1	119.6	—	Inferior	Sweet (1973); Sweet and Wells (1974)
	2	105.8	—		
	3	109.2	90.2		
	5	102.9	81.2		
8[b]	5	111.8	—	Superior	Arnold and Gleed (1985)
	10	103.6	104.9		
	11	102.8	105.9		
	13	102.0	105.4		
8[b]	11	95.9	~ 100.0	Superior	Arnold and Gleed (1985)
	13	96.0	~ 100.0		
	15	98.2	~ 100.0		
11–14	5	92.0	72.1	Similar	Tufuor and Libby (1973)
17	1	98.9	—	Inferior	Sweet (1973); Sweet and Wells (1974)
	2	87.8	—		
	3	93.0	72.4		
	5	92.6	64.7		

[a]Table compiled by Roger Arnold, Former Graduate Research Assistant, Department of Forestry, North Carolina State University, Raleigh.
[b]Data are from the same trial. The first data listed are from a destructive stem analysis sample. The second data listed are from a permanent sample plot.

with one exception to be mentioned later, the current knowledge of field growth of tissue culture propagules for loblolly pine and many other forest species is limited to those produced using the early method, referred to here as the cotyledon method.

Numerous differences have been reported between loblolly pine cotyledon method propagules and seedlings growing under field conditions (McKeand and Frampton 1984; Amerson et al. 1985, 1988; Frampton et al. 1985; McKeand 1985; Frampton 1986; Wisniewski et al. 1986; Frampton and Jett 1989). In eight field trials established in 1981, the early height and diameter growth of the cotyledon-method propagules was less than that of seedlings grown from the same open-pollinated families (Frampton 1986; Amerson et al. 1988). The 4th-year height increment was similar between the two propagule types despite a considerable difference in total height. Additionally, the cotyledon-method propagules exhibited a greater resistance to fusiform rust [caused by *Cronartium quercuum* (Berk.) Miyabe ex Shirai f. sp. *fusiforme*], displayed more mature shoot morphology (McKeand 1985) yet had a similar juvenile wood-specific gravity (Frampton and Jett 1989) relative to seedlings. Generally, loblolly pine propagules produced via this method of tissue culture appear to have by-passed a portion of their juvenile growth cycle in a manner similar to rooted cuttings of Monterey pine derived from adolescent ortets.

To better understand these differences in field development, two studies were established containing loblolly pine seedlings and cotyledon-method propagules as well as two additional propagule types produced in vitro: excised embryos and rooted hypocotyls (Frampton and Isik 1987). Excised embryos were simply young seedlings that had initially been germinated and grown in vitro without gametophytic tissue and thus, had a natural seedling root and shoot system. The rooted hypocotyls were similar to the excised embryos except that they had been cut at mid-hypocotyl and adventitiously re-rooted in vitro. After 3.5 growing seasons in the field, as in other trials, the cotyledon-method propagules were shorter in height, had less fusiform rust incidence, and displayed more mature morphology relative to the seedlings. Despite their smaller planting size, by 1.5 years the excised embryos and rooted hypocotyls performed similarly to the seedlings in every respect. Comparisons among these four propagule types suggested that the origin of differences between cotyledon-method propagules and seedlings resulted from the method of in vitro shoot production rather than from the tissue culture environment or rooting procedures. Similar results were observed in a reciprocal grafting study involving all four combinations of rootstock and scion from cotyledon-method propagules and seedlings (L.J. Frampton Jr. and S.E. McKeand, unpubl. data 1985).

By 1987, reliable techniques to micropropagate loblolly pine in vitro from the axillary and/or fascicular buds of greenhouse-grown seedling epicotyls were available (Amerson et al. 1987). Greenhouse and 1st-year field comparisons have indicated that unlike cotyledon-method propagules, those produced via this latter method do not display early reduced growth or morphological differences from seedlings (L.J. Frampton Jr. and H.V. Amerson, unpubl. data 1989). Further, propagules produced from the axillary and/or fascicular buds of cotyledon-method shoots do display these differences from seedlings. Apparently, these differences are inherent to the cotyledon method of propagation and, while they may be passed on, they do not necessarily occur as a result of other propagation methods. Two potential causes of the differences between the growth of cotyledon-method and other propagules are being examined: (1) an effect of the concentration and/or length of exposure to cytokinin during the adventitious shoot initiation process and (2) an inherent difference between adventitious and other types of shoots in loblolly pine.

From the above discussion it is clear that the in vitro methods employed influence the subsequent ex vitro growth and development of the tissue culture propagules produced. Recently, techniques to propagate loblolly pine via axillary and/or fascicular buds of mature shoot explants and via somatic embryogenesis using immature zygotic embryos as a starting source have become available and will undoubtedly be used to establish field trials of the future. The discovery of the same and/or unprecedented differences between these propagules and seedlings should not be surprising.

Other Examples. Additional literature dealing with forest tree species and (1) comparing field growth of vegetative propagules and seedlings or (2) assessing the effect of ortet age on growth are referenced below. The reader is encouraged to critically analyze the methods and conclusions reached in these reports in light of the above discussions.

Field trial comparisons between rooted cuttings (or air layers) and seedlings have been reported for *Eucalyptus* spp. (mainly *E. grandis* Hillex Maiden) (Zobel et al. 1984; Zobel 1985), *Picea* spp. (Roulund 1974; Kleinschmit and Schmidt 1977; Kleinschmit 1985), slash pine (*Pinus elliottii* var. *elliottii* Engelman) (Schultz 1972a, b; Bailey 1984), eastern white pine (*Pinus strobus* L.) (Thomas and Riker 1950; Patton and Riker 1954; Struve et al. 1984), and loblolly pine (Foster et al. 1987; Foster 1988).

Studies directly comparing the effect of ortet age on field performance have been reported for Monterey pine (Sweet 1973; Sweet and Wells 1974; Nicholls et al. 1977), loblolly pine (Foster et al. 1987), slash pine (Hoekstra and Johansen 1957; Franklin 1969) and Norway spruce [*Picea abies* (L.) Karst.] (von Wühlisch 1984). Additionally, results from field trials investigating the usefulness of hedging in Monterey pine (Libby et al. 1972; Libby and Hood 1976; Hood and Libby 1978; Bolstad and Libby 1982; Dodd 1988; Dodd and Walker 1988) and serial propagation in Norway spruce (Kleinschmit 1985; St. Clair et al. 1985) in maintaining juvenility have been reported.

Less is currently known about the field performance of tissue culture propagules relative to seedlings than that of rooted cuttings. Results from field trials of tissue culture propagules adventitiously derived from cotyledon explants have been reported for Monterey pine (Smith et al. 1985) and Douglas-fir (*Psuedotsuga menziesii* (Mirb.) Francol (Ritchie and Long 1986). Results of field trials containing tissue culture propagules derived from axillary buds for two *Eucalyptus* spp. (*E. tereticornis* Sm. and *E. toreliana* F. Muell.), teak (*Tectona grandis* L.) and the tooth-brush tree (*Salvadora persica* L.) in India have also been reported (Mascarenhas et al. 1988).

Although field testing of tissue culture propagules of forest species is relatively new, it is rapidly increasing, and knowledge of their performance will undoubtedly accrue as the operational feasibility of these techniques progresses. Recent successes with somatic embryogenesis technology will greatly facilitate field testing and the early testing stages have, in fact, commenced for somatic seedlings of several forest tree species.

2.2 Clonal Evaluation

Testing the relative worth of clones is the primary function of field testing once a program becomes operational. Reports of this type of information obtained from clonal testing programs have been made for Monterey pine (Burdon 1971; Shelbourne and Thulin 1974; Wilcox et al. 1976), Norway spruce (Dietrichson and Kierulf 1982, Kleinschmit 1985) (Figure 1), eucalypt species (Campinhos and Ikemori 1977, 1980; Mikkola 1982), and eastern cottonwood (*Populus deltoides* Bartr.) (Wilcox and Farmer 1968; Mohn et al. 1970; Ying and Bagley 1976) (Figure 2).

Aspects of testing and deployment for clonal forestry purposes have been discussed elsewhere (e.g., Zobel and Talbert 1984; Libby 1987). A four-step approach to testing clones has been suggested (Libby 1987). These steps are: (1) initial screening of a very large number of noncloned genotypes, (2) candidacy testing of a large number of genotypes which are cloned a few times, (3) clonal performance testing of a moderately small number of genotypes which are cloned many times, and (4) compatibility trials of a very small number of well-tested and highly desirable clones. While such extensive testing is ideal, in practice some of this testing is likely to be consolidated in time and space to reduce costs.

Facets of performance repeatability and test design associated with evaluating clones are discussed below.

2.2.1 Repeatability of Clonal Performance

Clonal testing is important since the performance of an ortet in natural stands, plantations, or seedling progeny trials is not necessarily indicative of the performance of its ramets. This discrepancy is in part due to the problems associated with vegetative propagation (Sect. 2.1) and in part due to the degree of environmental variation at the location where the selection is made. Estimates of individual-tree or family-mean correlations between seedling ortets and cutting ramets have generally been positive but range from low to moderately high (maximum r = 0.80) in value (Nicholls and Brown 1971; Schultz 1972a; Nicholls et al. 1976, 1977; Dietrichson 1981; Skrøppa 1981; Dietrichson and Kierulf 1982; Skrøppa and Dietrichson 1986; Stonecypher and McCullough 1986; Foster et al. 1987; Foster 1988). Correlations based on family means are generally higher than those based on clonal means (Skrøppa and Dietrichson 1986; Stonecypher and McCullough 1986) pesumably due to better estimates of seedling performance as a result of replication and randomization across environments. Generally, the more juvenile the seedling ortet or the closer the measurement ages of the seedling ortets and cutting ramets, the higher these correlations are likely to be. As expected, relative ranking of clones in field trials initially change, but with time eventually stabilize (Kleinschmit 1985; Foster et al. 1987). This stabilization is more rapid for cuttings from older ortets than from younger ortets (Skrøppa 1981). One disturbing phenomenon reported for clonal testing programs for both Monterey pine (Shelbourne and Thulin 1974)

Fig. 1. Eight-year-old Norway spruce rooted cutting clonal test. Note uniformity in growth habit

Fig. 2. Five-tree row plot of a fast-growing eastern cottonwood clone found in a 9-year-old clonal test

and Norway spruce (Dietrichson and Kierulf 1982; Kleinschmit 1985) is the poor correlation of clone means among propagation cycles.

2.2.2 Design Considerations

Several researchers have contributed to the theoretical basis of designing clonal field trials. The effects of test design alternatives on genetic gain estimates from selection within the tested population have been examined. Two of these studies have considered test design for a pedigreed population (Libby 1969; Shaw and Hood 1985), and one study has evaluated a simplified model that included only clones (Russell and Libby 1986).

The first study (Libby 1969) incorporated multiple ramets into an efficiency formula developed to optimize genetic gain from selection in a nested mating design. The revised formula requires estimates of heritability, number of male parents, number of females per male parent and number of ramets per cloned seedling. Various parameters of the model were varied to compare design efficiencies. In all cases, the optimum number of ramets per clone was one.

A more general model incorporating multiple ramets per cloned individual and at the same time accounting for the use of a variety of mating designs was later developed (Shaw and Hood 1985). Total test size and number of families were fixed in this model and several cases were examined where the number of clones, ramets per clone, heritability, and ratio of additive to dominance variance were varied. In this study, coancestry controls were imposed and optimum genetic gain served as the criterion for choosing the appropriate allocation strategy. The optimum solutions utilized from one to six ramets per clone under the conditions tested.

Three simulation models, each examining the effect of multiple ramets on gain from clonal selection, have been reported (Russell and Libby 1986). These models assumed no family structure for the clones. The model which seemed to be most appropriate for clonal forestry utilized a fixed number of selected clones and a fixed total number of trees. Several cases were investigated where the number of ramets per clone, the number of candidate clones, selection intensity and heritability varied. With broad-sense heritabilities of 0.6 or greater, maximum gain accrued when using a single ramet per clone and testing the maximum possible number of clones. For heritabilities less than 0.6, maximum gain generally resulted from the use of two to six ramets per clone.

General results from these theoretical studies agree that the best strategy for clonal testing programs is to maximize the total number of clones tested while utilizing one, or relatively few, ramet(s) at a given site. Future clonal testing programs should strongly consider the merits of such allocation, particularly since this reduces the propagation effort involved in clonal testing. The previously cited theoretical studies considered only surviving trees. Predicted survival must also be considered and more ramets included in the clonal tests to ensure the desired number of ramets after mortality. These considerations have been used in designing the field testing phase of at least one existing tree improvement program (Foster and Shaw 1987). In this program, clones of loblolly pine rooted cuttings have been established across three sites with two ramets per clone at each site. The program has family structure such that every family is represented in each of the six replications per site; however, a single ramet of each clone is only placed in two of the six replications at each site.

2.3 Assessment of Clone-by-Environment Interactions

The study of clone-by-environment interactions is essential to a clonal evaluation program. Knowledge of the extent of rank changes among clones across various sites is necessary to efficiently deploy clonal material. A further knowledge of site factors most responsible for these clonal rank changes is desirable to divide the land base into appropriate compartments that can be regenerated with common groups of clones.

When considering rankings, it is appropriate to keep in mind that changes in the middle of a group, where absolute differences are small, may be inconsequential for practical purposes. Conversely, large shifts where rank changes occur between quartiles, or from the selected versus cull groups in truncation selection, are critical to detect, since they may adversely effect the results of clonal deployment decisions.

Clone-by-site interactions have been investigated for Monterey pine in New Zealand (Burdon 1971; Burdon and Harris 1973). Considerable changes in clonal ranks among four sites were found for 12-year-old height, diameter, form, branching frequency, and stem straightness. The status of phosphorus at the site appeared to be the most important factor responsible for the interaction. A statistically significant clone-by-site interaction was also found for specific gravity; however, its magnitude was strikingly low relative to the clone and site main effects, so that selection at a single site would be sufficient for this characteristic.

Clone-by-environment interaction studies have also been reported for Norway spruce in northern Germany (Kleinschmit 1985; St. Clair and Kleinschmit 1986). This accounted for 8 to 50% of the total height variation in a series of clonal tests. In this case, elevation appeared to be the most obvious environmental factor contributing to the interaction, although other factors undoubtedly influence it.

Several reports of clone-by-environment interactions have been made for eastern cottonwood in the southern United States (Bridgwater 1972; Randall and Mohn 1969, Farmer 1970; Cooper and Randall 1973; Mohn and Randall 1973). Important interactions have been found for height, diameter, stem volume, stem dry weight; and in some cases, the variance associated with this interaction was similar in magnitude to, or exceeded, the genetic variance. The interaction variance for wood specific gravity was relatively small. Soil drainage and moisture appeared responsible for clone-by-site interaction in this species while clone-by-planting-year interactions appeared to be small (Mohn and Randall 1973).

2.4 Estimation of Quantitative Genetic Parameters

In the absence of appreciable C effects (Sect. 2.5), genetic parameters useful in predicting the amount of gain to be expected from utilizing clonal propagation can be estimated from clonal field trials. The variation among clones is commonly used as an estimate of total genetic variation and to calculate the degree of genetic control (broad-sense heritability) for a particular trait (Wilcox and Farmer 1968; Bridgwater 1972; Ying and Bagley 1976; Foster and Shaw 1988; Hood and Libby 1980; Gunion et al. 1982; Cannell et al. 1988; Foster 1988). Similarly, the among-clone covariance estimates between traits and/or ages can be utilized to estimate total genetic correlations between traits and/or ages (Foster 1986).

Theoretically, a clone is unique in its lack of genetic segregation, hence no genetic difference among its ramets. However, somaclonal variation is well known, especially for tissue culture propagules (e.g., Lester and Berbee 1977), so researchers must verify the fidelity of the clone prior to clonal testing. If the assumption of no within-clone genetic variation is valid, any phenotypic variation within a clone is then totally environmental in nature. Through an analysis of variance, pure estimates of genetic and environmental variance can be obtained from data of clonal trials, a feature which is precluded with data of seedling trials.

The use of family (open-pollinated, half- or full-sib) structure in clonal tests allows for further partitioning of the total genetic variance into that due to additive and nonadditive gene effects. In seedling tests, it is algebraically necessary to assume that epistatic effects are negligible when estimating additive and dominance genetic components of variation. In clonal tests where the ortets of the clones being tested are produced in a balanced mating design (i.e., factorial or diallel), then the nonadditive genetic variance component can be further partitioned into components approximating the variance due to dominance (interaction between genes at the same locus) and to epistatic (interactions among genes at different loci) effects (Foster and Shaw 1988). [Estimation of the epistatic genetic variance algebraically requires that non-environmental among-clone variation (C effects, Sect. 2.5) is assumed negligible.]

Numerous alternative combinations of mating and testing designs are available. For simplicity, four mating designs which have appeared in the clonal forestry literature will be examined as examples of variance component estimation. Each example has a randomized complete block design and clonal replication. All effects in these designs are considered to be random, and typical assumptions of the genetic models are used (Foster and Shaw 1988) including that of negligible C effects (Sect. 2.5).

NOTE: comprehension of the remaining portion of Section 2.4 requires a general knowledge of quantitative genetics techniques. Readers interested in clonal test design and not familiar with these techniques are encouraged to enroll in a course or consult a book (e.g., Falconer 1980; Becker 1984) on introductory quantitative genetics.

2.4.1 Design 1

In this design, multiple ramets (n) of clones (c) with no family structure are tested across multiple locations (1) with several complete blocks (b) per location. The form of the analysis of variance for this design is given in Table 2 and the

Table 2. Form of the analysis of variance of a clonal trial planted across multiple locations with no family structure (Design 1)

Source	Expected mean squares
Clones (C)	$\sigma^2 + n\sigma^2_{CB(L)} + nb\sigma^2_{CL} + nbl\sigma^2_C$
Locations (L)	$\sigma^2 + n\sigma^2_{CB(L)} + nc\sigma^2_{B(L)} + nb\sigma^2_{CL} + ncb\sigma^2_L$
C × L	$\sigma^2 + n\sigma^2_{CB(L)} + nb\sigma^2_{CL}$
Blocks (B)/L	$\sigma^2 + n\sigma^2_{CB(L)} + nc\sigma^2_{B(L)}$
C × B/L	$\sigma^2 + n\sigma^2_{CB(L)}$
Ramets/plot	σ^2

σ^2_C = variance among clones.
σ^2_L = variance among locations.
σ^2_{CL} = variance due to interaction of clones and locations.
$\sigma^2_{B(L)}$ = variance due to the interaction of clones and blocks in locations.
$\sigma^2_{CB(L)}$ = variance among blocks in locations.
σ^2 = variance among ramets in a plot.

experimental variance components have the following genetic expectations:

$$\sigma^2_C = V_A + V_D + V_{AA} + V_{AD} + V_{DD} + \cdots, \quad (2)$$

where V_A = additive genetic variance, V_D = dominance genetic variance, V_{AA}, V_{AD}, V_{DD} = epistatic genetic variance due to additive × additive effects, additive × dominance effects and dominance × dominance effects. (Total epistatic genetic variance, $V_I = V_{AA} + V_{DD} + V_{AD} + \cdots$).

Total genetic (V_G) and phenotypic (V_P) variances are estimated as:

$$V_G = \sigma^2_C \quad (3)$$

$$V_P = V_G + \sigma^2_{CL} + \sigma^2_{CB(L)} + \sigma^2. \quad (4)$$

2.4.2 Design 2

In this design, multiple ramets (n) of clones (c) from open-pollinated families (h) are tested across multiple locations (l) with several complete blocks (b) per location. The form of the analysis of variance for this type of design is given in Table 3. The experimental variance components (Table 3) have the following genetic expectations:

$$\sigma^2_H = COV_{HS} = 1/4V_A + 1/16V_{AA} + \cdots \quad (5)$$

$$\sigma^2_{C(H)} = 3/4V_A + V_D + 15/16V_{AA} + V_{AD} + V_{DD} + \cdots, \quad (6)$$

where COV_{HS} = covariance among half-sibs.

These genetic expectations assume that all individuals within an open-pollinated family are half-sibs. This assumption and appropriate adjustments to the genetic expectations if the assumption is not valid have been discussed (Squillace 1974).

Table 3. Form of the analysis of variance of a field trial of cloned individuals from half-sib families or single pair matings and planted across multiple locations (Designs 2 and 3)

Source	Expected mean squares[a]
Families (H)	$\sigma^2 + n\sigma^2_{B(L)C(H)} + nc\sigma^2_{HB(L)} + nb\sigma^2_{LC(H)} + nbl\sigma^2_{C(H)} + nbc\sigma^2_{HL} + nbcl\sigma^2_{H}$
Locations	$\sigma^2 + n\sigma^2_{B(L)C(H)} + nc\sigma^2_{HB(L)} + nch\sigma^2_{B(L)} + nb\sigma^2_{LC(H)} + nbc\sigma^2_{HL} + nbch\sigma^2_{L}$
H × L	$\sigma^2 + n\sigma^2_{B(L)C(H)} + nc\sigma^2_{HB(L)} + nbc\sigma^2_{HL}$
Clones (C)/H	$\sigma^2 + n\sigma^2_{B(L)C(H)} + nb\sigma^2_{LC(H)} + nabl\sigma^2_{C(H)}$
L × C/H	$\sigma^2 + n\sigma^2_{B(L)C(H)} + nb\sigma^2_{LC(H)}$
Blocks (B)/L	$\sigma^2 + n\sigma^2_{B(L)C(H)} + nc\sigma^2_{HB(L)} + ncb\sigma^2_{B(L)}$
H × B/L	$\sigma^2 + n\sigma^2_{B(L)C(H)} + nc\sigma^2_{HB(L)}$
B/L × C/H	$\sigma^2 + n\sigma^2_{B(L)C(H)}$
Ramets/plot	σ^2

[a]Terms common to Tables 2 and 3 have similar definitions.

σ^2_H = variance among families.
σ^2_{HL} = variance due to the interaction of families and locations.
$\sigma^2_{C(H)}$ = variance among clones in families.
$\sigma^2_{LC(H)}$ = variance due to the interaction of locations and clones in families.
$\sigma^2_{HB(L)}$ = variance due to the interaction of families and blocks in location.
$\sigma^2_{B(L)C(H)}$ = variance due to the interaction of blocks in locations and clones in families.

Estimates of genetic parameters are then calculated as:

$$V_{A'} = 4(\sigma^2_H) \quad (7)$$

$$V_D + V_{I'} = \sigma^2_{C(H)} - 3(\sigma^2_H) \quad (8)$$

$$V_G = \sigma^2_H + \sigma^2_{C(H)} \quad (9)$$

$$V_P = V_G + \sigma^2_{HL} + \sigma^2_{LC(H)} + \sigma^2_{HB(L)} + \sigma^2_{B(L)C(H)} + \sigma^2, \quad (10)$$

where $V_{A'}$ includes V_A and $1/4V_{AA} + \cdots$ and $V_D + V_{I'}$ includes only $3/4V_{AA} + \cdots$ rather than $1V_{AA} + \cdots$.

Using Eqs. (5) through (10), genetic parameters have been estimated from data of a single location of an eastern cottonwood study with a similar design (Foster 1985). Due to the use of a single location, these estimates were confounded with the environment-by-family and environment-by-clone interactions (Foster and Shaw 1988). Height, in this case, was mainly conditioned by additive gene effects, while dominance and epistatic gene interactions were more important for controlling diameter at breast height, survival, and stem volume.

2.4.3 Design 3

In this design, multiple ramets (n) of clones (c) from families (h) created by single-pair matings are tested across multiple locations (l) with several complete blocks (b) per location. The form of the analysis of variance for this type of design is the same as for Design 2 (Table 3); however, the genetic expectations of the experimental variance components (Table 3) change as follows:

$$\sigma^2_H = COV_{FS} = 1/2V_A + 1/4V_D + 1/4V_{AA} + 1/8V_{AD} + 1/16V_{DD} + \cdots \quad (11)$$

$$\sigma^2_{C(H)} = 1/2V_A + 3/4V_D + 3/4V_{AA} + 7/8V_{AD} + 15/16V_{DD} + \cdots. \quad (12)$$

Estimates of genetic parameters are then calculated as:

$$V_{A''} = 2\sigma_{H}^{2} = V_A + 1/2V_D + 1/2V_{AA} + 1/4V_{AD} + 1/8V_{DD} + \cdots \quad (13)$$

$$V_D + V_{I''} = 2(\sigma_{C(H)}^{2} - \sigma_{H}^{2}) = V_D + V_{AA} + 3/2V_{AD} + 7/4V_{DD} + \cdots, \quad (14)$$

where V_G and V_P are the same as in Eqs. (9) and (10), respectively.

Note that each of the calculated genetic parameters is only an approximation of the parameters of interest.

This model has been used to analyze data from a loblolly pine rooted cutting trial at a single location (Foster 1988). Again, in the case of a single location, estimates are confounded with the environment-by-family and environment-by-clone interactions. Additive gene effects increased in importance for height from age one to three while dominance and epistatic effects decreased in importance. Stem diameter, number of growth cycles per year and crown width were generally controlled by dominance and epistatic effects. Stem taper was totally controlled by additive gene effects.

2.4.4 Design 4

In this design, multiple ramets (n) of clones (c) from half-sib and full-sib families are created by mating male (m) and female (f) parents in a factorial mating design and tested across multiple locations (l) with several complete blocks (b) per location. The form of the analysis of variance for this type of design is given in Table 4. The experimental variance components (Table 4) have the following

Table 4. Form of the analysis of variance and expected mean squares for a factorial mating design with clonally replicated individuals within full-sib families and planted across multiple locations (Design 4). (After Foster and Shaw 1988).

Source	Expected mean squares
Location (L)	$\sigma^2 + n\sigma_{B(L)C(MF)}^2 + nc\sigma_{MFB(L)}^2 + nmc\sigma_{FB(L)}^2 + nfc\sigma_{MB(L)}^2 + nb\sigma_{LC(MF)}^2$ $+ nbc\sigma_{LMF}^2 + nmbc\sigma_{LF}^2 + nfbc\sigma_{LM}^2 + nfmc\sigma_{B(L)}^2 + nfmcb\sigma_{L}^2$
Blocks (B)/L	$\sigma^2 + n\sigma_{B(L)C(MF)}^2 + nc\sigma_{MFB(L)}^2 + nmc\sigma_{FB(L)}^2 + nfc\sigma_{MB(L)}^2 + nfmc\sigma_{B(L)}^2$
Males (M)	$\sigma^2 + n\sigma_{B(L)C(MF)}^2 + nc\sigma_{MFB(L)}^2 + nbc\sigma_{MB(L)}^2 + nbc\sigma_{LC(MF)}^2 + nbc\sigma_{LMF}^2$ $+ nfbc\sigma_{LM}^2 + nlb\sigma_{C(MF)}^2 + nlbc\sigma_{MF}^2 + nlbcf\sigma_{M}^2$
Females (F)	$\sigma + n\sigma_{B(L)C(MF)}^2 + nc\sigma_{MFB(L)}^2 + nmc\sigma_{FB(L)}^2 + nb\sigma_{LC(MF)}^2 + nbc\sigma_{LMF}^2$ $+ nmbc\sigma_{LF}^2 + nlb\sigma_{C(MF)}^2 + nlbc\sigma_{MF}^2 + nlbcm\sigma_{F}^2$
M × F	$\sigma^2 + n\sigma_{B(L)C(MF)}^2 + nc\sigma_{MFB(L)}^2 + nb\sigma_{LC(MF)}^2 + nbc\sigma_{LMF}^2 + nlb\sigma_{C(MF)}^2 + nlbc\sigma_{MF}^2$
Clones (C)/(MF)	$\sigma^2 + n\sigma_{B(L)C(MF)}^2 + nb\sigma_{LC(MF)}^2 + nlb\sigma_{C(MF)}^2$
L × M	$\sigma^2 + n\sigma_{B(L)C(MF)}^2 + nc\sigma_{MFB(L)}^2 + nfc\sigma_{MB(L)}^2 + nb\sigma_{LC(MF)}^2 + nbc\sigma_{LMF}^2$ $+ nfbc\sigma_{LM}^2$
L × F	$\sigma^2 + n\sigma_{B(L)C(MF)}^2 + nc\sigma_{MFB(L)}^2 + nmc\sigma_{FB(L)}^2 + nb\sigma_{LC(MF)}^2 + nbc\sigma_{LMF}^2$ $+ nmbc\sigma_{LF}^2$
L × M × F	$\sigma^2 + n\sigma_{B(L)C(MF)}^2 + nb\sigma_{LC(MF)}^2 + nbc\sigma_{LMF}^2$
L × C/(MF)	$\sigma^2 + n\sigma_{B(L)C(MF)}^2 + nb\sigma_{LC(MF)}^2$
M × B/L	$\sigma^2 + n\sigma_{B(L)C(MF)}^2 + nc\sigma_{MFB(L)}^2 + nfc\sigma_{MB(L)}^2$
F × B/L	$\sigma^2 + n\sigma_{B(L)C(MF)}^2 + nc\sigma_{MFB(L)}^2 + nmc\sigma_{FB(L)}^2$
M × F × B/L	$\sigma^2 + n\sigma_{B(L)C(MF)}^2 + nc\sigma_{MFB(L)}^2$
(B/L)(C/(MF))	$\sigma^2 + n\sigma_{B(L)C(MF)}^2$
Among ramets/plot	σ^2

expectations:

$$\sigma^2_{M} = \sigma^2_{F} = 1/4V_A + 1/16V_{AA} + \cdots \tag{15}$$

$$\sigma^2_{MF} = 1/4V_D + 1/8V_{AA} + 1/8V_{AD} + 1/16V_{DD} \tag{16}$$

$$\sigma^2_{C(MF)} = 1/2V_A + 3/4V_D + 3/4V_{AA} + 7/8V_{AD} + 15/16V_{DD} + \cdots. \tag{17}$$

Estimates of genetic parameters are then calculated as follows:

$$V_{A'} = 2(\sigma^2_{M} + \sigma^2_{F}) = V_A + 1/4V_{AA} + \cdots \tag{18}$$

$$V_{D'} = 4(\sigma^2_{MF}) = V_D + 1/2V_{AA} + 1/2V_{AD} + 1/4V_{DD} + \cdots \tag{19}$$

$$V_{I'''} = \sigma^2_{C(MF)} - (\sigma^2_{M} + \sigma^2_{F}) - 3\sigma^2_{MF} = 1/4V_{AA} + 1/2V_{AD} + 3/4V_{DD} + \cdots. \tag{20}$$

Note that each of the calculated genetic parameters is only an approximation of the parameters of interest.

Total genetic variance (V_G) and phenotypic variance (V_P) are then calculated as:

$$V_G = \sigma^2_{M} + \sigma^2_{F} + \sigma^2_{MF} + \sigma^2_{C(MF)} \tag{21}$$

$$V_P = V_G + \sigma^2_{LM} + \sigma^2_{LF} + \sigma^2_{LMF} + \sigma^2_{LC(MF)} + \sigma^2_{MB(L)} + \sigma^2_{FB(L)} + \sigma^2_{MFB(L)} + \sigma^2_{B(L)C(MF)} + \sigma^2 \cdots. \tag{22}$$

This model with the analysis of variance of a single location has been utilized to estimate genetic parameters for an eastern cottonwood study (Foster and Shaw 1988). The genetic parameter estimates are again confounded with environment-by-genetic entry (male parent, female parent, and clone) interactions. In this study, the importance of additive gene effects for height increased from age 1 to 8 while dominance, even though it was important at age 1, became unimportant. For diameter at breast height, additive variance increased from age 3 to 4 and then declined slightly at age 8 while the epistatic variance increased rapidly from age 1 to 8. Epistasis, in this case, was more important than additive gene effects at age height. Volume displayed a similar pattern to diameter at breast height except that dominance control was negligible.

In a similar fashion, a diallel mating scheme utilizing cloned individuals within full-sib families has been used to estimate genetic parameters in Douglas-fir (Stonecypher and McCullough 1986).

2.5 Estimation of C Effects

Due to the previously discussed influences of cyclophysis, topophysis, and periphysis, propagules derived from the same genotype may perform differently when established under field conditions. These factors not only contribute to within-clone variation and differences among types of propagules but, if common to members of a clone, can bias estimates of the true clonal performance. Such nongenetic factors common to members of a group of relatives such as a clone or family are referred to as C effects (Lerner 1958). In a general sense, the previously discussed differences between types of vegetative propagules or among propagules from ortets of different ages are the result of C effects (in cases where genetically similar groups were compared).

Quantitative geneticists are particularly concerned that C effects may bias estimates of total genetic variance and other genetic parameter estimates such as heritabilities, correlations among traits and genetic gains (Libby and Jund 1962; Burdon and Shelbourne 1974). Several researchers have therefore, designed studies to determine the amount of bias that specific C effect sources have on variance component estimates. These studies have most commonly investigated rooting characteristics or shoot growth prior to field planting (Wilcox and Farmer 1968; Foster et al. 1984, 1985; Power et al. 1988) and have concluded that C effects cause considerable upward bias in heritability and genetic gain estimates.

Clonal differences in rooting and early growth often result in clonal differences in planting size which are one form of C effects. In Monterey pine, the influence of initial mean size on clonal evaluation and genetic parameter estimates has been critically assessed (Burdon and Sweet 1976). Four approaches were compared: (1) actual height and diameter measurements, (2) height and diameter measurements adjusted for initial height via covariance analysis, (3) height increments and (4) relative rates of height growth. None of the adjusted or derived measures of growth allowed evaluation of young clones more reliably than the actual height and diameter measurements due to the strong and positive relationship between initial clonal size and subsequent growth. However, the variation due to initial starting size generally constituted greater than 50% of the estimated variance among clones. Consequently, heritability and genetic gain estimates were biased considerably upward.

One report has examined the longer-term influence of potential sources of C effects. In this report (Cannell et al. 1988), the bias to the clonal mean variance due to two-cycle cloning where the first generation was grown in two contrasting environments was estimated for Sitka spruce [*Picea sitchensis* (Bong.) Carr.] and lodgepole pine (*Pinus contorta* Dougl.). Second-cycle clones propagated from the 14- to 15-year-old first-cycle cuttings were grown at a common location. Despite a twofold difference in height at the first-cycle environments, this effect contributed less than 1% to the total variation of 5th-year height in second-cycle tests for each species.

Although the influence of C effects appears to lessen over time, researchers should take every precaution in producing propagules for field testing to equitably treat clones as much as is practical, particularly since early selection in clonal trials is often desirable. The use of secondary and higher cloning cycles is usually necessary to rapidly multiply clones and is highly recommended in order to reduce bias to clonal means (Libby and Jund 1962). In every stage of propagule production, ramets of each clone of each cycle should be properly randomized and replicated.

2.6 Demonstration or Promotion

Undoubtedly, numerous unreported field plantings exist that demonstrate or promote the use of clones in forestry. Promotional plantings are ideally established in convenient and easily accessed locations such as near company headquarters, a research station, nursery or university. These tests should be intensively site prepared and managed with strict weed and pest control. Permanent monumentation in the planting that is both concise and informative is recommended. Planting objectives might include the demonstration of differences between seedling and clonal planting stock, differences among clones or the effect of ortet age on propagule growth and development. Such plantings can be shown to visiting researchers, the general public, or to management and operational personnel internal to the organization. The establishment of promotional plantings is encouraged and likely to result in both tangible and intangible benefits to researchers and operational personnel working with vegetative propagation.

3 Conclusion

Field testing vegetative propagules is a multi-objective and multi-stage task necessary to help optimize the operational use of clonal propagules for reforestation. The objectives, design, and implementation of field tests have not always been given appropriate attention in the past. Nevertheless, considerable knowledge of the behavior of rooted cuttings in the field has accumulated for some forest species. In some cases, this knowledge has resulted in revisions of the expectations and intended methods of operationally applying rooted cutting technology (Kleinschmit 1985; Shelbourne 1986). In other cases, this knowledge has led to the fluent integration of vegetative propagation into intensive forestry operations resulting in phenomenal yield improvements (e.g., Zobel et al. 1984; Zobel 1985).

Less information is available about the field performance of tissue culture propagules relative to rooted cuttings but will undoubtedly accrue as the operational feasibility of these techniques increases. The future use of molecular genetic technology to insert gene sequences or otherwise alter tree genomes is certain to add new biological and political challenges to field testing vegetative propagules. At the same time, these techniques will offer tremendous new opportunities to better explore and understand fundamental genetic and developmental questions of forest tree species in order to enhance their productivity.

References

Abo El-Nil MM (1982) Method for asexual reproduction of coniferous trees. US Patent 4, 353, 184

Amerson HV, Frampton LJ Jr, McKeand SE, Mott RL, Weir RJ (1985) Loblolly pine tissue culture: laboratory, greenhouse and field studies. pp 271–295 In: Henke RR, Hughes KW, Constantine MJ, Hollander A (eds) Tissue culture in forestry and silviculture. Univ TN Symp Proc, Sep 1984. Plenum, New York, 390 pp

Amerson HV, Frampton LJ Jr, Mott RL, Spaine PC (1988) Tissue culture of conifers using loblolly pine as a model. In: Hanover J, Keathley D (eds) Genetic manipulation of woody plants. Plenum, New York, 519 pp

Arnold R, Gleed JA (1985) Raising and managing radiata pine cuttings for production forests. Aust For 48:199–206

Bailey JK (1984) Juvenile cutting production techniques and early field performance of cuttings versus seedlings of slash pine. PhD Diss, TX A&M Univ, Coll Stn

Becker WA (1984) Manual of quantitative genetics, 4th edn. Academic Enterprises, Pullman WA, 190 pp

Berlyn GP, Beck RC, Renfroe MH (1986) Tissue culture and the propagation and genetic improvement of conifers: problems and possibilities. Tree Physiol 1:227–240

Bolstad PV, Libby WJ (1982) Comparisons of radiata pine cuttings of hedge and tree-form origin after seven growing seasons. Silv Genet 31:9–13

Bonga JM (1982) Vegetative propagation in relation to juvenility, maturity and rejuvenation. In: Bonga JM, Durzan DJ (eds) Tissue culture and forestry. Nijhoff, Boston, 420 pp

Bridgwater FE Jr (1972) Multiple trait selection in a population of eastern cottonwood. Thesis, OK State Univ, Stillwater

Brix H, van den Driesche R (1977) Use of rooted cuttings in reforestation: a review of opportunities, problems and activities. BC For Serv/Can For Ser Joint Rep 6, 16 pp

Burdon RD (1971) Clonal repeatabilities and clone-site interactions in *Pinus radiata.* Silv Genet 20:33–39

Burdon RD, Harris JM (1973) Wood density in radiata pine clones on four different sites. NZ J For Sci 3:286–303

Burdon RD, Shelbourne CJA (1974) The use of vegetative propagules for obtaining genetic information. NZ J For Sci 4:418–425

Burdon RD, Sweet GB (1976) The problems of interpreting inherent differences in tree growth shortly after planting. pp. 483–502. In: Cannell MGR, Last FT (eds) Tree physiology and yield improvement. Academic Press, New York London, 567 pp

Campinhos E Jr, Ikemori YR (1977) Tree improvement program of *Eucalyptus* spp. Preliminary results. pp 717–738. In: Third World Consult Forest tree breeding, Canberra, Aust

Campinhos E Jr, Ikemori YR (1980) Mass-production of *Eucalyptus* spp by rooted cuttings. In: Symp Genetic improvement and production of fast-growing trees, Sao Pedro, Sao Paulo, Brazil 16 pp

Cannell MGR, Sheppard LJ, Cahalan CM (1988) C-effects and second generation clone performance in *Picea sitchensis* and *Pinus contorta.* Silv Genet 37:15–19

Cooper DT, Randall WF (1973) Genetic differences in height growth and survival of cottonwood full-sib families. In: Proc 12th So For Tree Improv Conf LSU, Baton Rouge LA, pp 206–212 (352 pp)

Dietrichson J (1981) Ortet selection of Norway spruce at age 7 and ramet performance after six propagation cycles. In: Symp Clonal forestry. Dep For Genet Res Note, Swed Univ Agric Sci, Uppsala, pp 57–64 (313 pp)

Dietrichson J, Kierulf C (1982) Selection of eight-year-old spruce [*Picea abies* (L) Karst] plants in a progeny trial and mass propagation by cuttings. Norw For Res Inst Rep 38(1):1–28

Dodd RS (1988) Cyclophytic effects on wood structure in *Pinus radiata* D Don II Tracheid morphology. Can J For Res 18:413–420

Dodd RS, Power AB (1988) Clarification of the term topophysis. Silv Genet 37:14–15

Dodd RS, Walker NK (1988) Cyclophytic effects on wood structure in *Pinus radiata* D Don I densitometry and grain angle. Can J For Res 18:406–412

Eldridge KG (1981) Cuttings from seedlings of selected radiata pine families. Aust for Counc Res Working Group 1, Traralgon, Victoria

Falconer DS (1980) Introduction to quantitative genetics, 2nd edn. Longman, London New York

Farmer RE Jr (1970) Variation and inheritance of eastern cottonwood growth and wood properties under two moisture regimes. Silv Genet 19:5–8

Fielding JM (1970) Trees grown from cuttings compared with trees grown from seed (*Pinus radiata* D Don). Silv Genet 19:54–63

Foster GS (1985) Genetic parameters for two eastern cottonwood populations in the lower Mississippi Valley. In: Proc 18th So For Tree Improv Conf, Long Beach MS, pp 258–266 (382 pp)

Foster GS (1986) Provenance variation of eastern cottonwood in the lower Mississippi Valley. Silv Genet 35:32–38

Foster GS (1988) Growth and morphology of rooted cuttings and seedlings of loblolly pine and their genetic analysis. pp 67–78 In: Proc 10th N Am Forest biology Worksh, July 1988, Vancouver, Univ B C

Foster GS, Shaw DV (1987) A tree improvement program to develop clones of loblolly pine for reforestation. In: Proc 19th So For Tree Improv Conf, TX For Serv; and TX Agric Exp Stn, Coll Stn, pp 17–21 (456 pp)

Foster GS, Shaw DV (1988) Using clonal replicates to explore genetic variation in a perennial plant species. Theor Appl Genet 76:788–794

Foster GS, Campbell RK, Adams WT (1984) Heritability, gain and C effects in rooting of western hemlock cuttings. Can J For Res 14:628–638

Foster GS, Campbell RK, Adams WT (1985) Clonal selection prospects in western hemlock combining rooting traits with juvenile height growth. Can J For Res 15:488–493

Foster GS, Lambeth CC, Greenwood MS (1987) Growth of loblolly pine rooted cuttings compared with seedlings. Can J For Res 17:157–164

Fouret Y, Arnaud Y, Larrieu C, Miginiac E (1986) *Sequoia sempervirens* as an in vitro rejuvenation model. NZ J For Sci: 16(3) 319–327

Framption LJ Jr (1986) Field performance of loblolly pine tissue culture plantlets. In: IUFRO Conf Proc Joint Meet Work Parties Breeding theory progeny testing and seed orchards, Williamsburg VA. Hosted by NC State Univ; Ind Tree Improv Prog, Raleigh NC, pp 547–553 (673 pp)

Frampton LJ Jr, Isik K (1987) Comparison of field growth among loblolly pine seedlings and three plant types produced in vitro. TAPPI J 7:119–213

Frampton LJ Jr, Jett JB (1989) Juvenile wood specific gravity of loblolly pine tissue culture plantlets and seedlings. Can J For Res 19:1347–1350

Frampton LJ Jr, Mott RL, Amerson HV (1985) Field performance of loblolly pine tissue culture plantlets. In: Proc 13th So For Tree Improv Conf, Long Beach MS, pp 113–144 (262 pp)

Franclet A (1985) Rejuvenation: theory and practical experiences in clonal silviculture. In: Clonal forestry: its impact on tree improvement and our future forests. Proc 19th Meet Can Tree Improv Assoc, Aug 1983, Toronto, pp 96–134 (235 pp)

Franklin EC (1969) Ortet age has strong influence on growth of vegetative propagules of *Pinus elliotti* Engelm. In: 2nd World Consult Forest tree breeding, Washington, pp 815–827

Gunion M, Hood JV, Libby WJ (1982) A clonal study of intraspecific variability in radiata pine II growth and form. Aust For Res 12:191–210

Hackett WP (1976) Control of phase change in woody plants. Acta Hortic 56:143–154

Hallé F, Oldemann RAA, Tomlinson PB (1978) Tropical trees and forests. An architectural analysis. Springer, Berlin Heidelberg New York

Hartney VJ (1980) Vegetative propagation of the eucalyptus. Aust For Res 10:191–211

Hoekstra PE, Johansen RW (1957) Growth of planted slash pine air-layers. J For 55:146

Hood JV, Libby WJ (1978) Continuing effects of maturation state in radiata pine and a general maturation model. In: Hughes KW, Henke R, Constantin M (eds) Propagation of higher plants through tissue culture: a bridge between research and application. Conf 7804111 Natl Tech Inf Serv USDE, Springfield VA, pp 220–232

Hood JV, Libby WJ (1980) A clonal study of intraspecific variability in radiata pine I Cold and animal damage. Aust For Res 10:9–20

Kleinschmit J (1985) Concepts and experiences in clonal plantations of conifers. In: Clonal forestry: its impact on tree improvement and our future forests. Proc 19th Can Tree Improv Assoc, Aug 1983, Toronto, pp 26–56 (235 pp)

Kleinschmit J, Schmidt J (1977) Experiences with *Picea abies* cuttings propagation in Germany and problems connected with large-scale application. Silv Genet 26:197–203

Klomp BK, Hong SO (1985) Performance of *Pinus radiata* seedlings and cuttings to age 15 years. NZ J For Sci 15:281–297

Kramer PJ, Kozlowski TT (1979) Physiology of woody plants. Academic Press, New York London, 811 pp

Larkin PJ, Scowcroft WR (1981) Somaclonal variation – a novel source of variability from cell cultures for plant improvement. Theor Appl Genet 60:197–214

Lerner IM (1958) The genetic basis of selection. John Wiley & Sons, New York

Lester DT, Berbee JG (1977) Within-clone variation among black poplar trees derived from callus culture. For Sci 23:122–131

Libby WJ (1969) Some possibilities of the clone in forest genetics research. In: Bogart R (ed) Genetic lectures. Oregon State Univ Press, pp 121–136 Corvallis

Libby WJ (1983) The clonal option. Norw For Res Inst 32 pp

Libby WJ (1987) Testing for clonal forestry. Ann For 13(1):69–75

Libby WJ, Hood JV (1976) Juvenility in hedged radiata pine. Acta Hortic 56:91–98

Libby WJ, Jund E (1962) Variance associated with cloning. Heredity 17:533–540

Libby WJ, Brown AG, Fielding JM (1972) Effects of hedging radiata pine on production, rooting, and early growth of cuttings. NZ J For Sci 2:263–283

Mascarenhas AF, Khuspe SS, Nadgauda RS, Gupta PK, Khan BM (1988) Potential of cell cultures in plantation forestry programs. In: Keathley JW, Hanover DE (eds) Genetic manipulation of woody plants. Plenum, New York, pp 391–412 (519 pp)

McKeand SE (1985) Expression of mature characteristics by tissue culture plantlets derived from embryos of loblolly pine. J Am Soc Hortic Sci 110(6):619–623

McKeand SE (1988) Optimum age for family selection for growth in genetic tests of loblolly pine. For Sci 34:400–411

McKeand SE, Frampton LJ Jr (1984) Performance of tissue culture plantlets of loblolly pine in vivo. In: Proc Int Symp Recent advances in forest biotechnology, Traverse City MI, pp 82–91 (194 pp)

Mehra-Palta A, Smeltzer RH, Mott RL (1977) Hormonal control of induced organogenesis: experiments with excised plant parts of loblolly pine. TAPPI 61:37–40

Mikkola L (1982) Results from a trial of *Eucalyptus grandis* cuttings assessed after seven years. Div For Res, Kitwe, Zambia, Res Note 29:12 pp

Mohn CA, Randall WK (1973) Interaction of cottonwood clones with site and planting year. Can J For Res 3:329–332

Mohn CA, Randall WK, McKnight JS (1970) Fourteen cottonwood clones selected for midsouth timber production. USDA For Serv So For Exp Stn Res Pap So- p 6217

Mott RL (1981) Trees. In: Conger BV (ed) Cloning agricultural plants via in vitro techniques. CRC, Boca Raton, pp 217–254 (273 pp)

Mott RL, Amerson HV (1981) A tissue culture process for the clonal production of loblolly pine plantlets. N C Agric Res Ser Tech Bull 271

Nicholls JWP, Brown AG (1971) The ortet-ramet relationship in wood characteristics of *Pinus radiata*. APPITA 25:200–209

Nicholls JWP, Brown AG, Pedrick LA (1974) Wood characteristics of sexually and vegetatively reproduced *Pinus radiata*. Aust J Bot 22:73–79

Nicholls JWP, Pawsey CK, Brown AG (1976) Further studies on the ortet-ramet relationship in wood characteristics of *Pinus radiata*. Silv Genet 25:73–79

Nicholls JWP, Pedrick LA, Brown AG (1977) A summary of the ortet-ramet relationship in wood characteristics of *Pinus radiata*. APPITA 30:496–502

Olesen P (1978) On cyclophysis and topophysis. Silv Genet 27:173–178

Patton RF, Riker AJ (1954) Top growth and root development of rooted white pine cuttings. J For 52:675–677

Pawsey CK (1971) Comparisons of vegetatively propagated and seedling trees of *Pinus radiata*. Aust For Res 5(3):47–57

Power AB, Dodd RS (1984) Early differential susceptibility of juvenile seedlings and more mature stecklings of *Pinus radiata* to *Dothistroma pini*. NZ J For Sci 14:223–228

Power AB, Dodd RS, Libby WJ (1988) Cyclophysis and topophysis in coast redwood stecklings I rooting and nursery performance. Silv Genet 37:8–14

Randall WK, Mohn CA (1969) Clone-site interaction of eastern cottonwood. In: Proc 10th So For Tree Improv Conf, pp 89–91 Houston, TX

Ritchie GA, Long AJ (1986) Field performance of micropropagated Douglas-fir. NZ J For Sci 16:343–356

Roulund H (1974) Comparative study of characteristics of seedlings and clonal cuttings. NZ J For Sci 4:378–386

Russell JH, Libby WJ (1986) Clonal testing efficiency: the trade-off between clones tested and ramets per clone. Can J For Res 16:925–930

Schultz RP (1972a) Inherent vigor influences growth of slash pine more than intensive cultural treatments. Silv Genet 21:26–28

Schultz RP (1972b) Root development of intensively cultivated slash pine. Soil Sci Soc Am Proc 36:158–162

Shaw DV, Hood JV (1985) Maximizing gain per effort by using clonal replicates in genetic tests. Theor Appl Genet 71:392–399

Shaw DV, Hellburg A, Foster GS, Bentzer B (1988) The effect of damage on components of variance for fifth-year height in Norway spruce. Silv Genet 37:19–22

Shelbourne CJA (1986) New approaches in breeding radiata pine. In: IUFRO Conf Proc Joint Meet Work Parties Breeding theory, progeny testing and seed orchards, Williamsburg VA. Hosted by NC State Univ; Ind Tree Improv Prog, Raleigh NC, pp 269–284 (673 pp)

Shelbourne CJA, Thulin IJ (1974) Early results from a clonal selection and testing programme with radiata pine. NZ J For Sci 4:387–398

Skrøppa T (1981) Some results from a 20-year-old cuttings experiment with Norway spruce. In: Symp Clonal Forestry. Dep For Genet Res Note, Swed Univ Agric Sci, Uppsala, pp 105–116 (131 pp)

Skrøppa T, Dietrichson J (1986) Genetic variation and ortet/ramet relationships in a clonal test with *Picea abies*. Scand J For Res 1:323–332

Smith DR, Shelbourne CJA, Dibley MJ, Carswell T (1985) Field performance of micropropagated *Pinus radiata*. In: Abstr 3rd Meet Int Conifer Tissue Culture Work Group, For Res Inst NZ For Serv, Rotorua, p 23 (37 pp)

Squillace AE (1974) Average genetic correlations among offspring from open-pollinated forest trees. Silv Genet 23:149–156

St. Clair JB Jr, Kleinschmit J (1986) Genotype-environment interaction and genotypic stability in ten-year height growth of Norway spruce clones (*Picea abies* Karst). Silv Genet 35:177–186

St. Clair JB Jr, Kleinschmit J, Svolba J (1985) Juvenility and vegetative propagation of Norway spruce clones (*Picea abies* Karst). Silv Genet 34:42–48

Stonecypher RW, McCullough RB (1986) Estimates of additive and non-additive genetic variances from a clonal diallel of Douglas-fir *Pseudotsuga menziesii* (Mirb) Franco. In: IUFRO Conf Proc Joint Meet Work Parties Breeding theory progeny testing and seed orchards, Williamsburg VA. Hosted by NC State Univ; Ind Tree Improv Prog, Raleigh NC, pp 211–227 (673 pp)

Struve DK, Talbert JT, McKeand SE (1984) Growth of rooted cuttings and seedlings in a 40-year-old plantation of eastern white pine. Can J For Res 14:462–464

Sussex I (1976) Phase change: physiological and genetic aspects. Acta Hortic 56:275–280

Sweet GB (1973) The effect of maturation on the growth and form of vegetative propagules of radiata pine. NZ J For 3:191–210

Sweet GB, Wells LG (1974) Comparison of the growth of vegetative propagules and seedlings of *Pinus radiata*. NZ J For Sci 4:399–409

Thomas JE, Riker AJ (1950) Progress on rooting of cuttings of white pine. J For 48:474–480

Tufuor K, Libby WJ (1973) First-lift pruning times of radiata pine seedlings and rooted cuttings in a small California experiment. NZ J For Sci 18:124–134

von Wühlisch G (1984) Propagation of Norway spruce cuttings free of topophysis and cyclophysis effects. Silv Genet 33:215–219

West CG (1984) Establishment requirements of *Pinus radiata* cuttings and seedlings compared. NZ J Sci 14:41–52

Wilcox JR, Farmer RE (1968) Heritability and C effects in early root growth of eastern cottonwood cuttings. Heredity 23:239–245

Wilcox MD, Thulin IJ, Vincent TG (1976) Selection of *Pinus radiata* clones in New Zealand for planting from cuttings. NZ J For Sci 21:239–247

Wisniewski LA, Frampton JR LJ, McKeand SE (1986) Early shoot and root quality effects on nursery and field development of tissue-cultured loblolly pine. Hortic Sci 21:1185–1186

Ying C, Bagley WT (1976) Genetic variation of eastern cottonwood in an Eastern Nebraska provenance study. Silv Genet 25:67–73

Zobel BJ (1985) Vegetative propagation in Eucalyptus. In: Clonal forestry: its impact on tree improvement and our future forests. Proc 19th Meet Can Tree Improv Assoc, Aug 1983, Toronto, pp 136–144 (235 pp)

Zobel BJ, Talbert JT (1984) Applied forest tree improvement, Chap 10: Vegetative propagation. John Wiley & Sons, New York, pp 309–344 (505 pp)

Zobel BJ, Brandao LG, Ikemori YK, Campinhos E Jr (1984) The new eucalypt forest. In: Marcus Wallenburg Found Symp Proc, Sep 1984, Falun, Sw 62 pp

10 Biotechnology and Clonal Forestry

M.R. Ahuja[1]

1 Introduction

Traditionally, forests have been regenerated from seedlings derived from bulked seed collected in nature, or more recently from seeds collected from randomly pollinated plus-trees. In most of these forests there is a large variation in growth, form, and vigor. Besides, there are forest tree species that are characterized by poor and irregular seed set, while in others the seed may be prone to genetic damage or rapid loss of viability. Clonal forestry has been practiced in Japan for centuries to establish Sugi (*Cryptomeria japonica*) forests (Ohba, Chap. 4, Vol. 2), and in recent times in Germany and Sweden to establish Norway spruce (*Picea abies*) forests (Kleinschmit et al., Chap. 2; Bentzer, Chap. 6, Vol. 2), Eucalypts (*Eucalyptus* spp) in Brazil (Zobel, Chap. 7, Vol. 2), and radiata pine (*Pinus radiata*) in New Zealand (Gleed, Chap. 8, Vol. 2).

By employing conventional methods of vegetative propagation, the numbers of plants (stecklings) that can be propagated from a tree species in a growing season often are relatively small. The limiting factors for a mass scale utilization of selected materials are restricted availability of improved genotype as planting material and the available space. Furthermore, in many tree species, such as *Quercus*, *Fagus*, *Eucalyptus*, and in most conifers, woody cuttings from mature trees are generally difficult to root, or the rooting frequency may be rather low. In still others, such as European aspen (*Populus tremula*) and quaking aspen (*Populus tremuloides*), root initials may be lacking and difficult to induce in the woody cuttings. To overcome these problems, tissue culture/biotechnological approaches offer opportunities not only for mass cloning of selected genotypes throughout the year, but also for the genetic modification of tree species. In this respect, biotechnology can serve as an adjunct to the time-honored genetic and breeding programs for tree improvement.

In this chapter I present an overview of biotechnological approaches to mass cloning and genetic improvement of forest tree species.

[1]Institute of Forest Genetics, Federal Research Centre for Forestry and Forest Products, Sieker Landstrasse 2, W-2070 Grosshansdorf, FRG

Clonal Forestry I, Genetics and Biotechnology
Ed. by M.R. Ahuja and W.J. Libby

2 Vegetative Propagation

Clonal propagation may be achieved by macropropagation or micropropagation. Vegetative propagation by macropropagation involves conventional methods, while micropropagation is carried out with the help of tissue-culture technology.

2.1 Macropropagation

Serious consideration to genetic improvement of forest tree species has been given in the past few decades. This involves hybridization between superior trees and establishment of seed orchards. In order to achieve genetic gain in forest tree species, it would be necessary to breed selected trees for at least a few generations, each requiring anywhere from 15 to 50 years. One way to capture genetic gain more readily is to clonally propagate material by vegetative means. In addition to its utility in conservation of the genotype of the donor plant in its propagules (barring mutations), vegetative propagation also offers opportunities for selection and maintenance of both additive and nonadditive gene effects. Nonadditive gene effects arising from gene interactions are not normally transmitted through sexual reproductions. However, nonadditive gene effects can give rise to exceptional individuals, within superior families and elsewhere, and in most tree species these genetic effects can be best captured by vegetative propagation. The alternative strategy of producing homozygotes, progeny testing for specific favorable combinations, and then operationally producing heterotic F_1 families is very difficult with long generation tree species.

The conventional methods of macropropagation involve vegetative propagation by rooted cuttings, rooted needle fascicles (in pines), grafting, or root suckers (in aspens). Macropropagation has been carried out in a majority of cases from juvenile trees, while vegetative propagation from mature trees is still difficult, and usually employs grafting as the method of choice. However, several approaches to maintenance of juvenility in trees involving hedging and serial propagation have been discussed by Talbert et al. (Chap. 11, Vol. 1). By employing conventional methods of vegetative propagation, small to large-scale clonal orchards have been established for a number of tree species in different countries of the world (Talbert et al., Chap. 11, Vol. 1).

2.2 Micropropagation

Plant propagation through tissue culture may be accomplished by employing callus, organ, cell, and protoplast cultures. Although tissue explants from tree species are generally difficult to grow and differentiate in vitro, the first types of cultures have been experimentally employed with varying degrees of success for micropropagation of a number of tree species (Bonga and Durzan 1987; Ahuja 1988a). Although initially callus cultures were employed for plantlet regeneration, now mostly organ cultures are employed for clonal propagation. Cell suspension and protoplast cultures have not been successfully employed for large-scale plant regeneration in clonal forestry. Long-term callus cultures exhibit a relatively high degree of genetic variation. Organ cultures, particularly meristem cultures, on the other hand, are relatively more stable, and involve less genetic risk for clonal propagation. Regeneration problems and genetic variation in plant tissue-cultures are discussed by Chen and Ahuja (Chap. 7, Vol. 1).

In more than 50 tree species, plantlets have been regenerated by employing in vitro techniques (Boulay 1984; Dunstan and Thorpe 1986). However, only in a handful of forest tree species have the techniques progressed far enough for production of large numbers of plants for evaluation of morphologic and genetic stability of the regenerants under field conditions (Ahuja and Muhs 1985; Talbert et al., Chap. 11, Vol. 1). In most of these studies, organ cultures were employed for micropropagation. In some of the angiospermic tree species, tissues from both juvenile and mature trees have been successfully employed for clonal propagation (Gupta et al. 1980; Cornu and Chaix 1981; Ahuja 1983, 1986a, 1987; Ahuja et al., 1988). On the other hand, in the gymnospermic tree species, mainly juvenile explants have been cultured for micropropagation. Organogenesis has been observed on the mature tissues of a number of conifer tree species, but large-scale clonal propagation from mature conifers is still very difficult by tissue culture (Talbert et al., Chap. 11, Vol. 1).

2.2.1 Somatic Embryogenesis

Micropropagation by somatic embryogenesis is another promising avenue for clonal propagation in plants. Just as zygotic embryos, the somatic embryos also develop in a bipolar fashion having both a shoot and a root pole. Somatic embryos develop from embryogenically competent somatic cells in vitro. In contrast to organogenesis, where shoots and roots differentiate sequentially on different media, somatic embryogenesis is a one-step process. That is not to say that one does not need different media for the induction, developmental and maturation phases of somatic embryos in vitro. Rather, the developmental genetic program in the embryogenically competent somatic cells, under the influence of specific phytohormones, proceeds in the same or closely similar fashion as it would in a zygote for the development of a zygote embryo. Somatic embryos have been produced in a number of forest tree species, both softwoods and hardwoods (Redenbaugh and Ruzin 1989; Becwar, Chap. 13, Vol. 1). It has been estimated that 1 l of liquid conifer embryogenic culture may contain about 100000 somatic embryos (Becwar et al. 1988). Since somatic embryos can be grown in liquid media, it should be possible to scale-up somatic embryo production in bioreactors for mass cloning of elite genotypes. In order to develop an efficient delivery system for planting, somatic embryos have been encapsulated to produce artificial or "synthetic" seeds (Redenbaugh and Ruzin 1989). In spite of all this excitement in the area of somatic embryogenesis, some basic problems need further attention (Chen and Ahuja, Chap. 7, Vol. 1). In addition to the difficulty of induction of somatic embryos in some species/genotypes, there are problems with maturation and conversion (germination) of somatic embryos and development of viable somatic seedlings in a number of forest tree species. Morphologic and genetic stability of somatic embryos, somatic seedlings and somatic plants needs to be fully examined.

2.2.2 Maturation and Rejuvenation

Forest trees have long generation cycles, with an extended vegetative phase. Although juvenile and vegetative phases are thought to be the same in herbaceous plants, and in the early stages of tree development, it seems that after about 10 years the vegetative phase is no longer fully juvenile in many tree species. A certain degree of maturation sets in the trees much before the flowering phase, which used to be considered diagnostic for the onset of maturation. In this context, by maturation it is implied that, among other maturation-sensitive traits, the rootability of woody cuttings declines with advancing age of a tree (Bonga and von Aderkas, Chap. 12, Vol. 1). And reduction in the degree of rootability starts in most tree species much before the onset of flowering. Not only is there reduction in rootability, but also the tissues (meristems from buds) from older trees are much less responsive to in vitro induction of organogenesis. For this reason, in most studies explants from juvenile tissues, such as embryos, cotyledons, seedlings, or young plants, have been employed for clonal propagation. Micropropagation from juvenile material, although useful for certain studies, has a disadvantage in that it is difficult to predict how a seedling would perform on reaching maturity. Of course, it would be ideal if biochemical or molecular markers were available in the juvenile stages that are correlated with yield traits. In that event, mature micropropagation might become superfluous! However, we are far from that goal.

It is thought that the aging process starts in the nucleus and permeates to the cytoplasm. Activation and inactivation of genes at different stages of growth and differentiation result in synthesis or abolition of specific proteins. Aging may involve selective and progressive inactivation of genes during development. Some of these genes may be essential for cell division/replacement and household proteins. Mechanisms of gene inactivation may involve methylation of chromosomal DNA (Holliday and Pugh 1975; Hood and Libby 1978; Razin and Riggs 1980), euchromatization (Schaffner and Nagl 1979), heterochromatization (Kolata 1979), or changes in the level of chlorophyll-ab (Cab) binding proteins (Greenwood and Hutchison, Chap. 3, Vol. 1). An alternative hypothesis could be proposed that invokes the activation of maturation-specific genes leading to gradual accumulation of aging-specific proteins, if any. Impairment of gene function by genetic changes in the genome (caused by mutational events or transposable elements) may be yet another possibility in the aging process.

The role of mitochondrial DNA (in plants and animals) and chloroplast DNA (in plants) in the aging process has not been fully investigated. Inactivation or impairment of mitochondrial genes might lead to loss of energy (ATP) and respiration potential of the cells. Genetic and physiological aging in plants may, in part, be regulated by chloroplast DNA, even in the mature trees. Tissue-culture techniques may allow selective expression of those totipotent cells from mature trees for clonal propagation.

3 Genetic Modification of the Genome

3.1 The Genome

Knowledge about the structure of genome is a prerequisite for genetic modification of forest trees and their subsequent clonal multiplication. Recent molecular biology studies on forest tree species have led to new insights into the structure of the tree genome (see Cheliak and Yapa 1988; Hällgren 1988; Kriebel, Chap. 14, Vol. 1). For one thing, the genome size is much larger in conifers than in most other living organisms, including angiospermic tree species (Dhillon 1987). For example, the number of base pairs in loblolly pine (*Pinus taeda*) is about ten times that of poplar, *Populus deltoides*. By employing restriction fragment length polymorphisms (RFLPs), a high-density genetic map is being developed for loblolly pine (Neale et al. 1989a). The chloroplast genome has been mapped by RFLP technology in Douglas-fir (*Pseudotsuga menziesii*) and radiata pine (*Pinus radiata*) (Howe et al. 1988). By using heterologous probes, conifer alcohol dehydrogenase and aldolase genes (Kinlaw et al. 1988) and photosynthesis genes (Lindholm et al. 1988) have been isolated and characterized.

The inheritance of chloroplast DNA (cpDNA) is predominantly maternal in higher plants, although bipaternal inheritance occurs in a few plant species. By employing RFLPs, it has been demonstrated in a number of conifer species that the inheritance of cpDNA is paternal, and not maternal as in many other plant species (Neale et al. 1988). The inheritance of mitochondrial DNA (mtDNA) is, by and large, maternal in plant species. However, In *Sequoia sempervirens* the mtDNA is paternally inherited (Neale et al. 1989b); the cpDNA is also paternally inherited in this species. Further molecular biology studies on a number of forest tree species (both gymnosperms and angiosperms) are necessary for a better understanding of the genome structure of trees, and determining the mode of inheritance of chloroplast and mitochondrial DNA, so that economic traits governed by nuclear and organelle genes are, by selecting the appropriate donor, incorporated in clonal forestry breeding programs.

3.2 Gene Transfer

Recent advances in molecular biology and cell-culture technology have made it possible to attempt genetic engineering in plants. At one point it was considered that protoplasts are essential for gene transfer and genetic modification of plants. However, by employing *Agrobacterium* as a vehicle for transfer of alien genes into plant cells, protoplasts may have lost some of their importance in genetic engineering. Furthermore, protoplasts from a number of economically important agricultural crops and forest tree species are still very difficult to grow and differentiate in vitro. For these reasons, the future of protoplast research is undergoing critical evalutation for its utility in genetic improvement of commercial crops. It is also not clear whether plants derived from protoplasts are true-to-type and remain genetically stable afterwards. After all, not every one is always looking for somaclonal variation. Relative genetic stability is essential for recovery of clonal material from genetically modified cells/tissues and plants.

Gene transfer in plants may be accomplished by (1) *Agrobacterium* mediation, (2) electroporation, (3) microinjection, (4) DNA-uptake, and (5) microprojectile bombardment. Of these, *Agrobacterium*-mediated transfer of foreign genes into plants has been the most widely and successfully used approach. *Agrobacterium* transfer requires wounded tissues of the host-genotype that can be infected with this bacteria. Gene transfer is achieved by: (1) insertion of the desired gene into the tumorigenic or disarmed Ti plasmid, (2) transport or reconstituted Ti plasmid by *Agrobacterium* through the wound into a plant cell, and (3) integration of the foreign gene sequence into the plant genome. By employing an *Agrobacterium*-mediated gene-transfer system, transgenic plants have been produced in a number of plant species, including woody plants, for example poplar (see Ahuja 1988b, c). However, some questions remain regarding the genetic stability of genetically engineered plants (Ahuja 1988c, d).

Direct DNA transfer (without the involvement of *Agrobacterium*) may be carried out by electroporation, microinjection, or uptake of DNA from a solution. In such studies mostly protoplasts have been employed for gene transfer. However, rapidly growing cells with very thin cell walls may also be employed for direct transfer of genes into plants. The latest method for gene transfer involves microprojectile bombardment of plant cells/tissues with DNA-coated tungsten bullets (see Sederoff and Stomp, Chap. 15, Vol. 1).

Although transgenic plants have been regenerated in a couple of hardwoods, including poplar (Fillatti et al. 1987), and walnut (McGranaham et al. 1988), the conifers have remained recalcitrant in this regard. This may have been partly because of lack of appropriate cell culture/regeneration systems following genetic transformation. However, with the availability of somatic embryogenesis systems in several conifers (Becwar, Chap. 13, Vol. 1), gene transfer and production of transgenic plants in this group of commercially important forest tree species may become a real possibility.

While studies on genetic engineering of tree species are progressing, we should not neglect to exploit the enormous existing genetic variability present in the populations of most forest tree species for recovering newer genotypes. Both types of programs, involving breeding and selection, on the one hand, and genetic engineering, on the other hand, can provide exceptional material for clonal forestry.

4 Preservation of Germplasm

Forest decline is proceeding at an alarming rate worldwide. In addition to deforestation, a number of other diverse factors are also involved in forest decline. These include the biotic factors, the abiotic factors, and the man-made factors, the so-called anthropogenic factors. The last category includes gaseous, metallic, and acid pollutants that are spewed into the atmosphere from factories and transportation sector. These pollutants are believed to be involved in the

"Waldschaden" (forest damage) and the "Waldsterben" (dying forests). Air pollution, in particular acid rain, may have initiated the process, but overall Waldsterben is most likely an end result of a multi-factor, multi-stage phenomenon. Some other factors that may be involved in forest decline may be diverse and varied. These might include microbes, pests, drought, soil acidity, radiation, ozone, and perhaps viruses. In other words, it is a cumulative effect of several harmful agents acting synergistically to produce the current state of Waldsterben (Ahuja 1991a).

The present situation forces us to focus on the necessity of planning forest conservation strategies for future exploitation. Various approaches for conservation of forest tree germplasm have been proposed (Melchior et al. 1986; Ledig 1988; Ahuja 1991b; Ahuja and Muhs 1989; Millar, Chap. 3, Vol. 2). Broadly speaking, preservation may be carried out by in situ or ex situ measures. The in situ measures involve preservation on site of the original stand by natural regeneration by seed or vegetative propagation. While the ex situ strategies involve preservation away from the site of the original stand, either under field conditions or under controlled conditions. This latter category includes storage of seed, pollen, plants and plant parts (buds, meristems, embryos, cells) at low temperatures (Ahuja, 1988e, 1989).

Seeds of forest tree species are generally stored at low temperatures from 4 to − 20°C. At these low temperatures, the seeds can be stored, in a majority of cases, only for a limited period. In most cases, the seeds begin to lose viability after a few years. Seeds of oak (*Quercus robur*, *Q. petraea*, *Q. rubra*), beech (*Fagus sylvatica*) and silver fir (*Abies alba*) can be stored only for 1 to 4 years at such low temperatures.

In recent years, the cryopreservation potential (storage in liquid nitrogen at − 196°C) of seeds and dormant buds for preservation of forest tree germplasm has been explored. Seeds from silver fir (*Abies alba*), Norway spruce (*Picea abies*), scots pine (*Pinus sylvestris*), larch (*Larix decidua*), beech (*Fagus sylvatica*), oak (*Quercus robur*, *Q. petraea*, *Q. rubra*), and hybrid aspen (*Populus tremula* × *P. tremuloides*) were stored for upto 6 days in liquid nitrogen. Seed germination data, after retrieval from liquid nitrogen, showed that there was no loss of viability in the seeds of these forest tree species, except those of *Quercus* (oak) and *Fagus* (beech) species (Ahuja 1986b, 1989). Beech and oak seeds did not germinate after storage in liquid nitrogen. It appears that relatively large seeds from beech and oak are more prone to freezing-thawing injury as compared to the relatively small seeds of silver fir, Norway spruce, larch, Scots pine, and aspen. Refinements of cryopreservation techniques, combined with tissue culture, are being carried out at the Institute of Forest Genetics, Grosshansdorf, Germany, for long-term preservation of beech and oak seeds (Ahuja, unpubl.).

Winter-hardy twigs from several tree species, for example willow, poplar, and apple were stored in liquid nitrogen and were found to be viable following storage (Sakai 1986). Dormant buds from a number of aspen and hybrid aspen clones were stored in liquid nitrogen, and bud explants from the retrieved buds were found viable and able to differentiate microshoots (Ahuja 1989, and unpubl.).

Thus, it is evident that cryopreservation can be effectively employed for preservation of forest tree germplasm. Preservation of seed is important for maintenance of genetic variability of a tree population. The storage of vegetative parts, such as dormant buds, on the other hand, is useful for preservation of superior genotypes. Preservation of both seeds and vegetative parts would provide valuable material for clonal forestry programs.

Other adaptations of technology to biology are also coming on line. For example, very accurate microcalorimetry developed to investigate cold fusion is now being applied for early selection of productive genotypes of forest trees (Anekonda et al. 1990).

5 Concluding Statement

Biotechnology offers prospects for rapid clonal propagation, germplasm preservation, and genetic engineering of forest tree species. However, there are limitations. The basic parameters of growth, differentiation, and regeneration are not well understood. Many tree species are still recalcitrant and do not respond well to in vitro conditions. Mature tissues from most forest tree species, in particular gymnosperms, are still very difficult to regenerate in vitro. Therefore, further physiological, biochemical, and molecular studies are necessary on growth and differentiation of woody plants. The molecular structure of the forest tree genome is currently under active investigation, and isolation and characterization of genes and RFLP genetic maps seem necessary for genetic engineering of forest tree species. Clonal forestry will greatly benefit from recent advances in biotechnology.

References

Ahuja MR (1983) Somatic cell genetics and rapid clonal propagation of aspen. Silv Genet 32:131–135

Ahuja MR (1986a) Aspen. In: Evans DE, Sharp WR, Ammirato PJ (eds) Handbook of plant cell culture, vol 4. Macmillan, New York, pp 626–651

Ahuja MR (1986b) Storage of forest tree germplasm in liquid nitrogen (– 196°C). Silva Genet 35:249–251

Ahuja MR (1987) In vitro propagation of poplar and aspen. In: Bonga JM, Durzan DJ (eds) Cell and tissue culture in forestry, vol 3. (Nijhoff, Dordrecht, pp 207–223

Ahuja MR (ed) (1988a) Somatic cell genetics of woody plants. Kluwer, Dordrecht, 225 pp.

Ahuja MR (1988b) Gene transfer in forest trees. In: Hanover JW, Keathley DE (eds) Genetic manipulation of woody plants. Plenum, New York, pp 25–41

Ahuja MR (1988c) Gene transfer in woody plants: perspectives and limitations. In: Ahuja MR (ed) Somatic cell genetics of woody plants. Kluwer, Dordrecht, pp 83–101

Ahuja MR (1988d) Molecular genetics of transgenic plants. In: Hällgren, JE (ed) Molecular genetics of forest trees. Swed Univ Agric Sci, Umeå, pp 127–145

Ahuja MR (1988e) Differential growth of aspen clones stored at sub-zero temperatures. In: Ahuja MR (ed) Somatic cell genetics of woody plants. Kluwer, Dordrecht, pp 173–180

Ahuja MR (1989) Storage of forest tree germplasm at sub-zero temperatures. In: Dhawan V (ed) Application of biotechnology in forestry and horticulture. Plenum, New York, pp 215–228

Ahuja MR (1991a) Air pollution and mammal resistance genes. In: MacNabb HS (ed) Genetics and molecular biology of interactions between harmful agents and trees. Springer, Berlin Heidelberg New York (submitted)

Ahuja MR (1991b) Application of biotechnology to preservation of forest tree germplasm. In: Ahuja MR (ed) Woody plant biotechnology. Plenum, New York London pp 307–313

Ahuja MR, Muhs HJ (1985) In vitro techniques in clonal propagation of forest tree species. In: Schäfer-Menuhr A (ed) In vitro techniques–propagation and long term storage. Nijhoff, Dordrecht, pp 41–49

Ahuja MR, Muhs HJ (1989) Biotechnologische Verfahren zur besseren Evaluierung, Erhaltung and Nutzung von forstlichen Gen-ressourcen. In: Biotechnologie in der Agrar- und Ernährungswirtschaft. Berichte über Landwirtschaft. Spec vol 201 Parey, Hamburg, pp 414–422

Ahuja MR, Krusche D, Melchior GH (1988) Determination of plantlet regeneration capacity of selected aspen clones in vitro. In: Ahuja MR (ed) Somatic cell genetics of woody plants. Kluwer, Dordrecht, pp 127–135

Anekonda ST, Criddle RS, Libby WJ (1990) The use of micro-calorimetry for early culling. In: Joint Meet WFGA; IUFRO WP S2.02-05, 06, 12 and 14, Olympia, 20–24 Aug 1990. Weyerhaeuser, Tacoma, pp Sec 6.1, (11 pp)

Becwar MR, Wann SR, Johnson MA, Verhagen VA, Feirer RP, Nagmani R (1988) Development and characterization of in vitro embryogenic systems in conifers. In: Ahuja MR (ed) Somatic cell genetics of woody plants. Kluwer, Dordrecht, pp 1–18

Bonga JM, Durzan DJ (eds) (1987) Cell and tissue culture in forestry, vol 3. Nijhoff, Dordrecht

Boulay M (1984) Some special aspects and application of the micropropagation of forest trees. In: Proc Int Symp In vitro propagation of forest tree species, Bologna, pp 51–81

Cheliak WM, Yapa AC (eds) (1988) Molecular genetics of forest trees. Petawawa Nat For Inst, Chalk River

Cornu D, Chaix C (1981) Multiplication par culture in vitro de mérisières adultes (*Prunus avium*)—application à un large éventail de clones. In: Coll Int Culture "in vitro" des essences forestières. AFOCEL, Nangis, pp 71–79

Dhillon SS (1987) DNA in tree species. In: Bonga JM, Durzan DJ (eds) Cell and tissue culture in forestry. vol 1. Nijhoff, Dordrecht, pp 298–313

Dunstan DI, Thorpe TA (1986) Regeneration in forest trees. In: Vasil IK (ed) Cell culture and somatic cell genetics of plants, vol 3. Academic Press, New York London, pp 223–241

Fillatti JJ, Selmer J, McCown B, Haissig B, Comai L (1987) *Agrobacterium*-mediated transformation and regeneration in *Populus*. Mol Gen Genet 206: 192–199

Gupta PK, Nadgir AL, Mascarenhas AF, Jagannathan V (1980) Tissue culture of forest trees: clonal multiplication of *Tectona grandis* L. (teak) by tissue culture. Plant Sci Lett 17: 259–268

Hällgren JE (ed) (1988) Molecular genetics of forest trees. Univ Press, Swed Univ Agric Sc, Umeå

Holliday R, Pugh JE (1975) DNA modification mechanism and gene activity during development. Science 187: 226–232

Hood JV, Libby WJ (1978) Continuing effect of maturation state in radiata pine and a general maturation model. In: Hughes KW, Henke R, Constantin M (eds) Propagation of higher plants through tissue culture. Proc. Symp Univ Tennessee, Knoxville, pp 220–232

Howe GT, Strauss SH, Doerksen AH, Palmer JD (1988) Gene and restriction site maps of Douglas-fir [*Pseudotsuga menziesii* (Mirb.) Franco] and radiata pine (*Pinus radiata* D. Don) chloroplasts genomes. In: Cheliak WM, Yapa AC (eds) Molecular genetics of forest trees. Petawawa Natl For Inst, Chalk River, pp 54–66

Kinlaw CS, Harry DE, Sleeter DD, Sederoff RR (1988) Use of heterologous probes to isolate and characterize conifer genes. In: Cheliak WM, Yapa AC (eds) Molecular genetics of forest trees. Petawawa Nat For Inst, Chalk River, pp 9–18

Kolata GB (1979) Developmental biology; where is it going? Science 206: 315–316

Ledig FT (1988) The conservation of diversity in forest trees. BioScience 38: 471–479

Lindholm J, Szmidt A, Gustafsson P (1988) Lack of one rDNA repeat and duplication of *psb*A in the chloroplast genome of lodgepole pine (*Pinus contorta* Dougl.) In: Cheliak WM, Yapa AC (eds) Molecular genetics of forest trees. Petawawa Natl For Inst, Chalk River pp 67–74

McGranaham GH, Leslie CA, Uratsu SL, Martin LA, Dandekar AM (1988) *Agrobacterium*-mediated transformation of walnut somatic embryos and regeneration of transgenic plants. Bio/Technology 6: 800–804

Melchior GH, Muhs HJ, Stephan BR (1986) Tactics for the conservation of forest gene resources in the Federal Republic of Germany. For Ecol Manag 17: 73–81

Neale DB, Marshall KA, Sederoff RR (1988) Inheritance of chloroplast and mitochondrial DNA in conifers. In: Hällgren JE (ed) Molecular genetics of forest trees. Univ Press Swedish Univ Agric Sci, Umeå, pp 89–100

Neale DB, Tauer CG, Gorzo DM, Jermstad KD (1989a) Restriction fragment length polymorphism mapping of loblolly pine: methods, applications and limitations. In: Proc 20th Southern forest tree improvement Conf. Charleston, pp 363–372

Neale DB, Marshall KA, Sederoff RR (1989b) Chloroplast and mitochondrial DNA are paternally inherited in *Sequoia sempervirens* D. Don Endl. Proc Natl Acad Sci USA 86:9347–9349

Razin A, Riggs AD (1980) DNA methylation and gene function. Science 201:604–610

Redenbaugh K, Ruzin SE (1989) Artificial seed production and forestry. In: Dhawan V (ed) Application of biotechnology in forestry and horticulture. Plenum, New York, pp 57–71

Sakai A (1986) Cryopreservation of germplasm of woody plants. In: Bajaj YPS (ed) Biotechnology in agriculture and forestry. vol 1. Trees. Springer, Berlin Heidelberg New York Tokyo, pp 113–129

Schaffner KH, Nagl W (1979) Differetial DNA replication involved in transition from juvenile to adult phase in *Hedera helix* (Araliaceae). Plant Syst Evol Suppl 2:105–110

11 Conifer Vegetative Propagation: an Overview from a Commercialization Perspective

C.B. Talbert, G.A. Ritchie, and P. Gupta[1]

1 Vegetative Propagation – A Revolution in Commercial Conifer Forestry...Or Is It?

> "Clonal forestry is one of the most exciting – and controversial – issues facing the industry today. There is no doubt about clonal forestry not working; rather, the issue is about a proper and responsible assessment."
>
> *Anon, Australian Forest Grower, Summer 1987*

Words like these, from a prominent forestry trade journal in Australia, reflect the degree of interest and awareness about vegetative propagation in forest trees around the world as we move into the 1990s, and they must be welcome words to those diligent pioneers who initiated the research from which later technological advances have been built. In the last two decades the world has witnessed an explosion of interest and research effort in vegetative propagation of commercial timber species. As this volume illustrates, many key technical breakthroughs have been made enabling the application of vegetative propagation in forestry on a fairly large scale. It would appear that we are in the midst of a technological revolution in forestry, one with profound impacts on a par with the first introduction of modern plantation management in the early decades of the 20th century.

The measure of this revolution will be in the changes brought about in the value realized from operational forestry. Are these technologies making a difference? This chapter will address this question, and will attempt to provide some insights into approaches which are working and those which are not.

First we will overview some of the major considerations in providing the genetic material to sustain a vegetative propagation system, and review the genetic improvement approaches of some of the organizations which are leading the way in the commercialization of vegetative propagation in forestry. Then we will discuss the propagation technologies themselves, not only in terms of a summary of technical developments, but also in terms of the methods and strategies being used by leading-edge organizations. In the third and final section we will present our views on the priorities for this next, critical stage in the commercialization process for vegetative propagation. Because there is very little published literature on operational production, we base much of the

[1] Weyerhaeuser Company Research and Development, Tacoma, WA, USA

Clonal Forestry I, Genetics and Biotechnology
Ed. by M.R. Ahuja and W.J. Libby

Table 1. Worldwide annual conifer rooted cutting production in 1989[a]

Country	Organization	Species	Annual production
			(thousands of plants)
Australia	S. Austr. Woods and Forests Dept., Mt. Gambier	*Pinus radiata*	300
	A.P.M. Forests, Ply. Ltd., Morwell, Victoria	*Pinus radiata*	4000
	Radiata Contractors of Austr. Mgt., Ltd.	*Pinus radiata*	3000
	Dept. of Cons. Forests & Lands, Victoria	*Pinus radiata*	500
	Egalme Nursery, Tumbarumba NSW	*Pinus radiata*	500
Belgium	Combined Federal & Private	*Picea abies*	140
Canada	Northern Clonal Forestry Center, Ontario	*Picea mariana*	1000
	Nova Scotia Dept. Lands and Forests	*Picea mariana*	3000
	J.D. Irving, Ltd., New Brunswick	*Picea mariana*	100
	Ontario Ministry Nat. Res., Orono	*Picea abies*	100
	British Columbia Min. For., Victoria	*Chamaecyparis nootkatensis*	650
Czechoslovakia	Numerous State Nurseries	*Picea abies*	250
	Forest Tree Breeding Sta., Tyniste n. Orlici	*Picea abies*	100
Denmark	Nat. For. & Nature Agency, Tree Imp. Sta.	*Picea abies*	400
	Nat. For. & Nature Agency, Tree Imp. Sta.	*Picea sitchensis*	100
Finland	Foundation for Tree Breeding, Helsinki	*Picea abies*	150
France	AFOCEL	*Picea abies*	1000
Germany	For. Res. Inst., Lower Saxony	*Picea abies*	700
	Inst. for Tree Breeding, Reinhardshagen	*Picea abies*	350
Great Britain	Brit. For. Comm., Nurseries, England	*Picea sitchensis*	1700
	Brit. For. Comm., Midlothian, Scotland	*Picea sitchensis*	130
	Elite Trees, Inverness, Scotland	*Picea sitchensis*	1500
	Tilhill Nurseries, Farnham	*Picea sitchensis*	250
	Banff & Buchan Nurs., Banff, Scotland	*Picea sitchensis*	120
Ireland	Coillte Teo., Dublin	*Picea sitchensis*	150
Japan	Private Forestry (summed figures)	*Cryptomeria japonica*	27700
	Forestry Agency National Forest	*Cryptomeria japonica*	3700
New Zealand	Tasman Forestry Ltd., Rotorua	*Pinus radiata*	1250
	New Zealand Forest Products	*Pinus radiata*	500
	N.Z. For. Res. Inst., Rotorua	*Pinus radiata*	100
Norway	Biri Plantskole, Biri	*Picea abies*	500
Sweden	Nat. Board of For., State For. Serv.	*Picea abies*	2000
	Hilleshog, Landskrona	*Picea abies*	4000

Table 1. (continued)

Country	Organization	Species	Annual production
			(thousands of plants)
	Inst. for Forest Impr.	*Picea abies*	300
	National For. Enterprise of Sweden	*Picea abies*	1000
United States	Int'l For. Seed Co., Alabama	*Pinus taeda*	100
	Weyerhaeuser Co., Washington	*Pseudotsuga menziesii*	1000
USSR	Latvian For. Expt. Sta.	*Picea abies*	50
	Ivantejevka Nursery, VNIILM Institute	*Picea abies*	100
Yugoslavia	Inst. za gozdno in Lesno	*Picea abies*	10

[a]Information derived from a survey completed April 4, 1989.

discussion on responses to a questionnaire which was mailed to 50 organizations around the world, incorporating questions on the size of programs, the species used, the techniques and systems of greatest promise in propagation and deployment, and the respondent's view of the steps or breakthroughs required for successful commercialization. (The organizations who responded to the questionnaire are listed in Table 1.)

2 Status of Conifer Vegetative Propagation: a Survey of Leading Programs

2.1 Supplying Elite Germplasm for Propagation

> "Clonal forestry is widely but erroneously regarded as a breeding method which represents the ultimate refinement in tree breeding...whatever its merits, clonal forestry is not a breeding method in itself but is rather a component of the total system of genetic management."
>
> *Burdon (1988a)*

2.1.1 Important Considerations in Genetic Population Development

Much like sophisticated data analysis software, the quality of the output from a vegetative propagation system depends heavily on the quality of the input. In the case of the software, the input is experimental data. In the case of the vegetative propagation system, the input is the population of genetic material to be propagated. This section will provide a brief overview of some of the major considerations or decisions involved in generating the genetic material to sustain a vegetative propagation program.

Definition of the Population to be Propagated. Genetic populations to be vegetatively propagated can take many forms, from species, to provenances, to hybrids, to families or individuals within provenances. In some cases, species or provenances are highly valued in themselves, but production from seed is difficult or costly. In the case of hybrids, several notable cases exist in which hybrids between provenances or species are strongly superior to the parental stocks and are more reliably propagated vegetatively than from seed (Burdon 1986; Haines and Nikles 1987; Bentzer Chap. 6, Vol. 2). In other situations, there is sufficient interest in a single species and/or provenance that organizations choose to select and propagate outstanding families or individuals within the species or provenance. In some of these situations, vegetative propagation becomes a logical production approach because seed production from orchards is inefficient or unworkable. Most of the time, however, the choice between generative and vegetative propagation is based on a weighing of the benefits and tradeoffs of the two approaches against the constraints and objectives of a given program. Some discussion of these benefits and tradeoffs is given in Section 2.6 below.

Scope of Genetic Improvement Efforts. A genetic improvement effort can be as simple as a single selection step, or as complex as multiple-generation recurrent selection and breeding. The difference between them is the level of effort required, and commensurately, the genetic gain which is expected. Recurrent programs, in which trees are selected and intermated in repeated cycles, generate new and better genotypes through recombination each cycle. A one-step program can only exploit the genetic variation which exists at the time of selection. It is possible for a single cycle of selection, testing and cloning to provide similar gains in a similar timeframe as a two-cycle program based on wind-pollinated orchard production (Cameron and Cotterill 1988). However, once the single-cycle program has "wrung" all the genetic potential from the population, there is no potential to develop and propagate new and even better genotypes.

In summary, an organization need not embark on a recurrent improvement program to justify vegetative propagation. The choice to be made is the level of improvement desired and the level of investment which is possible. Significant gains are possible in a single cycle, but the greatest long-term gains are available from recurrent improvement.

Population Size and Structure. At least two important objectives drive the population size for genetic improvement programs. One is to provide sufficient genetic variation to maximize long-term genetic gain potential. The second is to provide sufficient allelic variation to allow for future response to new environments or economic value functions.

The first consideration may have received more emphasis over the years than it deserves, judging from the sustained progress observed from mass selection for oil and protein content over 76 generations in maize:

> "There appears to be little concern about exhausting genetic variability for grain yield since, with gene frequencies between 0.25 and 0.5, only from 50–200 loci are required to allow a gain of 20σ. One would expect that there are more loci segregating for yield in most random-mating populations than

> for percent oil or protein. Thus, so far as quantitative traits are concerned, the most pressing problems may not be those of obtaining more genetic variability, but those of learning how to most efficiently concentrate the favorable alleles we now have and then proceeding to do it."
>
> *Dudley (1976)*

A locus-level simulation model of a tree breeding program supported this conclusion, projecting steady increases in gain through 14 generations of intensive selection and assortative mating with a population census number of 48 (Mahalovich and Bridgwater 1989).

Population size and within-population genetic diversity may be most important as a means of buffering against, and enabling effective response to, changes in environment and economic goals in space and time. This will be an important consideration for production populations as well as long-term breeding populations. The expectation is that population size and diversity increases the probability that alleles will be sampled which may be selectively neutral today but which will be important to adaptability or value in another environment or at some future time. The probability of sampling an allele rapidly approaches unity as the effective population size exceeds the inverse of the allele frequency (Burdon 1988b). Effective population sizes on the order of 50 to 200 should maintain unlinked neutral alleles with initial frequencies of 0.01–0.05 for 30 generations (Kang 1979), while alleles linked to traits under selection will require larger populations for acceptable rates of loss (Namkoong and Roberds 1982). Presumably, if initial program selections are taken from a broad environmental range, or if a concerted effort is made to include selections which exhibit a range for traits which are currently selectively neutral, the probability of inclusion of rare neutral alleles would be increased for a given population size.

Broad vs. Focused Improvement Goals. The degree of focus of improvement efforts must be defined in terms both of value traits and adaptation to environments. From the perspective of value traits, it is possible to direct selection and breeding efforts to produce a single population which shows at least some degree of improvement for a range of possible economic futures. In contrast, it is possible to direct two or more populations in different directions, supporting different value functions. The situation is analogous for single populations which are acceptably well adapted to a broad array of environments vs. multiple populations developed for strong performance on discrete environments. The choice of focus becomes more pressing with vegetative propagation due to the greater degree of production flexibility and greater genetic control available through mass propagation of controlled crosses or clones. The case for each direction is one of logistics, cost, uncertainty, and organizational goals.

In terms of cost and logistics, multiple populations are more complex to plan, but their simplicity and cost of implementation relative to a single-population approach depends on the number of traits or environments and the total number of trees to be measured in the two cases. From a population size and diversity point of view, it should not matter whether a given number of trees is subdivided into multiple populations or kept as one, larger population (Namkoong et al. 1988). In fact, genetic variability may be greater for multiple-population options, for a given population size, if selection is practiced along divergent paths.

If multiple populations are actually to be carried through to field establishment and beyond, the greatest logistical consideration is likely to be in forest management. Systems must be in place to ensure that site-optimized material goes to the right environment. Even more challenging are multiple economic populations, which must be tracked through the rotation cycle, managed, and processed appropriately for the economic objective.

On the other hand, multiple populations may serve primarily as buffer against the risk that environments or economic objectives will change over the course of an improvement cycle, and only the best suited population at the time of production may actually be deployed in the field. (This, of course, requires some production-system flexibility).

In summary, if the environments and/or economic objectives of interest are homogeneous, or if a single genetic population can be identified which provides good, stable progress across the range of objectives of interest, then a single population approach by merit of its simplicity is likely to be the way to go. However, it is often the case that important environments or economic objectives are at least to some extent mutually exclusive. Even more often, environments and economic futures are highly uncertain.

> "If breeders can determine traits that will affect future values, but cannot predict the directions in which to breed for trait combinations, then the utility of future populations can be increased by increasing the variation among populations. Extremal sets of trait values...could be selected in separate populations. With different kinds of future forest economic scenarios, an index type of selection system could be generated for each of several possible futures...to diversify populations."
>
> *Namkoong et al. (1988)*

Vegetative propagation simplifies the production of multiple populations, by making it possible to produce operational quantities of planting stock of diverse types produced through controlled crossing or cloning. In fact, through vegetative propagation, diverse types can be selected and propagated even from a single breeding population. The size and cost of traditional wind-pollinated orchards make multiple-population production much more difficult.

Approach to Recurrent Improvement. The most common recurrent improvement approach in forestry relies on testing of and selection for outstanding general combining ability (GCA). Because vegetative propagation enables the capture of specific combining ability (SCA) along with GCA, alternative improvement approaches have been proposed which seek to increase SCA and GCA in the population. Such approaches also are ideal for improving the performance of population hybrids. The most commonly discussed is Reciprocal Recurrent Selection, and a variation called Reciprocal Full-Sib Selection. In these systems, selection within a population is on the basis of performance of intercrosses with another population. Over repeated cycles, gene frequencies in the paired populations move in directions which most favor the intercross. The result is increased SCA, and better gain than recurrent selection for GCA, for traits controlled by loci at which overdominance or epistasis are important (Comstock et al. 1949). Such systems would produce results similar to selection for GCA for cases of zero or partial dominance.

For many species and traits it appears that SCA effects, and therefore, dominance effects, are much less important than GCA effects, and may in fact disappear with time (Stonecypher et al. 1973; Byram and Lowe 1986; Stonecypher and McCullough 1986). Where this is the case, it is probably better for an organization to focus its efforts on GCA improvement, directing any multiple-population efforts toward increasing economic or environmental flexibility, rather than to pursue improvement of SCA.

Selection and Testing Approaches. Selection effectiveness hinges on two elements: the selection intensity, and the ability to perceive the genetic value of the genetic entries to be screened.

The first element, selection intensity, is a simple concept in theory: the more entries that are screened, the greater the expected genetic superiority of a given number of selected entries from the upper tail of the population distribution. The pursuit of high selection intensities is often difficult in practice, however. Scarce resources often limit the number of entries which can be screened, especially when a number of trees must be evaluated to establish a mean value for a given entry, and testing costs are high. In vegetative propagation programs, selection intensity may be even more constrained by an inability to propagate some or many of the selected genotypes. Also, for a given test size, clonal replication tends to reduce the number of genotypes which can be tested, and therefore, the selection intensity.

The second key to effective selection is the ability to separate genetic from environmental effects in the observed value of different genetic entries, so that only the best genotypes are selected. This is fairly straightforward in the cases of species, provenance, or family selection, where wind-pollinated or control-pollinated seedlots of the different entries can be collected and planted in replicated trials across environments, allowing environmental effects to be partitioned out. When the units of selection are individual genotypes, this becomes more complex because, at least at the outset, a given genotype is represented by only a single seed. Of course, if vegetative propagation is feasible, then the actual candidate genotypes may be replicated across environments. This approach provides the highest selection effectiveness, and the greatest gain for a given number of candidates tested. Alternatively, siblings or progeny of the candidates can be produced and the resulting families planted in replicated trials over environments. In this option, the *genotype* of interest is not replicated specifically, but the *genes* comprising the genotype are.

To illustrate the importance of genetic replication to selection effectiveness, relative gains were predicted for individual selection based on phenotype alone, based on a family-individual combined index, and based on a clonally replicated test, using genetic and phenotypic parameter estimates for average height from 8-year-old Douglas-fir diallel tests (Stonecypher, Weyerhaeuser Company, unpublished data). These comparisons are given in Table 2. As expected, selection on phenotype alone was only half as efficient as selection based on a combined sib-individual index (5.9 vs. 13.3%). The incremental value of clonal replication over the best family-individual index approach is much less clear cut, however.

Table 2. Predicted percent gain for selection of ten individual genotypes using (A) mass selection; (B) selection based on family and individual data with no clonal replication; and (C) selection based on family and clone mean data with clonal replication, for two test size alternatives. Predictions are for average height, based on unpublished genetic and phenotypic parameters from 8-year-old Douglas-fir genetic tests

	Gain (%)	No. of trees tested[a]	Gain/1000 trees tested
A) Mass selection	5.9 +/− 5.4	14400	0.4%
B) Combined index (no clonal replication)	13.3 +/− 1.5	14400	0.9
C) Combined index (with clonal replication)			
C1) Same no. of test trees as A and B[b]	14.4 +/− 0.7	14400	1.0
C2) Same no. of tested genotypes as A, B	19.2 +/− 0.7	691200	0.03

[a]For cases A and B, number includes ten six-parent disconnected half diallel tests, planted with four trees per cross in each of eight replications on each of three sites. Cases C1 and C2 assume that each genotype is replicated two times in each of four reps on each of six sites.
[b]Represents 300 genotypes tested.

In the Douglas-fir example, clonal replication (48 total ramets per clone) provided an additional improvement in selection effectiveness of only 1 to 5% depending on the final selection intensity. In other, similar analyses, nearly 1.5 times the gain was predicted for selection based on clonal means, for traits of low to moderate heritability, compared to simple mass selection, assuming a fixed test size (Russell and Libby 1987). However, the incremental benefit of clonal replication was negligible when a more efficient family-plus-individual selection index approach was used as the base case for comparison (Shaw and Hood 1985). Clonal replication may be of greatest value as a tool for precise characterization of a smaller number of genotypes selected via family-individual index from conventional seedling tests. For this to be possible, however, the technology must exist to 'store' a high proportion of the original population of genotypes, in a condition from which the best may be propagated en masse later.

Since vegetative propagation is often used as a means to multiply controlled-cross families, an additional issue arises: is it necessary to create and test specific crosses, or is it possible to identify the best *crosses* by identifying the best *parents*? The low apparent magnitude of SCA effects for important species and traits suggests that there would be little to gain from the greater testing burden associated with crosses. In one comparison based on 42 Douglas-fir diallel tests, the gain from selecting and mating the best *parents*, and the set structure of these diallels did not even allow the best parents to be crossed (Stonecypher, Weyerhaeuser Company, unpubl. data).

Genotype by environment (G × E) interaction also deserves mention as it relates to vegetative propagation. A greater concern tends to be voiced about the magnitude of G × E which will be observed as the genetic entries to be deployed become more and more genetically homogeneous, and as we are able to exploit a greater range of genetic variability in a population, e.g., controlled-cross families or clones. Genotype by site interactions in clonal tests of Norway spruce and radiata pine have proven statistically significant, but results vary widely. In some of these tests G × E effects are only marginally significant and little rank change is observed among clones (St. Clair and Kleinschmit 1986; Bentzer et al. 1988, 1989). In other tests the G × E effects are strong and substantial rank change is observed (Burdon 1971; Kleinschmit and Schmit 1979). The jury is still out on the real importance that G × E effects will have; still, it is prudent to plan a testing program against the eventuality that such effects will be significant, and then to back off later if G × E proves unimportant.

From a testing perspective the spectre of significant G × E interaction can mean one of two things. If the objective of the program is to produce broadly adapted material, then the testing program should be designed to cover the range of expected environmental stress factors, to ensure that families or clones selected are stable. If the program is focused on production of multiple, narrowly adapted populations, then a single population might require only a narrow range of sites/stress factors for testing.

One last testing consideration to discuss relates to the timing of clonal testing. As proposed above, clonal tests may be initiated after conventional seedling tests have reached assessment age, so that only a subset of genotypes must be clonally tested. This sequential testing approach has a major disadvantage in time – even when early selection is possible in the seedling tests, it is necessary to wait while the seedling tests grow, and to wait still more while the selected subset is propagated to sufficient numbers for the clonal test. It may be possible to complete, or nearly complete, an additional cycle of breeding in the time it would have taken to conduct the sequential clonal test, accomplishing comparable gains without the cost of the clonal test. Early screening methods, which could reduce the time delay for sequential clonal testing, will not solve the problem if they involve significant compromise in selection effectiveness for the traits of ultimate interest.

An alternative is combined testing, in which clonal testing or some combination of clonal and seedling testing is used to accomplish all the testing objectives for the ongoing tree improvement program. The time delay of the sequential test is eliminated, in exchange for another major disadvantage. For the ongoing tree improvement program to achieve acceptable selection intensity, a large number of families and genotypes must be tested. For each of these to be clonally replicated, the testing program rapidly becomes unmanageable. In addition, certain families and genotypes will not propagate well, reducing the selection intensity and compromising gain from the ongoing program.

So what is the solution? It would appear that, under most circumstances, clonal testing should be carried out sequentially to some kind of seedling screening, in order to accomplish high selection intensities with a manageable testing load. Development of effective methods for early screening of clones should be of high priority. And finally, the traits to be focused on in the clonal testing and selection program should probably be traits which could not be as effectively improved via seedling testing and selection.

2.1.2 Approaches Used by Leading Programs

Two of the programs responding to the survey are using vegetative propagation to multiply valuable wild seed collections which are not part of recurrent improvement programs. The Nova Scotia Department of Lands and Forests is using rooted cutting technology to amplify supplies of black spruce seed of a selected provenance from the Cape Breton Island Highlands. The Institute for Forest Tree Breeding of the Hessian Research Station is using vegetative propagation to preserve genetic material which has become very rare due to forest decline.

In a majority of the programs responding to the survey, however, organizations pursuing vegetative propagation are doing so as an extension of recurrent selection, breeding, and testing programs. All of these programs are following forms of recurrent selection for general combining ability. Only one program advanced a multiple-population strategy, emphasizing a main, general purpose breed and small, specialty populations for disease resistance and length of internode (Carson 1987). Tree improvement efforts in Japan are actually de-emphasizing multiple, monoclonal varieties and are moving toward broad-based, single-population breeding efforts (see below). In some programs, the concept of assortative mating is advanced as a means to increase the frequency of extreme types for vegetative propagation (Foster 1986).

The focus of propagation efforts in each of these programs depends on the unique circumstances and goals of the program.

In some of these cases, orchard seed production is insufficient to meet planting requirements. This situation is common in western Europe and Scandinavia, where heavy emphasis has been placed on Norway spruce breeding and orchard programs, but orchards have produced low and sporadic seed crops. A variation of this situation is described by the Ontario Ministry of Forests at Orono. In this situation, a large geographic area is serviced by many small tree improvement programs which cannot support sufficient capital investment in production orchards to meet planting stock needs.

In other cases, the species of interest is difficult and costly to grow from seed, but grows well as a vegetative propagule. This situation exists for the British Columbia Ministry of Forests' breeding program for Alaska yellow-cedar, *Chamaecyparis nootkatensis.* This rare, high-value species is difficult to propagate from seed, but propagates readily and grows well as a cutting.

Other programs are using vegetative propagation to fill a gap in production of orchard families during the lag period after orchard establishment. J. D. Irving Company of New Brunswick, Canada, the British Forestry Commission's sitka spruce program, and the program at the Finnish Foundation for Forest Tree Breeding are organizations taking advantage of this approach.

Still other programs have well-established seed orchards, producing sufficient seed to meet planting stock requirements, and these programs are pursuing vegetative propagation as a means of increasing the genetic gains available through large-scale vegetative propagation of the very best crosses or clones. The radiata pine programs of the New Zealand Forest Research Institute (Shelbourne 1986; Carson 1986) and Tasman Forestry Ltd. (Arnold and Gleed 1985) are excellent examples.

Most programs responding to the survey stated the cloning of specific, high-value genotypes as an important long-term goal, but to date, only a few programs have realized this goal on a large scale. These include Tasman Forestry Ltd. with radiata pine, *Cryptomeria japonica* programs in Japan (described further by Ohba, Chap. 4, Vol. 2), and Norway spruce programs in Germany and Scandinavia (reviewed later in Volume 2 by Bentzer, Chap. 6). In addition, several organizations in Canada are proceeding rapidly with development of elite clonal stock, led by the B. C. Research Corporation, which is establishing clonal plantings of Sitka spruce from weevil-resistant parents.

Three of the programs involved in clonal production described detailed clonal testing and selection programs. In a Norway spruce program described by Kleinschmit (1983), over 2000 nursery seedlings are selected for height, cloned, and clonally tested in the nursery environment for 2 more years. At this point, one third of the original nursery selections are taken to the field. At ages 3, 9, 15, and 21, further reductions in the number of clones are made largely on the basis of volume and adaptability in field tests, so that by age 21 only 10% of the original selections remain. At each step, additional hedges of the selected genotypes are initiated so that, by the end, significant production is possible from the selected set. Another notable effort is the centuries-old cultivar development program used by foresters in Japan to develop preferred local varieties of *Cryptomeria japonica* (described by Ohba, Chap. 4, Vol. 2). Thirty two different cultivars have been developed over 5 centuries through repeated cycles of selection and field observation, selected for rooting ability and a wide range of growth, quality, appearance, and adaptive traits. However, as mentioned above, in recent years Japanese forestry agencies have attempted to reduce the dependence of foresters on highly selected monoclonal cultivars, for fear of future maladaptibility and catastrophic loss (Toda 1974). Instead, they propose to move toward more conventional breeding and orchard production, supplemented by cloning of the most desirable individuals. In the Tasman Forestry Ltd. program with radiata pine, selections are made from 5-year-old orchard-origin plantations, and hedges established to provide clones for deployment. Their intent is to make use of persistent maturation effects on stem and wood quality to improve value in the clonal stands.

2.2 Vegetative Propagation from Juvenile Tissue

Vegetative propagation from physiologically mature tissue (even as young as 1 year old in some species) tends to be quite difficult and results in important changes in the attributes of the resulting plants. Detailed discussion of these effects and their amelioration is provided by Greenwood and Hutchinson, by Bonga and von Ardekas, by Chen and Ahuja, and by Frampton and Foster, in other chapters of this volume. On the other hand, vegetative propagation from physiologically juvenile tissue (typically young seedlings) has been successful in a number of conifer species and appears to result in plants which are true to type. This phenomenon has led many organizations to focus on production of elite families through juvenile-tissue propagation in the near term, either for operational use or for clonal tests, while they continue to research methods of mature-tissue propagation.

Three methods of juvenile-tissue propagation have been widely explored as mass-propagation alternatives: rooted cuttings, organogenesis, and somatic embryogenesis.

2.2.1 Rooted Cuttings

Rooted cutting production programs with annual production approaching or exceeding 100000 propagules are listed in Table 1. Among the largest programs are those in Japan, eastern Canada, Sweden, France, the United Kingdom, and New Zealand.

Because of the size of the rooted cutting program in Japan, *Cryptomeria* is by far the most propagated conifer species. Behind *Cryptomeria* are the spruces and pines. The predominance of these species reflects at least two factors: (1) these species tend to propagate relatively easily, and (2) they have been the subjects of long-term selection and breeding programs.

Sources of Cuttings. The source of cutting material depends on several factors, including the objectives of the program, the breeding and propagation strategy, the species being propagated, the desired multiplication factor and rate, and levels of and sensitivity to costs.

Juvenile rooted cuttings are typically derived from one of three sources: seedling stock plants, seedling-origin hedges, or nursery seedlings.

Seedling stock plants are generally grown in pots in greenhouses under accelerated conditions for a fixed period of time, after which cuttings are harvested and the stock plants discarded. These conditions usually call for high fertility, constant temperatures in the range of 18 to 20°C, and extended photoperiod. Some programs grow seedling stock plants for only 1 year, while others require 2 years or more. This depends primarily on the growth rate of the stock plants and the rate of juvenility loss. Douglas-fir grows and matures rapidly with considerable branching so that 1-year growing regimes can be used successfully (Weyerhaeuser Company, unpublished data). In contrast, Sitka spruce grows and matures more slowly and so is typically raised for 2 years before cuttings are harvested (British Forestry Commission). Although stock plants are normally raised in greenhouses, results of trials by the British Forestry Commission in Scotland suggest that outdoor-grown stock plants, while taking longer to develop, are more economical to produce and provide a higher-quality cutting than indoor stock plants.

Hedges, in contrast to stock plants, are normally grown out of doors in soil, although they may also be maintained in pots. The more costly potted hedge approach is used to advantage by the Ontario Ministry of Forestry, where hedges are brought indoors in winter to force growth, which is then harvested in March as succulent cuttings. In general, hedges are sheared periodically to arrest maturation, and remain in production for several years, in contrast to stock plants which are usually discarded, field planted, or sold after cropping.

Some programs begin with cuttings taken from nursery seedlings before they are sent to the field. This practice is underway at the Hessian Forest Research Station, at the Ontario Ministry of Forests in Orono, at the Finnish Foundation for Forest Tree Breeding, and in the Lower Saxony program in Germany.

Rooting Procedures. Procedures for rooting cuttings have advanced remarkably little in the past decade. Critical factors remain condition of the cutting donor, date of collection, and quality of the cuttings. The environmental conditions which promote rooting are well known and generally include a porous rooting medium, intermittent mist, bottom heating, photoperiod extension, and relatively low light. Most rooting is done indoors in greenhouses, but some is done outdoors under cold frames.

Rooting media vary from gravel or sand to mixes of peat, vermiculite, perlite, and bark. Cuttings are normally rooted in containers but can also be rooted in open beds. Plastic tents are sometimes used to maintain high humidity around the cuttings. Rooting generally takes place over a 2- to 3-month period and is followed by a period of scheduled weaning and hardening off. This is a critical period because foliage developed in rooting tents is not adapted to low humidity and high light intensities. Without hardening it would quickly burn and die after field planting. Some cuttings are rooted in summer and over-wintered outdoors under snow before spring planting. Others are rooted over winter and planted out the following autumn.

The Norway spruce program at Biri Plantskolo, Norway provides a typical example. Cuttings are harvested from hedges from July to mid-August and are stuck immediately into containers holding a 70:30 peat-perlite mix. Containers are placed on a heated, concrete floor in a greenhouse and misted intermittently. After a 3-month rooting period they are hardened off under low temperature and short days. They are then overwintered outdoors under snow. The following spring and summer they are grown outdoors awaiting fall planting.

2.2.2 Organogenesis

Organogenesis involves the separate induction and growout of roots and shoots, originating either from unorganized callus, or from preformed or induced buds. Since 1934, and particularly in the 1970s, considerable research effort has been directed at vegetative propagation from seedlings via organogenesis. To date, plantlets have been produced from over 30 conifer species. However, as of this writing, large-scale production and field establishment has been achieved for only four species: coast redwood (*Sequoia sempervirens*), at AFOCEL in France (Poisonnier et al. 1980), radiata pine, by the FRI and Tasman Forestry Ltd. in New Zealand, loblolly pine, by North Carolina State University (Amerson et al. 1988), and Douglas-fir, by Weyerhaeuser Company in Washington (Ritchie and Long 1986).

Sources and Form of Explant. The greatest success in organogenesis has been achieved using juvenile explant tissue, in the form of embryos, cotyledons or seedling segments.

Shoot Induction, Multiplication, and Elongation. Recently, important advances have been made in the selection and pretreatment (with cytokinins) of explants before culturing (Thorpe and Biondi 1984).

Media improvements have provided important benefits for shoot induction, elongation, and multiplication in organogenic culture. Several basal media have been used with success, including modified MS (Murashige and Skoog 1962), Schenk and Hildebrandt (Schenk and Hildebrandt 1972), Brown and Lawrence

(Brown and Lawrence 1968) and DCR (Gupta and Durzan 1985a). Media have also been specifically designed for conifers, on the basis of seed analysis – examples are LM (Litvay et al. 1981) and P-6 (Teasdale et al. 1986).

Cytokinin, in the form of Benzyl aminopurine (BAP), kinetin, and isopentanyl adenine (2-ip), has been found to be essential for shoot bud induction. Addition of low concentrations of auxin (IAA, IBA, NAA) with the cytokinin improves response in some conifer species (Thorpe and Hasnain 1987).

A breakthrough in organogenesis automation potential has been made in radiata pine in the development of techniques to generate large numbers of meristematic nodules from embryos (Aitken-Christie et al. 1988). A single embryo produced 5480 meristematic nodules in 13.5 months, from which over 355000 elongated shoots could be produced. A semi-automated system for grinding and homogenizing plant tissue for organogenesis has been proposed (Levin 1985) and may offer a low-cost means of singulating meristematic nodules.

Activated charcoal has proven very beneficial to shoot elongation in most conifer species, likely due to absorption of excess hormones or inhibitory substances. It possibly plays some additional role, since activated charcoal of different origins appears to have different degrees of effectiveness. Conifer-derived charcoal was found to be best for shoot development in white and black spruce (Rumery and Thorpe 1984).

Use of a microporous membrane in the tissue culture vessel lid can increase gas exchange and enhance growth of tissue culture shoots (Aitken-Christie et al. 1990).

The use of liquid media over solid media has provided a means to simplify the freshening of media during subculture without tissue transfer (Aitken-Christie and Jones 1987). This improvement has important implications for cost reduction in tissue culture.

Rooting Procedures. Dramatic improvements have been made in recent years in rooting of shoots produced in vitro. In one example, different auxins used alone or in combination were provided in pulses of several hours to days to shoots in the dark, after which shoots were transferred to half-strength basal medium without hormone. This combination of treatments produced superior root development for many conifer species (Mohammed and Vidaver 1988). Rooting has also been achieved in non-sterile conditions. In one experiment, in vitro-grown shoots were transferred after an auxin pulse into peat-vermiculite, peat-sand, or peat-pumice (1:1 mixtures) under mist, with high light intensity and bottom heat, and rooting success ranged from 60–100% (Thorpe and Biondi 1984; Mohammed and Vidaver 1988).

2.2.3 Somatic Embryogenesis

Somatic embryogenesis is a process by which multiple embryos proliferate from somatic cells – typically suspensor cells – in culture. In contrast to organogenesis, somatic embryogenesis produces plants which develop in a bipolar fashion, as would natural zygotic embryos. An excellent review of progress in embryogenesis has been given by Dunstan (1988) and by Becwar, Chapter 13, Volume 1. We will offer a brief overview here.

Embryo-like structures have been produced in vitro from a number of conifer species since the mid 1970s (Abo El-Nil 1980; Durzan 1982). By the mid 1980s, understanding and repeatability of the process, and excitement over its implications, had advanced sufficiently that laboratories worldwide began research efforts in somatic embryogenesis (Hakman et al. 1985; Nagmani and Bonga 1985). To date, no laboratories have achieved large-scale operational production of conifer somatic embryos, although production in the hundreds was reported by researchers at the Institute of Paper Chemistry (Becwar et al. 1988).

Sources and Form of Explant. The first reports of conifer embryogenesis involved experiments using immature (just-fertilized) embryos as explants. Von Arnold and Hakman (1986), Gupta and Durzan (1986a,b), Krugstrup (1986) and Lelu et al. (1987) extended the technique to include mature-seed explants of Norway spruce and sugar pine as well. In other species, immature embryos continue to be the only consistent source of successful cultures. In fact, the time-lag after fertilization appears to be important to successful initiation, with embryos 4–5 weeks after fertilization being the best explants (Durzan and Gupta 1987).

Initiation, Multiplication, and Maturation. Initiation of embryogenic cultures typically takes place in the presence of high levels of the auxin 2,4-D and benzyladenine added to a selected basal medium. In contrast to bud induction for organogenesis, the auxin to cytokinin ratio for embryogenesis initiation is high. The tissue arising from initiation treatments has been termed Embryo Suspensor Mass or ESM by Gupta and Durzan (1987b). ESM is distinct from callus in that the tissue is inherently organized, in the form of early-stage "pro-embryos". These pro-embryos are made up of dense, compact embryo heads attached to long, thin suspensor cells. Pro-embryo numbers in culture can be multiplied through maintenance of ESM with five to ten times less hormone. Sustained multiplication requires regular renewal of media, every 10–12 days for cultures on solid media, or every 7–10 days for cultures in liquid suspension. Dunstan (1988) reports the maintenance of white spruce cultures in liquid suspension with multiplication rates of three to five times every 7 days. To date, ESM cultures have been maintained for up to 5 years with no change in multiplication rate (Gupta, unpubl. data).

Once cultures have been initiated, and perhaps multiplied, a new set of conditions is required to cause the pro-embryos to "mature" or develop into competent, germinable embryos. Many studies point to a key role for abscisic acid (ABA) in the maturation process (Durzan and Gupta 1987; Becwar et al. 1987b; von Arnold and Hakman 1988b; Hakman and von Arnold 1988), and the duration of the ABA treatment also appears to be critical.

Bioreactor systems have been proposed to provide controlled, optimum conditions to embryos for multiplication and synchronous development (Dunstan 1988). Such systems are, in fact, in use on a research scale with certain crop plants (Styer 1985).

2.3 Vegetative Propagation of Tested Genotypes

True cloning requires the ability to identify and mass-produce propagules from genetically superior individuals. Propagation of "mature" selected genotypes may focus on collection of tissue directly from the selected trees for mass propagation. Alternatively, tissue may be stored or maintained in a juvenile state while a clonal test of the same genotypes are conducted. Upon selection, mass-propagation may be carried out from the juvenile material of the genotypes which were selected from the clonal test.

2.3.1 Propagation from Physiologically Mature Tissue

Large-scale propagation from physiologically mature tissue has been the focus of much research over the past two decades, but this effort has been met with only modest success.

In some species, cuttings may be successfully produced without pretreatment from trees in the field of an age at which fairly reliable selections may be made. At Tasman Forestry Ltd. in New Zealand, cuttings are collected from phenotypic selections of radiata pine out of 3- to 5-year-old progeny tests or pedigreed plantations, and rooted directly. However, other authors (Shelbourne 1986; Johnson 1988) have backed away from such a practice in radiata pine due to negative growth rate influences of aging.

Pre-treatment of mature trees has made large-scale vegetative propagation possible in some species (Abo E-Nil 1982; Boulay 1988; Gupta and Durzan 1987a). Examples of pre-treatments are heavy pruning or sprays with growth regulators.

It has also been widely reported that repeated subculture of shoots from mature trees will result in more juvenile appearance and rooting of propagules from those trees (Franclet et al. 1987).

Initiation of embryogenesis from tissues of mature, tested trees has not yet been successful in conifers. The production of somatic embryos from seedling cotyledons of Norway spruce may be an important intermediate step to embryogenesis from mature needles or buds (Lelu et al. 1987). Alternatively, initiation from maternal tissue in developing seeds may offer promise. Nagmani and Bonga (1985) initiated embryogenic cultures from gametophyte tissue of *Larix decidua*, and recovered haploid plants. Successful initiation of embryogenesis has also been achieved for mango and citrus from nucellus tissue (Litz 1986; Rangan 1986).

2.3.2 Maintenance of Juvenility

Juvenility maintenance procedures attempt to maintain tissue in a juvenile condition so that clonal tests can be conducted and selected genotypes can be propagated effectively based on the clonal test results. Such procedures can also be useful in enabling a production system to be maintained at a high productivity level over a sustained period, through replacement of depleted active cultures with vigorous, juvenile cultures.

Four approaches have been explored extensively: serial propagation, hedging, repeated subculture, and tissue storage.

Serial propagation, the practice of repeated production of cuttings from cuttings, has been shown to arrest maturation effectively in some species (St. Clair and Kleinschmit 1986). Cuttings derived from serial propagation may themselves be used as final propagules, or they may be used as stock plants or hedges. The British Columbia Alaskan yellow-cedar program uses serial propagation as one method of producing cuttings, and in the Norway spruce program of Lower Saxony, Germany, it is the main method.

Like serial propagation, hedging has been shown to be an effective method of arresting maturation in a number of conifer species (Bolstad and Libby 1982). The British Columbia Ministry of Forests' rooted cutting program with Alaskan yellow-cedar and a number of programs involving radiata pine, loblolly pine and Sitka spruce are examples where hedging is being used successfully. However, in other species, hedging has not been successful as a means of preventing the negative effects of maturation. The Norway spruce cutting program in Finland has experienced maturation in hedges which has resulted in the abandonment of many clones. Similar situations exist for black spruce and Douglas-fir.

A third alternative, repeated subculture in vitro, may be used to maintain organogenic cultures in a juvenile state over extended periods, as should extended maintenance of Embryo Suspensor Masses under multiplication conditions. Upon completion of clonal tests, cultures of the best genotypes could be used to mass-propagate those genotypes in vitro, or if desired, to generate stock plants or hedges from which rooted cuttings could be derived.

Tissue storage methods can be broken down into three general categories: cold storage, which slows growth without freezing; cryopreservation; and desiccation.

The subject of cold storage has been discussed in detail in a recent review by Aitken-Christie and Singh (1987). They describe techniques which have been used to cold-store cultures of a range of species. They report that shoots of radiata pine have resumed growth after having been stored up to 5.5 years, although rooting was inhibited by storage after 17 months. Results have been highly variable among genotypes and species.

Cryogenic storage of Embryo Suspensor Masses in liquid nitrogen may offer another means of maintaining genetic material in a juvenile state in species with working embryogenesis systems. Gupta et al. (1987) and Kartha et al. (1988) demonstrated successful recovery of embryogenic cultures of loblolly pine, Norway spruce, and white spruce after storage in liquid nitrogen (– 196°C).

Desiccation has also been proposed as a method by which somatic embryos could be stored. Carrot and celery somatic embryos have been dried and coated with polyoxyethylene to form a wafer, and some of these embryos could be recovered in vitro after rehydration (Kitto and Janick 1985). Some success has been achieved with other species as well, and these systems are reviewed in detail by Fujii et al. (1987) and by Redenbaugh et al. (1988a,b,c).

2.4 Nursery Growing and Field Establishment of Vegetative Propagules

The availability of genetically superior vegetative propagules in themselves is of little commercial value if they cannot be delivered to the field cost-effectively. With most vegetative propagation systems, delivery to the field after rooting or embryo maturation is a multi-step process encompassing growout, acclimation or hardening, delivery to the field, and final field establishment.

The first step in the transition of vegetative propagules to the field from the carefully controlled conditions of the laboratory or the rooting tent is the acclimation or hardening step. This step may be carried out through movement of the plants or gradual changing of conditions while the plants are in the same containers in which rooting or germination first took place. Alternatively, this step can involve transplanting into new conditions.

Next, the propagules may either be grown out and delivered to the field in containers, or grown out in a nursery and delivered to the field as bareroot plants.

In Canada and Scandinavia, vegetative propagules are grown primarily in containers. In most cases, propagules go to the field in the containers in which they were rooted. Leach Tubes, Styroblocks, and Multipots are among the more popular systems in Canada, while the Lannen Paperpot and Ecopot systems are more popular in Scandinavia, especially in Finland. Sometimes the plugs are extracted from the containers before being shipped, but generally they are not.

In Germany and the UK, bareroot production predominates. Propagules are rooted or sown in flats, rather than in containers. They are hand-lifted and graded from the flats, then transplanted into the nursery, using a standard transplanting machine. Here they are allowed to grow an additional year or more before being lifted for field planting.

Container systems offer several advantages over bareroot systems for vegetative propagation. An important one is the opportunity for mechanization, which has been fully exploited by Hilleshog in Sweden. This can lead to lower cost. The major disadvantage of container systems is that plant quality may be compromised. Root restriction and dense spacing can lead to small plants which perform poorly in the field against heavy weed competition and animal browsing. In contrast, bareroot plants are typically larger than containerized plants. The decision to go with bareroot or container production for cuttings is often based on customers' size standards and often follows the most popular method of seedling production for a given geographic region.

Some organizations are experimenting with container-bareroot hybrid systems. In principle, such an approach could enable capture of the best features of both container and bareroot systems. In these hybrid systems, rooting or conversion of embryos to soil is done in a greenhouse in containers. Following rooting, the propagules are transplanted into a bareroot nursery where size can be controlled and root culturing operations, such as wrenching and undercutting, can be practiced. The Ecopot is especially well adapted to this operation. In another system, under development by Weyerhaeuser Company, cuttings are rooted in a greenhouse, in bottomless trays with individual cavity sizes of only 5 cc. Very efficient use of greenhouse space is achieved. Soon after rooting the cuttings are transplanted into a bareroot nursery using a specially designed, highly mechanized system. They are then grown out as bareroot stock for one or two growing seasons.

The subject of cost-effective handling, storage, and delivery to the nursery or field of fairly fragile propagules becomes a crucial one to the viability of a large-scale vegetative propagation system, particularly in commercial forestry where per-plant cost thresholds and values tend to be low. Artificial seed technology has arisen directly from recognition of the importance of these activities.

Three methods have been described for delivery of somatic embryos to the nursery or field: fluid drilling, polyoxyethylene tapes or sheets incorporating desiccated embryos, and artificial analogs of true seeds – artificial seeds (Redenbaugh et al. 1988a,c). In particular, the concept of an "artificial seed" delivery system to accompany somatic embryogenesis has received much interest, and was discussed in the forestry context at a recent symposium in Pakistan (Redenbaugh et al. 1988a).

Artificial seed research on species as diverse as alfalfa, asparagus, date palm, soybean, grapevine, and mulberry has resulted in successful embryo-to-plant conversion from artificial seeds, although successful conversion has largely been confined to in vitro or growth chamber conditions for most of these species. The focus of this research has been development of hydrated capsules of calcium or sodium alginate. The capsules may be formed through the mixing or insertion of embryos into alginate, after which droplets containing the embryos are hardened in a calcium nitrate complexing solution. Alternatively, the use of temperature-sensitive gels and gel molds have been proposed. (Fuji et al. 1987; Redenbaugh et al. 1988a,c). To facilitate handling of the artificial seeds, hydrophobic coatings have been developed. Coated capsules could actually be planted using a conventional seed sowing machine, and did not dry out excessively over a 1-day period (Redenbaugh et al. 1988c).

Most programs responding to our survey recognize the value of meshing their vegetative propagation systems with the effective and low-cost systems which are already in place in forest nurseries for large-scale seedling production and delivery. As we will discuss in the final section of this chapter, such integration will be important not only in achieving cost-effective production, but also in building acceptance of vegetatively propagated stock by field foresters.

2.5 Field Performance of Vegetative Propagules vs. Seedlings

Numerous field tests have been established with vegetative propagules of conifers around the world, and some of these tests have reached ages of 20 or more. The subject of field performance of vegetative propagules is covered in detail by Frampton and Foster in Chapter 9. We will touch on some highlights in the discussion to follow.

In general, height growth in cuttings of radiata pine has been similar to that of seedling-origin stock, for cuttings taken from trees under 10-15 years of age. Diameter growth has also been similar for cuttings and seedlings when cuttings were taken from juvenile trees; however, there appears to be a profound impact of maturation state on diameter growth of cuttings. A 0.60-0.70 ratio of cutting diameter to seedling diameter was observed in a 17-year-old radiata pine trial for which cuttings were taken from 7- and 8-year-old trees (Sweet 1973; Sweet and Wells 1974), versus ratios from 1.0 to 1.09 in 6-8-year-old radiata pine trials established from juvenile-origin cuttings (Shelbourne and Thulin 1974; Fielding 1970). However, one radiata pine trial showed equal or greater diameter growth for cuttings from 8-year-old trees than for seedling-origin stock (Arnold and Gleed 1985). It is important to note that comparisons in many of these trials are confounded by differing levels of genetic quality of seedling and cutting material.

Slower height and diameter growth was observed for loblolly pine cuttings compared to seedlings in a 6-year-old trial established with cuttings from 4-year-old trees (Foster et al. 1987), but growth rates were equivalent for the two stock types in another loblolly pine trial with cuttings from 1-year-old trees. In contrast to these results, rooted cuttings of Norway spruce have larger diameters for their height than do seedling-origin plants in long-term field trials (Kleinschmit and Schmit 1979).

Growth performance of tissue-culture-origin plants relative to seedling-origin plants has been particularly difficult to assess due to the typical small size of tissue-culture material at the time of test establishment. Frampton and Foster (Chap 9, Vol. 1) reviewed detailed results of several field experiments, which in general show greater height and diameter for seedling-origin plants of loblolly pine, radiata pine, and Douglas-fir over time than was observed for tissue-culture-origin plants. However, these experiments also suggest similar rates of height growth over time for the two stock types, a result which they suggest would cause height differences to become less and less important over time. In a field planting of over 1200 somatic seedlings derived from embryogenesis of interior spruce, growth of the somatic seedlings was similar to that of zygotic seedlings at the end of the first growing season.

It may be that the lower rates of growth often observed for vegetative propagules is reflective of imperfect technology – resulting in poorer quality root systems and vigor at establishment – rather than an intrinsically reduced growth potential for vegetatively propagated stock. This would be borne out by further work on radiata pine which has demonstrated that adjustments in nursery practice and culling methods have resulted in significant improvements in growth performance of rooted cuttings (Arnold, pers. commun.).

As indicated above, form and quality characteristics vary between vegetative propagules and seedling-origin plants of several conifer species. Particularly good information on this subject is available for *Pinus radiata*. On the basis of several studies, it appears that stem taper, bark thickness, sweep, and branch diameter are less on cutting-origin plants than on seedling-origin plants, when the cuttings are taken from trees 5 or more years of age (Thulin and Faulds 1968; Fielding 1970; Sweet and Wells 1974; Shelbourne and Thulin 1974; Arnold and Gleed 1985). In fact, stem straightness appears superior in cutting-origin stock even when the cuttings are taken from juvenile trees (Eldridge 1981). Stem taper and sweep have also been observed to be equal or better in tissue-culture plantlets than in seedling-origin stock (Smith et al. 1985). Incidence of juvenile-onset diseases such as fusiform rust in loblolly pine and *Dothistroma* needle blight in radiata pine appear to be reduced in stock of cutting or tissue culture origin (Power and Dodd 1984; McKeand 1985; Carson 1986), although *Armil-*

laria root rot and *Cyclaneusma* needle cast incidence and damage have actually been greater for cuttings than seedling-origin stock in radiata pine field trials (Carson 1986). Wood specific gravity and grain angle appear to be unfavorably impacted by vegetative propagation, with the impact being greatest for propagules from older trees (Nicholls et al. 1977; Sweet and Harris 1976).

2.6 Benefits, Costs and Returns from Vegetative Propagation

Many authors have described the benefits which could accrue to forestry from vegetative propagation. A recent and exhaustive review of potential benefits available from clonal forestry was made by Libby and Rauter (1984). Their review was challenged for radiata pine by Carson (1986) and Burdon (1988a), and some new benefits were presented for vegetative multiplication of controlled cross families. The arguments they present illustrate that the benefits and costs actually experienced for vegetative propagation will depend greatly upon the strategies and unique circumstances of individual species and programs. Experience among our survey respondents provides a valuable insight into the array of benefits which are being pursued and realized from large-scale vegetative propagation, and the costs which will be associated with those benefits in practice.

2.6.1 Projected and Realized Benefits

In this section we will review the benefits which are being used by leading organizations to justify their vegetative propagation efforts.

Benefit: large-scale production from high-value raw material which is difficult to produce from seed.

As we described in Section 2.1 above, vegetative propagation is enabling several organizations to meet planting needs with valuable genetic material which could not be produced in sufficient quantity from seed. This material takes the form of rare species and provenances, hybrids, or families from seed orchards which are producing poorly. In fact, a number of respondents to our survey cited this benefit as their primary justification for vegetative propagation.

Benefit: increased gain from mass production of control-cross families.

In theory, controlled crossing of selected genotypes can provide double or more the gain available from wind-pollinated orchards (Shelbourne 1986; Cameron and Cotterill 1988). Such increases in gain result primarily from the ability to put selection pressure on both the male and the female side of the pedigree. Also, in cases where crosses exhibiting favorable nonadditive genetic effects can be identified, the remake of these tested crosses increases the proportion of the production population expressing these effects. This benefit is receiving increasing attention, particularly among more advanced genetic improvement programs, and in some cases is the primary focus of vegetative propagation programs. In many cases, controlled crosses are costly to complete, and vegetative propagation allows large numbers of plants to be produced from a small amount of seed. However, for some species there may be other ways to achieve this benefit more cost-effectively. In *Pinus radiata*, innovative orchard designs and proper orchard siting can allow large-scale production of control-cross seedlings at a lower cost than vegetative multiplication of the same crosses via cuttings (Arnold 1990). Steady improvements have also been made in Supplemental Mass Pollination technology which could make large-scale controlled crossing quite economical (Bridgwater et al. 1987; Blush 1987).

Benefit: increased genetic gain from mass propagation of tested genotypes.

When individual genotypes can be tested, tracked, and propagated, the entire genetic potential of the selected individuals can be transferred to the production plants. In addition, when individual genotypes are propagated on a large scale, the selection intensity can be significantly increased for a given number of trees planted. For these reasons, the cloning of superior genotypes offers the greatest potential genetic gain within a generation. Matheson and Lindgren (1985) predicted 5 to 40% more genetic gain from cloning of the best 5% of tested genotypes, compared to orchard production of seed from the best 10% of tested individuals within the best 50% of families. The 5% figure corresponds to a case of zero nonadditive genetic variation, while the 40% figure represents non-additive variation equal to the additive genetic variation. Cameron and Cotterill (1988) compared an array of production options using a heritability of 0.1 and assumed zero nonadditive genetic variance. They predicted 14.4% gain for a rogued first-generation clonal seed orchard, 22.5% gain for crosses among the best progeny tested first-generation parents, and 25.7% gain from tested clones taken from the same crosses.

Benefit: significant decrease in the time elapsed before selected genetic material can be deployed in the field.

Among the respondents to the survey, this benefit is achieved through multiplication of seed orchard crops during the years before the orchard reaches full production. Simulation analyses with *Pinus radiata* and *Pinus sylvestris* suggest that vegetative propagation can reduce the time to full-scale production of selected stock by 4 to 7 years, resulting in significant increases in gain per unit time (Matheson and Lindgren 1985; Cameron and Cotterill 1988). However, this benefit is highly species-specific. In species which can produce large quantities of seed 2 to 4 years after grafting, this benefit would be far less, perhaps minimal.

Benefit: quality improvement through propagation from mature donor stock.

One organization responding to the survey, Tasman Forestry Ltd., is producing clonal stock of radiata pine via hedges produced from 5-year-old selections taken from plantations of orchard origin. Their objective is to improve stem and wood quality through operational exploitation of maturation effects. However, the majority of organizations answering the survey are actually working aggressively to *minimize* maturation effects in their planting stock, due to negative impacts of maturation on growth and propagation efficiency.

Benefit: providing for planting stock needs without the high cost of a traditional wind-pollinated seed orchard.

This benefit, pointed out by only one of the survey respondents, has high potential significance. Organizations which have been hesitant to pursue tree improvement due to a small landbase may find that they can support a program which does not depend on high-cost wind-pollinated orchards. Even programs with larger planting requirements could potentially reduce total tree improvement costs by replacing traditional seed orchards with small, lower-cost breeding orchards supported by vegetative multiplication. However, few programs have actually discarded wind-pollinated orchards in favor of vegetative propagation, most likely because of comfort with existing families coupled with uncertainty about performance and cost of vegetative propagation.

Benefit: rapid recovery of production potential after a disaster.

The Woods and Forests Department of South Australia lost all their seed orchards to brush fires in February of 1983, but they were able to recover rapidly to full-scale production of genetically select material through rooted cuttings.

Benefit: flexibility and responsiveness of genetic improvement program.

Orchards to produce small quantities of controlled-cross seed for vegetative propagation can be small, and can reach the necessary level of production much more quickly than traditional systems relying entirely on wind-pollinated seed production. Fewer genotypes are required, since it is not necessary to separate related individuals. For these reasons, new genetic material can be infused into the production population more quickly. It becomes just as straightforward to produce propagules of many different provenances or varieties than to produce the same number of crosses of a single variety. All of these factors increase the flexibility of a genetic improvement program.

Benefit: increased stand uniformity and predictability.

Reduced variability and increased predictability for stem dimensions and stem and wood quality would result in important harvesting, handling, processing and marketing advantages. Full-sib family and monoclonal block planting has been promoted as a means of achieving improved uniformity and predictability. However, for most commercially important traits, environmental effects play a more important role than genetics in observed value. For such traits, substantial gains in uniformity and predictability would require improved understanding and control of environmental variation. In addition, realization of this benefit would require establishment and tracking of families or clones as pure blocks.

Benefit: greater gain and simpler orchard design due to a reduced need for inbreeding avoidance in the production population.

Wind-pollinated orchards are typically designed with tight restrictions on the number of trees which can be selected from a given family in order to minimize related matings and inbreeding depression in the resulting production progeny.

Such restrictions can reduce the gain potential from production population selection by as much as one-third to one-half, since the best selections are usually concentrated in a small number of top families (Talbert, unpubl. data). Vegetative propagation makes it possible to produce operational quantities of planting stock from controlled pollinations. As a result, a production population need only include the number of unrelated parents required to produce the set of controlled crosses of interest, and the breeder has complete control over the level of inbreeding that will be allowed. Orchards could be simpler in design, and could be set up in clonal blocks to facilitate the controlled-crossing operation.

Benefit: new types will be identified through full-sib and clonal forestry which could fundamentally change the existing value system or forest management processes.

Clones, and even full-sib families, are likely to exhibit a broad array of characteristics, some of which might not have been noticed in genetically heterogeneous stands. It is quite possible that clonal and full-sib tests and plantation blocks will demonstrate the potential for dramatic improvements in traits which have not yet been exploited. Examples exist today of unique genotypes which were uncovered either serendipitously or through years of intensive, costly screening. Production and establishment of full-sib families or clones in blocks just improves our ability to identify such types.

2.6.2 Costs and Risks

Vegetative propagules have historically been more costly to produce than seedlings. This has been due in large measure to a greater number of handling steps in vegetative propagation, resulting in high labor inputs and costs. Rooted cutting production systems also tend to use lower planting densities in the nursery, increasing the growing costs for the stock.

Production costs for Norway spruce cuttings in three European programs range from 1.3 to 1.6 times the cost of seedlings (Bentzer, Chap. 6, this Vol. 2). Rooted cutting production costs for radiata pine are approximately five times higher than for comparable seedling stock (Arnold and Gleed 1985). However, in a separate analysis Johnson (1988) was able to show a lower cost for production of cuttings than for mass-production of controlled-cross seed. Costs are 1.8 to 3.6 times higher for rooted cuttings than for seedlings of Sitka spruce (Gill 1982).

Organogenesis systems, with their many transfer steps, costly facilities and materials, and high rates of mortality to the field, have proven to be very costly on a per-plant basis to date. Reported costs of production for radiata pine plantlets are ten times or more the cost of seedlings (Aitken-Christie and Jones 1987), and similar costs were also reported for plantlets of Douglas-fir (Timmis 1985).

An important element of attractiveness for somatic embryogenesis is the potential for automation and low labor input coupled with very high multiplication rates. Embryogenesis/artificial seed systems for a number of crop plants have resulted in costs per plant which were in fractions of a cent (Redenbaugh et al. 1988a). However, research with conifers has not yet progressed to the point where cost estimates have been reported.

One important reason for the higher cost of vegetative propagation systems may be the degree of refinement of large-scale production processes for vegetative propagation relative to seedling propagation. Methods and equipment for large-scale production and handling of seed and seedlings have been well refined through years of research and operational practice by many organizations worldwide. Vegetative propagation has, in contrast, been a research activity to date in most programs, with small production volumes and little or no effort expended on process efficiency and cost-reduction. This situation is changing rapidly, pioneered by organizations like the Forest Research Institute and Tasman Forestry Ltd. of New Zealand, and Hilleshog in Sweden. Automated or mechanized systems, cultural regimes which reduce labor requirements, and innovative applications of new technology such as in vitro hedging and artificial seeds will be certain to result in steadily lowering costs for vegetative propagules. In one case, technologies for in vitro hedging of shoots, automated liquid media replacement and other cost-saving innovations have been shown to reduce plantlet production costs by 70% (Aitken-Christie et al. 1988).

So far we have discussed only direct costs of production. In addition to direct costs, vegetative propagation programs can also be associated with significant indirect costs, and it is the total cost which must be justified in a decision to embark on such a program. Some of these are listed below:

- Research and development, including process engineering.
- Facility and equipment replacement or alteration.
- Clonal testing.
- Retraining of tree improvement, nursery and forestry personnel in new methods.
- Additional breeding cost for controlled crosses.
- Reduced rate of gain in ongoing breeding program:
  - due to reduced selection intensity because of clonal testing load,
  - due to delay in other tree improvement activities as result of implementation of vegetative propagation.

Finally, there is a third category of costs, termed "intangible" costs for this discussion, which may be incurred due to the application of vegetative propagation, and these must also be weighed against the benefits. One example of an "intangible" cost of vegetative propagation is in the greater complexity of forest management when many different genetic types must be managed and tracked. This could actually translate into greater indirect costs if additional personnel must be hired to handle the task.

A very important category of "intangible" costs relates to political complexity and regulatory risk. Around the world, forestry practices are increasingly under public scrutiny, and regulations are proposed daily which impact forestry. An entire chapter of this text is dedicated to the subject of regulations relating to clonal forestry. There should be little doubt that the more intensively selected are our production populations, and the greater the scale of operational use of small numbers of families or clones, the greater will be the political pressure to limit such activities. This political pressure comes from a widespread public perception that deployment of clonal, and even highly selected family, stock is associated with reduced genetic diversity and greater risk of catastrophic loss along the lines of the corn blight in the United States. Tree breeders are painfully aware that their improvement programs are far more conservative, and have maintained vastly more genetic diversity than most if not all agronomic crops. However, forest trees – rightly – are perceived as quite different from other crop plants, and political pressure on forestry can be expected to intensify. The extent of regulatory control over tree improvement and propagation will depend significantly on the ability of the leading organizations to develop responsible deployment strategies supported by sound data, and on their ability to communicate effectively and sensitively with the public on the subject of biological risk. This subject will be the focus of a discussion on Guidelines for Responsible Deployment, in Section 3 below.

2.6.3 Return on Investment

No conifer vegetative propagation programs have yet reached a stage in commercialization where it has been possible to evaluate a steady-state return on investment. However, several estimates have been made, and we will review them here. It is important to note that ROI analyses are highly dependent on the assumptions used, and these assumptions vary greatly from one investigator to another.

Some investigators begin with estimates of incremental genetic gain from vegetative propagation, and work backward to estimate the incremental cost at which their programs would reach a breakeven return compared to orchard seedling production. In one such analysis, an incremental cost of $305 per acre (about $0.6 per plant) could be justified from family bulking, while $874 per acre ($1.75 per plant) could be justified from propagation of selected individual genotypes (Frampton, unpubl. data 1983). These figures represented costs from four to ten times the cost of orchard seed. Timmis (1985) projected that cost increases of up to five to six times could be accepted for a constant net present value if vegetative propagation resulted in a doubling of genetic gain. In another analysis, the estimated value added by vegetative propagation was $0.005 to $0.4 per plant over seedlings, depending on the forest management scenario employed (Foster et al. 1981).

Other investigators have combined cost and gain information from their programs in a sensitivity analysis approach to evaluate the returns from vegetative propagation. In an analysis of a Sitka spruce program, the net present values from vegetative propagation went from strongly positive to strongly negative, depending on the multiplication rate of cuttings, the level of gain, and the productivity of the competing seed orchard technology (Gill 1982). Kleinschmit (1973) compared seed orchards, seed stands and vegetative propagation in terms of their net present value impacts, and found that *none* of the three had a positive net present value at discount rates about 2%, given the assumed costs and gains for the three methods.

As we mentioned in the section on Benefits above, radiata pine programs in New Zealand have found innovative ways to capture value from vegetative propagation. Despite rooted cutting costs reported to be five times greater than seedlings, and higher establishment costs due to larger stock, total forest management costs were reduced by approximately 25% through vegetative propagation (Arnold and Gleed 1985). Cost reductions were realized through lower initial stocking levels, which resulted in lower pruning and thinning costs. In another analysis for radiata pine which did not incorporate such forest management cost reductions, cutting propagation from controlled crosses resulted in net present value estimates which differed little from controlled-cross seedling production (Johnson 1988).

3 The Future: Making Vegetative Propagation a Commercial Success in Forestry

We began our chapter with a question: are the technological advances which have been made in vegetative propagation of conifers translating into meaningful changes in operational forestry practice? Based on the results of our survey, it is clear that a large number of programs have the technology in hand to create such change. Yet, with a few notable exceptions, vegetative propagation has been slow to integrate into operational practice for major commercial conifers. Even the most aggressive vegetative propagation programs with the most easily propagated conifer species produce only a fraction of their total planting stock by vegetative propagation, and in most countries the fraction is 5% or less, and, in general, it is only in countries with the most advanced programs that the discussion of vegetative propagation has begun to make its way into the language and the literature in forest management, silviculture and biometrics. It appears that we have yet to make the jump from an "exciting new technology" to an accepted way of doing business in forestry.

Clearly, there is significant promise in vegetative propagation, to increase realized value from forest lands, to enhance the uniformity and predictability of raw material and products, to improve flexibility and responsiveness in production of "varieties" which meet priority needs. As we have reviewed in the preceding discussion, technologies are available for a number of species to enable large-scale vegetative propagation resulting in vigorous plants of high genetic value, at costs which in many cases seem to provide excellent economic returns. Commercial acceptance should flow quickly and naturally...or should it?

We believe that the answer to this question is: *not necessarily*! Organizations experienced in commercialization (used here to mean full operational implementation) of new technologies suggest that it is a process which (1) must be given specific attention and focus to be successful; (2) can be described in terms of several elements which are important to success regardless of the specific nature of the technology; and (3) often requires activities, personalities, and skills which are quite different from those which typically characterize research activities. We believe that some new areas of focus will be critical to successful, widespread operational implementation of vegetative propagation in forestry – and that these new areas are at least as critical as further refinements in technology. These areas are:

- Technology transfer.
- Cost-effective production.
- Genetic improvement and propagation as part of a forest management system.
- Guidelines for responsible deployment and communication.

Technology Transfer. In a 1984 review in the Journal of Forestry, Robert Callaham reviewed several factors which have been identified as necessary for the adoption of technological innovations. The technology must work, and it must be credible, observable, and documented in terms easily understood by the ultimate user. The ultimate user, and any other groups who are part of the distribution chain, must value the technology and demonstrate commitment to it. This means that each group must understand and acknowledge the benefits the technology offers to their own operation, as well as any incremental costs to their operation. Adequate resources, personnel, facilities, funds, capacity, and training must be available for scale-up and implementation. Users must feel ownership, of the technology and the implementation process.

The keys here are communication, and an orientation toward satisfying the needs and concerns of the other people who must be involved in producing and using the vegetative propagules. Some of these other people or groups might be involved in selection and testing, in production of seed for vegetative multiplication, in the actual propagation operation, in growing of the propagules in the nursery, or in planting and culture of the vegetatively propagated stands.

Each of these people or groups perceives the value of vegetative propagation from a different perspective, yet each will be impacted and must value the benefits more than the costs – tangible or intangible – for the technology to be widely accepted in commercial practice. This principle will be important not only for large, vertically integrated organizations, but for small, single-function organizations as well. However, technology transfer generally proceeds most quickly and effectively in smaller organizations or in situations where few organizational boundaries exist between the developers of technology and the ultimate users of the technology.

Cost-Effective Production. Commercial forestry has historically been characterized by low per-plant value and a low per-plant cost threshold. This has in part been due to a lack of differentiation in the quality of planting stock. The introduction of technologies for genetic improvement has begun to change the basic "value equation" to reflect differentiation and improved genetic quality, and this change has been reflected in higher willingness to pay for genetically improved planting stock. However, cost thresholds for planting stock in forestry remain low, and are likely to continue to increase slowly relative to increases in genetic value, due to the long carrying period for establishment costs, the high risk of loss or value fluctuations for stumpage over the rotation, and many other factors. Low-cost production methods will be critical to the large-scale acceptance of vegetative propagation in commercial forestry.

A number of promising innovations have already been noted which have or will shortly greatly reduce costs of vegetative propagation. Automation, cultural regimes or media which necessitate less handling, and improved biological response and efficiency have already had an impact. However, only a few research laboratories as yet report active efforts in the area of automation, process engineering, and scale-up. Increased effort and focus in this area will be critical.

In our organization, experience has shown that "practice makes perfect". Gradual scale-up, and high involvement by operational nursery personnel, result in a broad base of experience and ideas which would not have come out on a small scale without such involvement.

Genetic Improvement and Propagation as Part of a Forest Management System. Programs which are achieving the most rapid scale-up and the most dramatic benefits from vegetative propagation are those in which there is tight integration from genetic selection through propagation and stand management to final stand harvest. An example of such integration can be found in the Tasman Forestry Ltd. program described by Arnold and Gleed (1985). Stem quality benefits of maturation are coupled with growth rate benefits through genetic selection in 5-year-old orchard-origin plantations. Rooted cuttings from these selections are grown in existing production nurseries, with modifications in nursery regimes, and are then delivered to the foresters. Based on an understanding of the stem quality and growth attributes of the stock, the foresters plant the trees at a wider than normal spacing, enabling them to reduce their forest management costs while retaining or enhancing final harvested value.

Advantages of such integration are twofold. First, the benefits which may be realized by any one group in the process (the breeder, the nursery grower, the forester) can have real synergy, as demonstrated in the example. Second, integration can lead to a strong sense of commitment and ownership for the new technology across each group in the process, and such commitment is the key to successful implementation.

Guidelines for Responsible Deployment and Communication. Through their approaches to development and deployment of genetic material, foresters possess the ability to integrate vegetative propagation into their operations on a significant scale with minimal impact on the level of biological risk their stands will face. Through our breeding programs, we can provide a broad base of diverse genetic material from which to select. Intensive testing programs allow us to expose candidate genotypes and families to a wide array of stresses so that we have a good understanding of the behavior of our material, and so that genotypes which are intolerant of important stress factors will not be included in production populations. Genetic diversity and adaptive performance for primary commercial species can be maintained through vegetative propagation far more effectively than it can through natural regeneration or planting of improved stock, since we understand far more about the material we are planting, and we know precisely where we are putting it. Through tracking of stands of known pedigree, we can observe long-term behavior of our genetic material, learn to manage it effectively, and stop planting particular families or genotypes if they show unacceptable characteristics over time. And, related to all the above, we can *choose* the sites where our material is planted, establishing blocks or mixtures of controlled size, and even interspersing unimproved stock or other species for added buffering if we find that such measures are important.

Yet, with the above we do not intend to be cavalier about the complexity of the systems in which we operate. The mechanisms of response of specific genotypes and mixtures of genotypes to the array of operational environments over a harvest rotation are highly complex. The capability to produce large quantities of stock of specific families and genotypes will often precede the accumulation of sufficient data to predict the behavior of that material against all the major types of stresses it is likely to face in space and time, in pure blocks or in controlled mixtures.

Given the above, much responsibility rests on the shoulders of forest managers and geneticists to develop and apply responsible guidelines for deployment of improved stock, whether vegetatively propagated or otherwise. The stakes are large, from more than one perspective. Collectively, the organizations involved in improvement activities with a species are stewards of that species' gene pool, and they share a joint responsibility to ensure the long-term biological fitness and productivity of the species. Forestry organizations also depend on the long-term fitness of their commercial species for sustained *financial* well-being, and the more that is invested in genetic improvement and other intensive management practices, the more important it becomes to protect those investments through prudent management of biological risk. And last but certainly not least, our ability to carry out what we perceive to be biologically and economically sound and productive forest management activities is likely to be severely constrained through public pressure and regulation unless we can demonstrate both the intent and the practice of responsible management of our forests.

We believe that certain elements will be very important:

1. Sound research and data. Topics like the relationships of genetic diversity with adaptive behavior and productivity, the critical criteria for measuring diversity, and the mechanisms of response to important environmental factors will be critical to development of sound deployment plans and guidelines for clonal stock. There can be no substitute for well-planned, statistically rigorous research to address these issues. Emphasis must be given not only to development of a strong database, but also to establishment of demonstration plantings which can help to increase public understanding of these topics. Given the complexity of these problems and the broad base of interest across forestry organizations worldwide, some type of cooperative approach to such research may be the most effective way to get it accomplished.

2. Gradual scale-up. Good research is essential to provide understanding of general relationships impacting deployment of clonal material. At the same time, understanding of the consequences of specific practices is often greatly enhanced through small-scale operational deployment, coupled with intensive tracking to observe any response of concern. Through such a program, only a small proportion of the total planting effort would be subject to any increased risk, but valuable information, experience and partnership with forestry personnel would be gained through the aggressive, small-scale trials.

The concept of gradual scale-up also applies to internal guidelines or policies developed by an organization to regulate its application of a technology like cloning. In the face of significant unknowns, guidelines for large-scale forestry operations should initially be conservative. At the same time, if small-scale trials are established to push the limits of likely gains and losses for various deployment options, then the scene is set or guidelines to be gradually refined over time, and relaxed or strengthened based on the data.

It is important to note that gradual scale-up also provides for a gradual increase in the comfort level of foresters and the general public with a new technology, plus an array of demonstration areas to be used for communication and education. As mentioned above, this is an important element of technology transfer.

3. Understanding of genotypes and sites. Historically, tree improvement has been based on development of a modest amount of information on a large and heterogeneous population of genetic material. With the availability of technology for large-scale production of control-cross families and clones, the emphasis must shift toward a more detailed understanding of a smaller population of material. Commensurately, there is also a greater need to understand the sites on which that material will be deployed. This is particularly true for organizations developing varieties which are optimized for specific site types or stress factors. However, even organizations pursuing broadly stable varieties will have a greater need to understand the types of stresses which are likely to be encountered on major site types. By so doing, procedures and guidelines can be put in place to minimize the likelihood that a clone or family will be subjected to conditions it cannot tolerate.

4. Self-regulation. The subject of internal policies and guidelines has now been mentioned more than once. We have said that it will be very important for organizations to demonstrate that they are willing and able to manage biological risk, and well-considered, clearly defined guidelines are an effective way to do this. The added benefit of internal guidelines rigorously adhered to, is a high level of consistency across even complex operations, which can be maintained through time even as scientists, managers, and operations personnel come and go.

5 Teamwork. Tree improvement organizations, and particularly those desiring to implement clonal forestry, must act in a coordinated fashion, with respect shown for differing objectives and pressures. Excessively restrictive guidelines fashioned and publicized by one organization may not apply well to the circumstances of other organizations, but are likely to be seized by the public and applied across the board. On the opposite extreme, irresponsible actions resulting in significant biological impacts are likely to lead to regulatory restrictions which are applied broadly.

6. Sensitivity to public concerns. Different groups of people and even different forestry organizations will often have different objectives. Sometimes these objectives will be mutually exclusive. In a large number of situations, however, a "middle ground" probably exists but is never sought, or is pursued but lost in the heat of polarized thinking, inefficient communication, and/or pursuit of political advantage. It is our view that responsible societies must dedicate some significant land area to be responsibly managed for sustained timber production to meet their long-term demands for wood products. Over time we believe that the public will recognize and accept this need. It is the definition of "responsible" which is at issue. Forestry organizations which are to survive these increasingly regulated times, and especially those who wish to implement new technologies, must demonstrate genuine interest in addressing and resolving issues or questions that the public has relative to their activities. Many questions will be resolved through education, and through research and sharing of results. Some issues may arise, however, which reflect significant differences in philosophy about how forest resources should be managed, and such issues will be resolved either through compromise or conflict. For public concerns to be resolved in a reasonable manner, an atmosphere of mutual trust and respect must be built and maintained. In the absence of such an atmosphere, the future will hold only tighter restrictions on the practice of forestry.

4 Summary

The subject of vegetative propagation in forestry has become a very broad one, rich in experience and diversity. Like the editor of the Australian Forest Grower, we believe that this technology certainly has the potential to create a revolution in forestry practice. However, its level of success largely remains to be established. Success no longer revolves solely around the technical capabilities of the scientists who initiated the technology, but now also requires new and diverse skills and a shift of focus toward the questions, needs, and concerns of the public, and of the practitioners who will be implementing the technology. The skills, know-how, and determination are definitely in place, and our survey indicates that many organizations have recognized the challenges of large-scale implementation and are proceeding ahead responsibly and effectively. We conclude this chapter with a great deal of respect for the efforts of those involved in vegetative propagation research around the world, and a strong sense of optimism and momentum for the future.

Acknowledgments. We very gratefully acknowledge the following individuals and organizations who responded to our survey.

Bengt Bentzer, Hilleshog AB, Sweden
Charles Blackwood, Forestry Commission, Cheshire, UK
David Boomsma, Woods and Forests Dept, South Australia
Michael J. Carson, New Zealand Forest Research Institute, Rotorua
R. Currie, UK
Lado Elersek, Institut za gozdno in lesno gospodarstvo Slovenije, Yugoslavia

Zdenek Fojtik, VcSL, Czechoslovakia
G. Sam Foster, International Forest Seed Co., U.S.A. (now with USFS)
D.P. Fowler, Canadian Forest Service-Maritimes, New Brunswick
Yoshiyuki Fugimoto, Kyushu Forest Tree Breeding Institute, Kujimoto Pref., Japan
John Gleed, Tasman Forestry Ltd., New Zealand
Gunnar Haug, Biri Plantskole, Norway
Bo Karlsson, Inst. for Forest Improvement, S. Ekebo District, Sweden
Jochen Kleinschmit, Lower Saxony Forest Research Inst., Escherode, FRG
Hans J. Kosmer, Dept. of Conservation, Forests & Lands, Melbourne, Australia
Martti Lepisto, Finnish Foundation for Tree Breeding, Finland
Martin Lindell, Skogsstyrelsen, National Board of Forestry, Sweden
Jan Matras, Instytut Badawczy Lesnictwa, Poland
William Mason, Northern Research Station, Roslin, Midlothian, UK
D. McDonald, Aberdeen Tree Nurseries, Aberdeenshire, UK
Glenn McLeod, Orono Ministry of Natural Resources, Ontario, Canada
Bev Nairn, Tasman Forestry Ltd., New Zealand
Alphonse Nanson, Station de Recherches Forestieres, Belgium
R.S.D. Ogilvy, Bogallan Nursery, Inverness, Scotland
Kihachiro Ohba, Inst. of Agriculture and Forestry, Univ. of Tsukuba, Japan
Pavel Radosta, Forestry Research Institute, Prague, Czechoslovakia
Deborah Rogers, Ontario Ministry of Natural Resources, Ontario, Canada
Hans Roulund, Royal Veterinary and Agricultural University, Denmark
John Russell, British Columbia Ministry of Forests, B.C., Canada
K.E. Scarisbrick, International Forest Seed Co., USA
Tore Scroppa, Div. of Forest Genetics and Tree Breeding, Norwegian FRI, Norway
Robin Sym, Tilhill Nurseries Division, Surrey, UK
W. Weisgerber, Inst. for Forest Tree Improvement, FRG

References

Abo El-Nil MM (1980) Embryogenesis of gymnosperm forest trees. US Patent 4,217,730

Abo El-Nil MM (1982) Method for asexual reproduction of coniferous trees. US Patent 4,353,184

Aitken-Christie J, Jones C (1987) Toward automation: radiata pine shoots hedge in vitro. Plant Cell Tissue Org Cult 8:185–196

Aitken-Christie J, Singh AP (1987) Cold storage of tissue cultures. In: Bonga JM, Durzan DJ (eds) Cell and tissue culture in forestry, vol 2. Nijhoff, Dordrecht, pp 285–305

Aitken-Christie J, Singh AP, Davies H (1988) Multiplication of meristematic tissue: a new tissue culture system for radiata pine. In: Hanover JW, Keathley DE (eds) Genetic manipulation of woody plants. Plenum, New York, pp 413–432

Aitken-Christie J, Davies HE, Kobata C, Kozai T (1990) Benefits of using a miroporous membrane in the tissue culture vessel for the micropropagation of *Pinus radiata*. In: Abstr 7th Int Congr Plant tissue culture, Amsterdam, June 24–29, p 86

Amerson HV, Frampton LJ, Mott RL, Spaine PC (1988) Tissue culture of conifers using loblolly pine as a model. In: Hanover JW, Keathley DE (eds) Genetic manipulation of woody plants. Plenum, New York, pp 117–138

Anon (1987) Exclusive-clonal forestry in Australia. Aust For Grower, Summer 1987:11–18

Arnold RJ (1990) Control pollinated radiata pine seed — a comparison of seedling and cutting options for large-scale deployment. NZ For 35(3):12–17

Arnold R, Gleed JA (1985) Raising and managing radiata pine cuttings for production forests. Aust For 48:199–206

Becwar MR, Noland TL, Wann SR (1987a) A method for quantification of the level of somatic embryogenesis among Norway spruce callus lines. Plant Cell Rep 6:35–38

Becwar MR, Noland TL, Wann SR (1987b) Somatic embryo development and plant regeneration from embryogenic Norway spruce callus. TAPPI J 70(4):155–160

Becwar MR, Wann SR, Johnson MA, Verhagen SA, Feiner RP, Nagmani R (1988) Development and characterization of in vitro embryogenic system in conifers. In: Ahuja MR (ed) Somatic cell genetics of woody plants. Kluwer, Dordrecht, pp 1–18

Bentzer BG, Foster GS, Hellberg AR, Podzorski AC (1988) Genotype × environment interaction in Norway spruce involving three levels of genetic control; seed source, clone mixture, and clone. Can J For Res 18:1172–1181

Bentzer BG, Foster GS, Hellberg AR, Podzorski AC (1989) Trends in genetic and environmental parameters, genetic correlations and response to indirect selection for 10-year volume in a Norway spruce clonal experiment. Can J For Res 19:897–903

Blush TD (1987) An operational trial of supplemental mass pollination in a loblolly pine seed orchard. In: Proc 19th SFTIC, Coll Stn TX, pp 240–246

Bolstad PV, Libby WJ (1982) Comparisons of radiata pine cuttings of hedge and tree-farm origin after seven growing seasons. Silv Genet 31:9–13

Boulay M (1988) In vitro propagation of tree species. In: Green CE, Somers DA, Hackett WA, Biesboer DD (eds) Plant tissue and cell culture. Liss, New York, pp 367–382

Bridgwater FE, Bramlett DL, Matthews FR (1987) Supplemental mass pollination is feasible on an operational scale. In: Proc 19th SFTIC, Coll Stn TX, pp 216–222

Brown CL, Lawrence RH (1968) Culture of pine callus on a defined medium. For Sci 14:62–65

Burdon RD (1971) Clonal repeatabilities and clone-site interactions in *Pinus radiata*. Silv Genet 20:33–39

Burdon RD (1986) Clonal forestry and breeding strategies: a perspective. In: Proc IUFRO Meet Breeding theory, progeny testing and seed orchards, Williamsburg VA, pp 645–659

Burdon RD (1988a) When is cloning on an operational scale appropriate? In: IUFRO Meet Breeding tropical trees, Pattaya, Thailand, Nov–Dec pp 9–27

Burdon RD (1988b) Recruitment for breeding populations: objectives, genetics and implementation. In: Wier, B.S, Eisen, E.J., Goodman, M.M., Namkoong, G (eds) Proc 2nd Int Conf Quant Genet. Raleigh, North Carolina pp 555–572

Byram TD, Lowe WJ (1986) General and specific combining ability estimates for growth in loblolly pine. In: proc IUFRO Breeding theory, progeny testing and seed orchards, Williamsburg VA, pp 352–360

Cameron JN, Cotterill PP (1988) Comparison of genetic gains and economic returns for the many packaging population options for *Pinus radiata*. In: Proc 10th Meet Aust RWG 1 Forest genetics, Gympie, Qld

Carson MJ (1986) Advantages of clonal forestry for *Pinus radiata* – real or imagined? NZ J For Sci 16(3):403–415

Carson MJ (1987) Improving log and wood quality: the role of the radiata pine improvement programme NZ For 31(4):26–30

Comstock RE, Robinson HF, Harvey PH (1949) A breeding procedure designed to make maximum use of both general and specific combining ability. Argon J 41:360–367

Dudley JW (1976) Generations of selection for oil and protein percentage in maize. In: Pollak E, Kempthorne O, and Bailey TB (eds) Proc 1st Int Conf Quant Genet. pp 459–474 Ames, Iowa.

Dunstan DI (1988) Prospects and progress in conifer biotechnology. Can J For Res 18(12):1497–1506

Durzan DJ (1982) Somatic embryogenesis and sphareoblasts in conifer cell suspensions. In: Fujiwara A (ed) Plant tissue culture 1982. Maruzen, Tokyo, pp 113–114

Durzan DJ, Gupta PK (1987) Somatic embryogenesis and polyembryogenesis in Douglas-fir cell suspension cultures. Plant Sci 52:229–235

Eldridge KG (1981) Cuttings from seedlings of selected radiata pine families. Aust For Counc Res Work Group 1. Traralgon, Victoria

Fielding JM (1970) Trees grown from cuttings compared with trees grown from seed (*Pinus radiata* D Don). Silv Genet 19:54–63

Foster GS (1986) Making clonal forestry pay: breeding and selection for extreme genotypes. In: Proc IUFRO Meet Breeding theory, progeny testing and seed orchards, Williamsburg VA, pp 582–590

Foster GS, Bridgwater FE, McKeand SE (1981) Mass vegetative propagation of loblolly pine – a revaluation of direction. In: 16th SFTIC, Blacksburg VA, pp 311–319

Foster GS, Lambeth CC, Greenwood MS (1987) Growth of loblolly pine rooted cuttings compared with seedlings. Can J For Res 17:157–164

Franclet A, Boulay M, Bekkaoui F, Fouret Y, Verschoove-Martouzet B, Walker N (1987) Rejuvenation. In: Bonga JM, Durzan DJ (eds) Cell and tissue in forestry, vol 1. Nijhoff, Dordrecht, pp 232–248

Fujii JA, Slade DT, Redenbaugh K, Walker KA (1987) Artificial seeds for plant propagation. Trends Biotechnol 5(12): 335–339

Gill JGS (1983) Comparisons of production costs and genetic benefits of transplants and rooted cuttings of *Picea sitchensis*. Forestry 56(1): 61–73

Gupta PK, Durzan DJ (1985a) Shoot multiplication from mature trees of Douglas-fir and sugar pine. Plant Cell Rep 4: 177–179

Gupta PK, Durzan DJ (1985b) Establishment and maintenance of mature and immature trees of Douglas-fir and sugar pine. In: Proc Intl Symp Eur Soc Hortic, 21–26, Sep, Belgium, p 16

Gupta PK, Durzan DJ (1986a) Plantlet regeneration via somatic embryogenesis from subcultured callus of mature embryos of *Picea abies* (Norway spruce). In Vitro Cell Dev Biol 22: 685–688

Gupta PK, Durzan DJ (1986b) Somatic polyembryogenesis from callus of mature sugar pine embryos. Bio/Technology 4: 643–645

Gupta PK, Durzan DJ (1987a) Micropropagation and phase specificity in mature, elite Douglas-fir. J Am Soc Hortic Sci 112: 969–971

Gupta PK, Durzan DJ (1987b) Biotechnology of somatic polyembryogenesis and plantlet regeneration in loblolly pine. Bio/Technology 5: 147–151

Gupta PK, Durzan DJ, Finkle BJ (1987) Somatic polyembryogenesis in embryonic cell masses of *Picea abies* and *Pinus taeda* after freezing in liquid nitrogen. Can J For Res 17: 1130–1134

Haines RJ, Nikles DG (1987) Mass propagation of superior families and individual genotypes of some conifers in Queensland. Aust For 50(2): 106–111

Hakman I, von Arnold S (1988) Somatic embryogenesis and plant regeneration from suspension cultures of *Picea glauca* (white spruce). Physiol Plant 72: 579–587

Hakman I, Fowke LC, von Arnold S, Eriksson T (1985) The development of somatic embryos in tissue culture initiated from immature embryos of *Picea abies* (Norway spruce). Plant Sci 38: 53–59

Johnson GR (1988) A look to the future: clonal forestry. In: Menzies MI, Aimers JP, Whitehouse LJ (eds) FRI Bulletin 135, Worksh Growing radiata pine from cuttings. pp 79–85 Rotorua, NZ.

Kang H (1979) Long-term tree breeding. In: Proc 15th SFTIC, Starkville MS, pp 66–72

Kartha KK, Fowke LC, Laing NL, Caswell KL, Hakman I (1988) Induction of somatic embryos and plantlets from cryopreserved cell cultures of white spruce (*Picea glauca*). J Plant Physiol 132: 529–539

Kitto S, Janick J (1985) Hardening treatments increase survival of synthetically coated asexual embryos of carrot. J Am Soc Hortic Sci 110: 283–286

Kleinschmit J (1974) A programme for large-scale cutting propagation of Norway spruce. NZ J For Sci 4: 359–366

Kleinschmit J (1983) Concepts and experiences in clonal plantations of conifers. In: Proc 19th Meet Can Tree Improv Assoc, Toronto, pp 26–56

Kleinschmit J, Schmit J (1979) Experiences with *Picea abies* cutting propagation in Germany and problems connected with large-scale application. Silv Genet 26: 197–203

Krugstrap P (1986) Embryo-like structures from cotyledons and ripe embryos of Norway spruce (*Picea abies*). Can J For Res 16: 664–668

Lelu M, Boulay M, Arnaud Y (1987) Formation of embryogenic calli from cotyledons of *Picea abies* (L) collected from 3- to 7-day-old seedlings. CR Acad Sci Paris 305 (Ser III): 105–109

Levin R (1985) Process for plant tissue culture propagation. EP Patent 0132414A2

Libby WJ, Rauter RM (1984) Advantages of clonal forestry. For Chron 60: 145–149

Litvay JD, Johnson MA, Verma D, Einspahr D, Weyrauch K (1981) Conifer Suspension culture medium development using analytical data from developing seeds. Inst Paper Chemistry, Appleton, WI. Tech. Paper. 115

Litz PE (1986) Mango. In: Evans DA, Sharp WR, Ammivato PV (eds) Handbook of plant cell culture, vol 4. MacMillan, New York, pp 612–625

Mahalovich MF, Bridgwater FE (1989) Modeling elite populations and positive assortative mating in recurrent selection programs for general combining ability. In: Proc 20th SFTIC, Charleston SC, pp 43–49

Matheson AC, Lindgren D (1985) Grains from the clonal and the clonal seed orchard options compared for tree breeding programs. TAG 71: 242–249

McKeand SE (1985) Expression of mature characteristics by tissue culture plantlets derived from embryos of loblolly pine. J Am Soc Hortic Sci, 110:619–623

Mohammed GH, Vidaver WE (1988) Root production and plantlet development in tissue-cultured conifers. Plant Cell Tissue Org Cult 14:137–160

Murashige T, Skoog F (1962) A revised medium for rapid growth and bioassay with tobacco tissue culture. Physiol Plant 15:473–497

Nagmani R, Bonga JM (1985) Embryogenesis in subcultured callus of *Larix decidua*. Can J For Res 15:1088–1091

Namkoong G, Roberds JH (1982) Short-term loss of neutral alleles in small population breeding. Silv Genet 31:1–6

Namkoong G, Kang HC, Brouard JS (1988) Tree breeding: principles and strategies. Springer, Berlin Heidelberg New York, 180 pp

Nicholls JWP, Pedrick LA, Brown AG (1977) A summary of the ortet-ramet relationship in wood characteristics of Pinus radiata. APPITA 30:496–502

Poisonier N, Franclet A, Dumant MJ, Gauntry JY (1980) Enracinement de tigelles in vitro de *Sequoia sempervirens*. Ann AFOCEL, pp 231–253

Power AB, Dodd RS (1984) Early differential susceptibility of juvenile seedlings and more mature stecklings of *Pinus radiata* to *Dothistroma pini*. NZ J For Sci 14:223–228

Rangan TS (1986) Somatic embryos of citrus. In: Vasil IK (ed) Cell culture and somatic cell genetics of plant, vol 1. Academic Press, New York London, pp 68–73

Redenbaugh K, Slade DT, Fujii JA (1988a) Artificial seeds. In: Proc Int Conf In vitro selection and propagation, Pakistan, 7 pp

Redenbaugh K, Paasch BD, Nichol JW, Kossler ME, Viss PR, Walker KA (1988b) Somatic seeds: encapsulation of asexual plant embryos. Bio/Technology 4:797–801

Redenbaugh K, Fujii JA, Slade DT (1988c) Encapsulated plant embryos. Biotechnol Agric 9:225–248

Reilly K, Washer J (1977) Vegetative propagation of radiata pine by tissue culture. NZ J For Sci 7:199–206

Ritchie GA, Long AJ (1986) Field performance of micropropagated Douglas-fir. NZ J For Sci 16:343–356

Rumery C, Thorpe TA (1984) Plantlet regeneration in black and white spruce. Can J For Res 14:10–16

Rumery C, Patel RK, Thorpe TA (1986) Plantlet formation in black and white spruce. Can J Bot 64:997–1002

Russell J, Libby WJ (1987) Clonal testing efficiency: the trade off between clones tested and ramets per clone. Can J For Res 16(5):925–930

Schenk RU, Hildebrandt AC (1972) Medium and techniques for induction and growth of monocotyledonous and dicotyledonous plant cell cultures. Can J Bot 50:199–204

Shaw DV, Hood JV (1985) Maximizing gain per effort by using clonal replicates in genetic tests. TAG 71:392–399

Shelbourne CJA (1986) New approaches in tree breeding with radiata pine in New Zealand. In: Proc IUFRO Meet Breeding theory, progeny testing and seed orchards, Williamsburg VA, pp 269–284

Shelbourne CJA, Thulin IJ (1974) Early results from a clonal section and testing programme with radiata pine. NZ J For Sci 4:387–398

Smith DR, Shelbourne CJA, Dibley MJ, Carswell T (1985) Field performance of micropropagated *Pinus radiata* p23. In: Abstr 3rd Meet Int Conifer tissue culture work group. For Res Inst NZ For Ser, Rotorua, 37 pp

St. Clair JB, Kleinschmit J (1986) Genotype-environmental interaction and stability in ten-year height growth of Norway spruce clones (*Picea abies* Karst). Silv Genet, 35:177–186

Stonecypher RW, McCullough RB (1986) Estimates of additive and non-additive genetic variances from a clonal diallel of Douglas-fir. In: Proc IUFRO Meet Breeding theory, progeny testing and seed orchards, Williamsburg VA, pp 211–227

Stonecypher RW, Zobel BJ, Blair R (1973) Inheritance patterns of loblolly pines from a nonselected natural population. Tech Bull 220, NC Agric Exp Stn, NC State Univ

Styer DJ (1985) Bioreactor technology for plant propagation. In: Henke RR, Hughes KW, Constantin MJ, Hollander A (eds) Tissue culture in forestry and agriculture. Plenum, New York, pp 117–130

Sweet GB (1973) The effect of maturation on the growth and form of vegetative propagules of radiata pine. NZ J For Sci 3: 191–210

Sweet GB, Harris JM (1976) Wood properties of *Pinus radiata*: seed-grown trees compared with grafts from different-aged ortets. NZ J For Sci 6(1): 114–121

Sweet GB, Wells LG (1974) Comparison of the growth of vegetative propagules and seedlings of *Pinus radiata*. NZ J For Sci 4: 399–409

Teasdale RD, Dowson PA, Woolhouse HW (1986) Mineral nutrition requirement of a loblolly pine (*Pinus taeda*) cell suspension culture. Plant Physiol 82: 942–945

Thorpe TA, Biondi S (1984) Conifers. In: Sharp WR, Evans DA, Ammirato P, Yamada Y (eds) MacMillan, New York, pp 435–470

Thorpe TA, Hasnain S (1987) Micropropagation of conifers. Method, opportunities and costs. Proc 21st Can Tree Improv Assoc Mtg: 68–84

Thulin IJ, Faulds T (1968) The use of cuttings in breeding and afforestation of *Pinus radiata*. NZ J For 13: 66–77

Timmis R (1985) Factors influencing the use of clonal material in commercial forestry. In: Tigerstedt PMS, Puttonen P (eds) Crop physiology of forest trees. Univ Press, Helsinki, pp 259–273

Timmis R, Ritchie GA (1984) Progress in douglas-fir tissue culture. In: Proc Int Symp Recent advances in forest biotechnology, Traverse City MI, June 10–13, pp 37–46

Toda R (1974) Vegetative propagation in relation to Japanese tree forest improvement. NZ J For Sci 4: 410–417

Von Arnold S, Hakman I (1986) Effect of sucrose on initiation of embryogenic callus from mature zygotic embryos of *Picea abies* (Norway spruce). J Plant Physiol 122: 261–265

Von Arnold S, Hakman I (1988a) Plantlet regeneration in vitro via adventitious buds and somatic embryos in Norway spruce (*Picea abies*). In: Hanover JW, Keathley DE (eds) Genetic manipulation of woody plants. Plenum, New York, pp 199–215

Von Arnold S, Hakman I (1988b) Regulation of somatic embryo development in *Picea abies* by abscisic acid (ABA). J Plant Physiol 132: 164–169

Von Arnold S, Woodward S (1988c) Organogenesis and embryogenesis in mature zygotic embryos of *Picea sitchensis*. Tree Physiol 4: 291–300

Wagley LM, Gladfelter HJ, Phillips GC (1987) De novo shoot organogenesis of *Pinus eldarica*. Plant Cell Rep 6: 163–166

Webb DT, Santiago OD (1983) Cytokinin induced bud formation on caribbean pine (*Pinus caribaea*) embryos. Plant Sci Lett 32: 17–21

Webster FB, Roberts DR, McInnis SM, Sutton BCS (1990) Propagation of interior spruce by embryogenesis. Can J For Res 20: 1759–1765

Winton L (1970) Shoot and tree production from aspen tissue culture. Am J Bot 57: 904–909

Winton LL, Verhagen SA (1977) Shoots from Douglas-fir cultures. Can J Bot 55: 1246–1250

Wochok ZS, Abo El-Nil MM (1977) Conifer tissue culture. Proc Int Plant Propagators Soc 27: 131–136

12 Rejuvenation of Tissues from Mature Conifers and its Implications for Propagation in Vitro

J.M. Bonga[1] and P. von Aderkas[2]

1 Introduction

1.1 Why Clonal Propagation of Mature Trees?

Clonal propagation of trees old enough to have expressed their desirable characteristics is an effective means of rapidly obtaining improved planting stock. However, cloning of mature trees of conifer species is often difficult or impossible with present-day techniques. Therefore, specimens are often selected for clonal propagation when they are still too young for proper assesment. For many species such assesment is, at present, not possible until the trees have reached about half their rotation age (Namkoong et al. 1980). Because of this predicament, maximum benefit of clonal propagation cannot be expected until we find the appropriate means either to clone older trees or to determine at a young age what the trees will be like at the end of their rotation.

Improvement of planting stock by cloning of mature elite trees has been effective for species in which the mature trees can provide juvenile tissues, i.e., stump sprouts, for relatively easy cloning (e.g., *Populus*, *Eucalyptus*, see Zsuffa et al., Chap. 5, and Zobel, Chap. 7, Vol. 2). In fact, where cloning is easy, it often produces better results in the short run (i.e., over several decades of planting) than can be obtained from the one or two generations of selection and breeding possible in that time. Cloning of elite trees would be particularly interesting if it could be combined with genetic engineering, for example, if genes for disease or other resistance could be incorporated in clones of elite trees that lack these characteristics.

1.2 Problems with Clonal Propagation of Mature Conifers

Unfortunately, conifers at an age old enough to allow proper evaluation of their growth potential are generally difficult to clone, except by grafting. Grafting can be carried out with scions from trees of almost any age, except that graft rejection may occur if the scion and rootstock are not properly matched. Cloning by rooting of cuttings is possible with young or sometimes adolescent

[1]Forestry Canada—Maritimes Region, P.O. Box 4000, Fredericton, N.B. E3B 5P7, Canada
[2]Department of Biology, University of Victoria, Victoria, B.C. V8W 2Y2, Canada

Clonal Forestry I, Genetics and Biotechnology
Ed. by M.R. Ahuja and W.J. Libby

trees, but generally not with sexually mature conifers. A problem with grafting or rooting of cuttings is that, except when the scions or cuttings are taken from trees not more than a few years old, grafted scions or rooted cuttings often fail to form trees resembling the donor of the scions or cuttings. The propagules are often plagiotropic, and even if they become orthotropic they may still retain some mature characteristics (see discussion of cyclophysis in Greenwood and Hutchison, Chap. 3, Vol. 1). Consequently, unless highly juvenile donors are used, grafting and rooting of cuttings results in propagules in which the juvenile growth phase is largely bypassed. This can be a disadvantage because the juvenile phase generally displays a fast growth rate. On the other hand, bypassing the juvenile phase has, on a few occasions, resulted in trees with improved characteristics (see Gleed, Chap. 8, Vol. 2).

Cloning by tissue culture is, for most conifers, only possible with explants from embryos or young seedlings, i.e., it can generally only be carried out with donors younger than the ones that still allow propagation by rooting of cuttings or grafting. However, tissue culture probably has a better potential for obtaining phenotypically true-to-type propagules. The reasons for this assumption are the following. (1) In tissue culture we can use very small pieces of tissue or, theoretically, even single cells. This allows selection of those cells which are the least physiologically mature and, thus, most capable of providing true-to-type propagation. By culturing these cells free from the correlative controls normally exerted by neighbouring tissues, they can express their total potential (Bonga 1987) (also see miniaturization in Section 4.3). (2) Because tissue culturing is carried out aseptically and in a strictly controlled environment, highly specific chemical and physical treatments can be applied easily at particular stages in development. Consequently, a wider range of precise experimental controls is possible in tissue-culture propagation than in rooting of cuttings and grafting.

The problems in conifer tissue culture are still numerous. Even propagules produced from explants from a donor as young as a mature embryo (cotyledon culture) do not always behave phenotypically true to type. Frequently, such propagules display slower initial growth rates and earlier appearance of mature characteristics than do sexually produced seedlings (Amerson et al. 1988; Lesney et al. 1988). Presumably this means that either the explant was not sufficiently juvenile for true-to-type organogenesis, or, more likely, that maturation progressed more rapidly during plantlet formation in vitro than during development of seedlings from seeds. Problems of this nature are progressively more severe as the explants are excised from progressively older donor plants. Clearly, maturation and our present inadequate ability to reverse it by rejuvenation are major obstacles hindering micropropagation of conifers. In the following, we will discuss these obstacles. Our main objective is to deal with rejuvenation, but obviously one cannot deal with that subject in isolation. Maturation, which is the force opposing rejuvenation (also see Greenwood and Hutchison, Chap. 3, Vol. 1), will have to be considered first.

2 Maturation

2.1 General Aspects

There is a clear difference between maturation and aging. Whereas aging implies gradual loss of vigor and, eventually, senescence and death, maturation refers to gradual changes in several maturation-sensitive traits, in particular to the attainment of a physiological state that allows flowering to be induced (Borchert 1976a; Hackett 1985). Therefore, maturity is not necessarily associated with loss in vigor. In fact, vigorous, older trees generally grow faster than young seedlings (Young and Hanover 1976).

Maturation does not occur at the same speed in all parts of the plant. For example, many tree species possess meristems that normally are perennially dormant and, therefore, mature more slowly than active ones. After release from dormancy these meristems often produce juvenile sprouts, the most prevalent being stump sprouts. Such stump sprouts are common in hardwood species, but are found in only a few conifers (Burrows 1987). Nevertheless, conifer shoots forced from branches in positions close to the trunk and near the bottom of the tree are more juvenile than those arising from other positions. This has led to the use of heavily pruned hedges as sources of juvenile cuttings for clonal propagation (Bonga 1982; Haines and Nikles 1987; also Talbert et al. Chap. 11, Vol. 1).

When a tree changes from the juvenile to the mature state, several morphological and cellular characteristics are altered in a typical fashion, either abruptly or gradually. Some of these changing characteristics have been listed previously (Bonga 1982). Lately some others have been noted. For example, juvenile conifers have flatter shoot meristems than mature ones (Monteuuis 1987). In eastern larch, juvenile plants have less chlorophyll than mature plants, the chlorophyll a/b binding protein gene may be expressed differently in tissue from young than in that from mature plants, and grafted scions of young plants branch more and have a different needle structure than those of older plants (Greenwood 1984; Greenwood and Hutchison Chap. 3, Vol. 1). Differences have also been noted in nucleotide content in shoot apices of juvenile and mature *Sequoiadendron* (Bon 1988).

2.2 Maturation Concepts

Fortanier and Jonkers (1976) have divided aging into chronological, ontogenetic, and physiological aging. In their description, chronological aging refers to the time elapsed since germination. Ontogenetic aging refers to the passing of the plant through successive phases of development, i.e., embryogenesis, germination, and the vegetative and sexual growth phase. Physiological aging, according to their definition, primarily refers to "the negative aspects of aging, such as loss of growth vigour, the increased susceptibility to adverse conditions, or deterioration in general". In our discussion we will largely adhere to these definitions, except that, in agreement with most authors who have dealt with the subject (Wareing and Frydman 1976; Greenwood 1987), we will use the term "maturation" in the case of ontogenetic aging. It is mostly maturation, not aging, that will be discussed.

2.3 Mechanisms of Maturation

There are several schools of thought regarding the mechanism of maturation. The main issue distinguishing these is whether cell development is primarily determined by nutritional-environmental or by intrinsic cellular factors. In the case of the former, meristems in a tree initiate juvenile or mature growth characteristics primarily in response to nutritional and hormonal inputs. In the case of the latter, even though physiological factors play a role in the behavior of the meristems, whether the meristems act in a juvenile or a mature fashion is primarily determined by a strong imprinting ("memory") in these meristems. These two views will be reviewed below (also see discussion by Greenwood and Hutchison, Chapter 3, Vol. 1).

2.3.1 Maturation Under Positional, Physiological Control

The main proponent of the idea that the behavior of meristems on the tree is determined primarily by physiological patterns that develop within the tree as it matures was Borchert (1976a, b). These patterns develop as the tree increases in size and complexity, with the meristems furthest away from the roots being the most stressed for water and nutrients. It is well known that stress often causes maturation. For example, reduced rootability of cuttings (Leakey 1985) and early flowering (Grime et al. 1986; Jones 1987) are often induced by poor nutrition or other stress. However, these stresses do not promote maturation sufficiently to initiate flowering in plants that are still in a fully juvenile physiological state at the time the stress is applied (Zimmerman 1973; Hackett 1985). Stress also controls the female to male flower ratio in some trees (Lev-Yadun and Liphschitz 1987, also see Greenwood and Hutchison, Chap. 3, Vol. 1). Stress, induced by keeping rooted cuttings of late-juvenile and early-adolescent douglas-fir, redwood and giant sequoia in small containers or in the lathhouse or in both, causes plagiotropic growth. However, this stress-induced plagiotropism is not permanent; orthotropic and otherwise juvenile or early-adolescent type growth occurs shortly after transfer of the plagiotropic stecklings to conditions providing full light and unrestricted space for the roots to grow in (personal communication W.J. Libby, 1989).

Rooting of cuttings or grafting low on the rootstock often reduces stress because it transfers scions to a location closer to the roots, thus creating an improved water and nutrient supply. This "reinvigoration" often results in a postponement of flowering, and in a gradual loss of such characteristics as plagiotropism (Allsop 1964; Borchert 1976a; Hackett 1985). For a few conifer species, repetitive grafting and rooting of cuttings of mature material has resulted in subsequent micropropagation of this material (Franclet et al. 1987). This suggests the possibility that some degree of rejuvenation has taken place in these species. However, for most conifers a change in the maturation state appears not, or barely, to occur upon such grafting or rooting of cuttings. Evidence of reinvigoration without a change in the maturity status has been provided by Tufuor (1973). He found that cuttings from the upper crown and basal sprouts of 200–400 year old redwoods took up to 4 years to root, and that the rooting percentage of basal-sprout cuttings was about four times as high as that of upper-crown cuttings. Cuttings taken from the ramets rooted much quicker (within 6 months) than the original ortet cuttings. However, the cuttings from the ramets of upper-crown origin and those from the ramets of basal-sprout origin still showed the same relative difference in rooting percentage as the upper-crown and basal-sprout cuttings of the ortet. The reduction in time needed for rooting (from 4 years to 6 months) suggests reinvigoration, but the unchanged difference in rooting percentage indicates that the relative maturation state did not change.

2.3.2 Determination

From the previous section it appears that in most conifers stress is only a minor factor in determining whether a propagule behaves, at least for some maturation-sensitive traits, in a juvenile or more mature fashion in vitro. Therefore, we have to consider additional mechanisms of maturation. Many authors think that meristems in trees are more or less strongly "determined", i.e., they have a "memory" of the age (cyclophysis) of the donor plant, and of their position (topophysis) in it (Bonga 1982, 1987; Durzan 1984; Stange 1984). In the following, we will consider what some of the mechanisms that control determination could be.

Why Determination? Meristems often act as autonomous organs; they maintain a specific growth program independent of factors arising from elsewhere within the plant or external stimuli (cf. Stange 1984), i.e., they are determined. Once determination is established, it is passed on clonally through successive cell divisions (Trewavas 1986), sometimes for the rest of the life cycle of the plant. When persisting throughout the life cycle, determination is reversible only in the sexual process (Bonga 1982, 1987). There are specific chemical markers that indicate determination in tissues (Raff et al. 1979), some of which appear long before differentiation occurs (Bruck and Walker 1986).

What is the function of this determination? It has been argued that plasticity is only required for short, specific periods during the life cycle of the plant; for example, before the embryo meristems are formed, and at moments when shoot meristems switch from one type of growth to another (Bonga 1982, 1987; Bonga and von Aderkas 1988). At other moments, it is presumably advantageous to have the meristems rather strongly programmed. Such rigidity would assure that exposure to minor environmental or endogenous physiological changes would not change the growth pattern of the meristems. For example, it can be argued that once the root, shoot, and cambial meristems have been formed in the embryo, the need for plasticity is diminished. From that point on, the root meristems will normally only form roots, the shoot meristems only shoots, and the cambium only vascular tissue.

Of course the strength of this determination varies greatly from species to species. In some angiosperm tree species this determination is less strong than in conifers. For example, adventitious shoot formation from roots is common in some tree species (Leakey 1985). The quiescent center of the root apex of one woody species, *Vanilla planifolia*, has the capability to transform itself directly into shoot meristemoids when cultured in vitro (Philip and Nainar 1988).

The question now arises, why is determination so much stronger in some conifer families than in others? McCown (1986) has divided the conifers into four groups of difficulty of micropropagation, with *Picea*, *Pseudotsuga*, *Abies*, and some *Pinus* species being the most difficult. In our own experience with the above group, *Abies* is the least responsive. At present we do not know why this should be so. One possible speculation is that the rigid architecture (the very regular branching pattern and shape) of *Abies* is somehow related to its nonresponsiveness in vitro. This rigid architecture, we speculate, requires strongly determined shoot meristems, which, in turn, could be the cause of the low morphogenetic plasticity of these meristems when cultured in vitro.

There are indications that determination is established early in embryogenesis. For example, it has been argued that the zygote of *Citrus* is already determined as a protodermal cell. To overcome this determination, internal cells of the young globular embryo have to lose, by dedifferentiation, their epidermal programming before they can assume other functions (Bruck and Walker 1986). For conifers, determination has been noted in embryos as young as those in the globular stage (Monnier and Norstog 1986).

Little is known about the mechanisms that control determination in conifers. Determination as a factor in the control of maturation has been studied in somewhat greater detail in nonconiferous tree species and in microbial and animal species. In the following, we will look at some of the possible mechanisms of determination.

Epigenesis. In nuclear transplant experiments (reviews, Bonga 1982; Raff and Kaufman 1983) with animals and microorganisms, it was shown that cytoplasm from a cell of a mature organism causes maturation of the nucleus from a juvenile cell (egg cell). Conversely, nuclei that have undergone a degree of maturation can be rejuvenated by juvenile cytoplasm. On the other hand, a juvenile nucleus does not rejuvenate a mature cytoplasm. This suggests that, at least in the animals and microorganisms investigated, juvenility and maturity are, to a degree, under cytoplasmic control. Nuclear transplant experiments have never been carried out with higher plants, and, therefore, we do not know how the nucleus and cytoplasm interact in the control of maturation and rejuvenation. Nevertheless, there is evidence that the cytoplasms of mature and juvenile tissues and cells are different, and that the cytoplasm may play some role in the control of maturation and rejuvenation.

The changes in cytoplasmic factors that cause loss of juvenility are persistent in cell division, but generally reversible in the sexual process. Because at present little is known about these factors, one can only speculate what their nature might be. Candidates discussed earlier (Bonga 1982, 1987) are persistent changes in organelles, cytoplasmic DNA, and cytoplasmic membranes. Lately, new information has come to light regarding the relationship between organelles and adventitious embryogenesis (embryogenesis indicates rejuvenation). During embryogenesis the organelles, in particular the plastids, appear to be unstructured. In a variety of herbaceous plant species, androgenesis occurred only in species with pollen with unstructured proplastids; in the nonandrogenic species the pollen contained amyloplasts (Sangwan and Sangwan-Norreel 1987). Similarly, plastids lacking organization are common during conifer embryogenesis (Becwar et al. 1988; Rohr et al. 1989). Bernard-Durix and Kovoor (1987) have observed differences in mitochondrial DNA in organogenetic and nonorganogenetic cultures of *Corylus avellana*. They suggested that different classes of mtDNA may be associated with stable changes in organogenetic potential during ontogeny.

In some plants, apomixis, i.e., naturally occurring somatic embryogenesis, appears to be under the control of a cytoplasmic gene, most likely, in a plasmid (Lehnhardt and Nitzsche 1988). In these plants, the cells involved in apomixis presumably are either kept continuously in a juvenile state by the cytoplasmic genes or are rejuvenated under their influence shortly before apomixis occurs.

One possible mechanism of determination residing in the cytoplasm is the formation of hormone receptors. As reviewed by Guern (1987), hormonal "imprinting may occur at various stages of development of the plant. Receptors induced by this imprinting can persist in the plant for a long time and thus control specific aspects of its behavior.

Nuclear Changes. Distinct nuclear changes take place during plant maturation. Some of these presumably are only little affected by the cytoplasm, others probably are under stronger cytoplasmic control, i.e., they are largely induced by changes in the cytoplasm.

The observed nuclear changes include DNA methylation and amplification. Since this subject has been reviewed extensively, we refer the reader to these reviews for details (reviews Walbot and Cullis 1985; Halperin 1986; Bonga 1987). It has also been suggested that the ability to pair during meiosis could be due to changed methylation of the DNA, and that this could be a part of the rejuvenation event at or just before meiosis (Hood and Libby 1978).

In some plants DNA is progressively lost with tissue aging, while amplification of repeated sequence occurs in tissues developed after the plants start flowering (Altamura et al. 1987). In these plants, the loss of DNA sequences was associated with an impaired morphogenetic ability in vitro.

DNA amplification is common in many plants and animals (Walbot and Cullis 1985). Sometimes amplified DNA appears as extra chromosomal fragments, which are implicated in cell differentiation (Walbot and Cullis 1985; Van't Hof 1987). These, however, are not always stable; they tend to disappear gradually in tissues that remain mitotic (Walbot and Cullis 1985). Recent evidence indicates that amplified DNA has a function in morphogenesis by acting as a proton sink (Nagl and Popp 1983; Chwirot et al. 1985). However, it has also been reported that in *Hedera helix* there is no phase-change-correlated alteration in DNA replication (König et al. 1987).

Finally, there is evidence that repositioning of intact chromosomes within the nucleus alters phenotypic expression at specific developmental stages (Bennett 1984). According to Bennett, "it is not unreasonable to suppose that peripheral chromosomes, which were less condensed at metaphase, had engaged in greater or more prolonged transcriptional activity in interphase than had the more condensed centrally located ones". He further suggested that "it is not difficult to envisage that some major stages of development could each have a unique intranuclear genome architecture with different chromosome arms or segments regularly near to, or far from, the nuclear membrane and/or each other, and that together these different structural features could help to modulate the patterns of gene activity characteristic of each development stage."

It is feasible that, because of persistent nuclear changes, totipotency has been lost in most tissues of many of the presently difficult-to-micropropagate plants. It is even possible that, in some of these plants, only those cells that will eventually enter into the sexual process have the capacity to return their nuclear DNA to the state of the DNA present in the zygote (cf. Holliday 1984). This does not necessarily mean that there is a germ-line in plants similar to the one in animals. In animals the germ-line is a cell-line set apart early in embryogenesis and undergoing very few divisions until maturation of the sexual cells. Many zoologists subscribe to the idea that germ-line cells are, because of their low division rate, somehow protected from accumulation of DNA errors and other detritus during maturation of the organism, thus maintaining genetic fidelity in the sexual cell-line. However, there is no evidence for a germ-line in plants comparable to the one found in animals, i.e., there is little indication that a group of cells is set aside early in embryogenesis and carried in an inactive state into the floral structures. Therefore, we must assume that mechanisms other than low division rates protect the totipotency of those cell-lines that eventually enter meiosis.

Several authors have suggested that maturation is in part influenced by a mechanism that "counts" the number of successive cell divisions in a cell line (Summerbell et al. 1973; Holliday and Pugh 1975; Sussex 1976; Greenwood 1984; Ununger and Ekberg 1987). This clock mechanism presumably is reset during the sexual process and could be controlled by specific genes, i.e., "chronogenes" (Edmunds and Adams 1981) that turn on and off other genes responsible for specific maturation characteristics at the appropriate time. At a future date, it may become possible, through specific environmental controls, to affect the chronogenes in cell cultures of mature conifers in such a fashion that they will reactivate the genes active in the juvenile phase. It is further conceivable that, by genetic engineering, genes can be introduced in juvenile or rejuvenated clones that will prevent the chronogenes from initiating the mature phase. This could result in propagules that will remain juvenile throughout their life span. In some species, this could have a beneficial effect on volume, crown shape and other growth characteristics.

3 Juvenility Retention

3.1 Positional Effects in Mature Trees

Maturation does not proceed at an equal pace throughout the tree. In the mature tree, there are some zones where the tissues are maturing at slower rates than in others (Bonga 1982; Hackett 1985). For example, many hardwood trees produce juvenile stump-sprouts or sometimes juvenile root or epicormic sprouts that can be used for cloning. With conifers, except for a few, e.g., redwood (Tufuor 1973), these options are generally not available. Nevertheless, tissues more juvenile than others, and thus more responsive in vitro, are available (Bonga 1987). For example, shoot cuttings obtained from positions close to the lower portion of the trunk root more easily than those obtained from other positions (Monteuuis 1984). On that basis one would expect that conifer buds removed from positions close to the lower part of the trunk would be more morphogenetic in vitro than buds taken from other positions. However, experiments carried out to show this have, so far, been inconclusive (von Arnold 1984; Dunstan et al. 1987; Evers 1987). Detached meristems (small, inhibited meristems which normally do not form vascular connections with the rest of the tree), which have been found in trees of several gymnosperm (Burrows 1987) and angiosperm (Ikemori 1987) species, could be a source of juvenile tissue. To our knowledge, culture of these has not been reported.

3.2 Delaying the maturation of Juvenile Material

There are techniques to delay maturation in juvenile material. The most commonly practiced of these is hedging, i.e., regular severe pruning of seedlings or stecklings (Bolstad and Libby 1982). *Pinus radiata* hedges have retained juvenility for more than 27 years (Libby, pers. commun. 1988). Young material has also been maintained in a prolonged juvenile state by serial rooting of cuttings; for example, after five serial rootings, every 3 years, Norway spruce stecklings showed no signs of maturation (St. Clair et al. 1985). *Pinus radiata* shoot cultures, obtained from cotyledons in vitro, were kept juvenile for up to 5 years by low-temperature storage (Aitken-Christie and Singh 1987). All these techniques were used to maintain valuable stock in a juvenile state while clonal propagules of the stock plants were being field-tested. A technique that has not yet been tested for that purpose is cryopreservation. Regeneration of plants from cryopreserved cells has been achieved for a number of conifer species (Gupta et al. 1987; Kartha et al. 1988).

4 Rejuvenation

4.1 Natural Rejuvenation

The most obvious natural rejuvenation is the one that occurs during the sexual process. It has been suggested earlier (Bonga 1982, 1987) that this rejuvenation involves a drastic restructuring of the cytoplasm during microsporogenesis and egg formation (cf. Li and Dickinson 1987; Pennell and Bell 1987a, b). Restructuring of the cytoplasm during microsporogenesis changes the cell from one programmed as a sporophyte cell to one programmed as a gametophyte cell (Dickinson 1987; Pennell and Bell 1987b). The cytoplasmic restructuring during egg formation presumably reprograms the cell from one determined as a gametophyte to one programmed for embryonic growth. In view of earlier comments (Section 2.3.2), it is tempting to speculate that these events rejuvenate the cytoplasm and thus indirectly the nucleus.

Natural reprogramming of cells is not restricted to sexual reproduction. During embryogenesis and at later stages in development, there are several moments when the nucleus-cytoplasm interactions are dramatically changed. One mechanism by which this is achieved is by unequal cell division. This results in an unequal distribution of cytoplasmic components in the two daughter cells (Bonga 1980; Trewavas 1985), and in the smaller of the two cells becoming a center of specialized activity (Sachs 1984). Unequal divisions are a prominent feature in some embryogenic conifer cultures (Hakman and Fowke 1987; von Aderkas and Bonga 1988).

There are well-defined, short periods during the life cycle of the tree when specific tissues or cells regain some of the capacity for organogenesis lost in the maturation process. For example, shoot regeneration from somatic tissues of female cones of *Larix decidua* is only possible for a 1-2-week period around the time of meiosis (Bonga and von Aderkas 1988). Shoot regeneration from shoot explants of mature conifers (Bonga and von Aderkas 1988), and somatic (Lu and Thorpe 1987: Becwar et al. 1988) and gametic (von Aderkas and Bonga 1988) embryogenesis in conifers, often occur more readily during short, specific developmental stages of the donor.

Short periods of increased morphogenetic plasticity, as described above, are not unique to trees but occur in other plants as well. Dickinson (1987) has ascribed the phenomenon to nucleus-cytoplasmic interactions. He states that the emission of nuclear signals decreases to a low level at certain stages during the life cycle of the plant and that, at these points, cells are more developmentally labile and receptive to outside stimuli.

The above examples of natural rejuvenation warrant further study. More information on rejuvenation mechanisms operating in situ will be useful in our search for methods to induce organogenesis and embryogenesis in tissues of mature conifers in vitro.

4.2 Rejuvenation by Horticultural Means

Grafting on juvenile rootstock, especially if it is done sequentially, is sometimes a means of achieving at least some degree of rejuvenation (Boulay 1987; Franclet et al. 1987; Monteuuis et al. 1986). However, such rejuvenation is generally not persistent; after all, grafting to establish breeding and seed orchards, and grafting of fruit trees, is based on the principle that rejuvenation, if occurring at all, is only temporary. Nevertheless, grafting and re-grafting can be useful. For example, vegetative buds from mature *Picea abies* trees, increased their capacity to form adventitious buds in vitro after each of several successive cycles of grafting (von Arnold 1984). Micrografting (in vitro), as discussed later (Sect. 4.3), has also been effective in a few cases.

Reappearance of juvenile morphological characteristics or improved rooting was also obtained by spraying with the plant growth regulator BAP (Dumas 1986) or GA (Hackett 1985), or after sequential rooting of cuttings or grafting

(Monteuuis 1984; Wühlisch 1984; Haines and Nikles 1987; Ununger and Ekberg 1987), or after annual crown-pruning (Copes 1983).

Because explants from material thus treated generally still are not capable of true-to-type micropropagation, it is legitimate to ask if we are dealing with true ontogenetic rejuvenation or not. According to Ununger and Ekberg (1987), rooting of cuttings of *Picea abies* results in true, partial rejuvenation, i.e., rejuvenation at the gene level. However, others consider the phenomenon to be mostly just an expression of phenotypic plasticity (Kaplan 1983), i.e., we are dealing with environmentally induced physiological changes and not with changed gene activity.

In those cases where the reappearance of juvenile characteristics is phenotypic rather than a true developmental-phase reversion, one can expect true-to-type cloning still to be difficult. Nevertheless, reinvigoration treatments are a good general practice because they ensure that the explants will be in top physiological (minimal stress) condition when excised. Only in a few instances has material rejuvenated by horticultural means improved subsequent micropropagation (Monteuuis 1984; Franclet et al. 1987)

4.3 Rejuvenation in Vitro

As was pointed out in previous sections, the first consideration when starting cultures of explants from mature trees is to select the most juvenile tissues in the tree, i.e., tissues excised from specific positions in the tree at specific times. Furthermore, these tissues should be in a vigorous physiological (reinvigorated) condition at the time of excision. Once in culture, further rejuvenation is often possible by serial subculture (Mullins et al. 1979; Sriskandarajah et al. 1982; Fouret et al. 1986a; Krogstrup 1987). Generally, the main beneficial result of this procedure is improved rooting of shoots (Economou and Read 1986). Partial rejuvenation has also been achieved by repetitive excision and subculture of apical meristems of micropropagated shoots each time these shoots reach a certain length. This has resulted in improved rooting of micropropagated shoots of a several-centuries-old *Sequoia sempervirens* tree, alas without a concurrent switch from plagiotropic to orthotropic shoot growth (Walker 1985). Presumably, in many cases the above effects are comparable to those achieved by serial grafting or rooting of cuttings as described in the previous section. If so, then here again, we may be dealing, at least in some cases, mostly with reinvigoration, i.e., an expression of phenotypic plasticity, rather than with a true reversal to an earlier ontogenetic stage (Fouret et al. 1986a).

One procedure for reinvigoration in vitro that has worked well for a number of species is single or serial micrografting in conjunction with other in vitro procedures. This has resulted in micropropagation of a number of adolescent and mature conifer trees (Misson and Giot-Wirgot 1984; Tranvan and David 1985; Monteuuis 1986).

It has been noticed that tissues from mature conifers, as well as tissues that have matured in culture, have chemical characteristics that are different from those of seedlings. Tissues from mature trees or old cultures have a high calcium to potassium ratio (de Champs 1986; Boulay 1987), and high levels of sodium (von Arnold and Roomans 1983), polyphenol (Monteuuis and Bon 1985), peroxidase (Boulay 1987), abscisic acid (ABA) (Fouret et al. 1986b), and cytokinin (Bekkaoui et al. 1986). It is not yet known if treatments that would reduce these levels would promote organogenesis or embryogenesis. Possibly such treatments would only reinvigorate rather than rejuvenate the cultures. In some laboratories cultures are washed once, or at regular intervals, in water (John and Webb 1987) or liquid nutrient medium (Aitken-Christie and Jones 1987) to improve micropropagation. It is conceivable that the effectiveness of these treatments is based on partial removal of some of the chemicals that have accumulated in excess in cultured tissues.

The most common explant used to initiate cultures of tissues from mature conifers is the excised shoot tip. Not all cells in these shoot tips may have the same organogenic or embryogenic potential. As stated earlier (Sect. 2.3.2), it is possible that only those cells that eventually become involved in the sexual process are totipotent. Furthermore, the cells best suited to form adventitious organs or embryos may be prevented from doing so by less morphogenetically plastic neighboring cells (correlative controls, see Barlow and Carr 1984). Therefore, the most morphogenic cells should, whenever possible, be separated from their less competent neighbors. This has led to miniaturization of the explant, i.e., excision of explants composed of only the dome of the apical meristem or the dome with only a few needle primordia. One problem with miniaturization is that, at least initially, it increases the stress level in the explants. In small explants the cut surface (wound) area is larger in relation to the volume of the explant than in large explants, and thus the damage is more severe. Wounding results in pronounced physiological changes in cells (Davies 1987; Thompson et al. 1987), including those of conifer explants (Monteuuis and Bon 1985). Nevertheless, miniaturization has improved organogenesis in several instances (Walker et al. 1984; Walker 1985; Krogstrup 1987) and may become even more effective once we learn how to counteract the damaging effects of wounding. Single-cell or protoplast culture is the ultimate in miniaturization.

5 Discussion

Maturation starts as soon as the zygote is formed. From that point on it progresses at a rate that is determined by many environmental and intrinsic factors. Study of this maturation early in the life cycle is relevant because it will not only give us insight into the maturation process in general, but may also suggest the means to reverse it (rejuvenation).

As was pointed out above, one important factor in maturation is stress. When the conifer zygote develops inside gametophyte, it does so in a well-protected, strictly regulated, continuously changing environment. When culturing tissues in vitro, it is difficult to replicate these precise conditions at the right moments and, therefore, various degrees of stress will probably be experienced during organogenesis or embryogenesis in vitro. For example, the commonly observed abnormalities in conifer somatic embryogenesis (Lu and

Thorpe 1987; Hakman and von Arnold 1988) could in part be the result of stress. In addition, stress is probably also a main contributor to the frequently observed premature appearance of mature characteristics in plantlets produced by organogenesis (Amerson et al. 1988). However, we cannot automatically equate all abnormalities with stress. In some cases, lack of continued, normal development of somatic embryos, or formation of abnormal plantlets from adventitious shoots, may have resulted from embryogenesis and organogenesis in tissues that were incompletely rejuvenated rather than stressed.

When trees are exposed to stress during the growing season, they respond by producing ABA (Johnson 1988) and by slowing down growth rates. ABA is also involved in growth-rate reduction not associated with environmental stress. For example, it suppresses premature germination in embryos in a gene-controlled fashion (Williamson and Quatrano 1988). Similar gene-regulated changes in ABA production may occur at other critical moments in embryogenesis to mediate the switch from the current to the next ontogenetic stage. One such moment may be when the proembryonal head has just been formed. In early zygotic embryogenesis, several conifers show cleavage of the proembryonal head (Singh 1978). This cleavage has to be kept under strict control because excessive, continuous cleavage would deny the cleavage products the chance to mature into a proper embryo. Presumably ABA has a function in this control.

In many of the embryogenic conifer cultures obtained to date, the embryogenic masses appear to be largely in a perpetual state of cleavage, i.e., most of the cells keep producing masses of proembryo-like cells that fail to mature sufficiently to allow normal ontogeny to proceed (cf. Gorst et al. 1987). Addition of ABA to the nutrient medium often stops this excessive cleavage, thus allowing normal embryogenesis to proceed (Durzan and Gupta 1987; Boulay et al. 1988). ABA also controls normal development at later stages of embryogenesis (Boulay et al. 1988; Hakman and von Arnold 1988).

Not all embryogenic cultures require exogenous ABA for proper embryo development. We have haploid embryogenic cultures of *Larix decidua* (von Aderkas and Bonga 1988) that have gone through numerous subcultures over a 4-year period. Some lines of these cultures are still producing large numbers of normally developing and germinating embryos without ABA ever having been added to the medium.

It has been suggested that, in some nonorganogenic cultures, a situation akin to the one in excessively cleaving embryogenic cultures may be present. Some callus could be a mass of proliferating primordia rather than a mass of noncommitted cells, these primordia being restrained from proper organogenesis by inappropriate conditions (Gorst et al. 1987). This again could be considered as a failure of proper maturation and growth rates may have to be slowed down before a proper response can be expected. Slowing down of the growth rate is often a prerequisite for cell differentiation and organogenesis to occur (Yeoman 1987). In this connection, it is noteworthy that slower-growing *Picea abies* genotypes were more responsive in vitro than faster-growing ones (von Arnold 1984).

From the foregoing, it is clear that, to obtain normal embryogenesis in potentially embryogenic cultures, factors that will induce proper maturation have to be applied to these cultures. However, when we are trying to initiate embryogenesis in cultures of tissues of mature trees, we are probably dealing with cultures that, to become potentially embryogenic, require rejuvenation rather than maturation. To achieve such rejuvenation, we probably need a far better understanding of how rejuvenation occurs during the sexual process.

Once this knowledge has been obtained, we can try to induce a similar rejuvenation process in somatic tissues in vitro. Another approach that may have some merit would be to transplant nuclei from somatic cells of mature trees into a juvenile cytoplasm. For example, nuclei from meristems of mature trees could be transferred to de-nucleated protoplasts obtained from embryos. Conceivably, the juvenile cytoplasm would rejuvenate the mature nucleus so that we would obtain protoplasts or cells in which both the cytoplasm and the nucleus are juvenile. At present such nuclear transfers are not possible, and whether such a transfer would result in viable juvenile cells is still very much an open question.

Acknowledgments. We thank Dr. D.J. Durzan, University of California, Davis, Dr. W.M. Cheliak, and Dr. K. Klimaszewska, Petawawa National Forestry Institute for their advice during the preparation of the manuscript. The helpful comments of the editors of this Volume were much appreciated.

References

Aitken-Christie J, Jones C (1987) Towards automation: radiata pine shoot hedges in vitro. Plant Cell Tissue Org Cult 8:185–196

Aitken-Christie J, Singh AP (1987) Cold storage of tissue cultures. In: Bonga JM, Durzan DJ (eds) Cell and tissue cultures in forestry, vol 2. Nijhoff, Dordrecht, pp 285–304

Allsop A (1964) Shoot morphogenesis. Ann Rev Plant Physiol, 15:225–254

Altamura MM, Bassi P, Cavallini A, Cionini G, Cremonini R, Monacelli B, Pasqua G, Sassoli O, Tran Thanh Van K, Cionini PG (1987) Nuclear DNA changes during plant development and the morphogenetic response in vitro of *Nicotiana tabacum* tissues. Plant Sci 53:73–79

Amerson HV, Frampton LJ, Jr. Mott RL, Spaine PC (1988) Tissue culture of conifers using lobolly pine as a model. In: Hanover JW, Keathley DE (eds) Genetic manipulation of woody plants. Plenum, New York, pp 117–137

Barlow PW, Carr DJ (eds) (1984) Positional controls in plant development. Univ Press, Cambridge

Becwar MR, Wann SR, Johnson MA, Verhagen SA, Feirer RP, Nagmani R (1988) Development and characterization of in vitro embryogenic systems in conifers. In: Ahuja MR (ed) Somatic cell genetics of woody plants. Kluwer, Dordrecht, pp 1–18

Bekkaoui F, Maldiney R, Pilate G, Boulay M, Franclet A (1986) Degénérescence clonale du douglas (*Pseudotsuga menziesii*) cultivé in vitro; relation avec la teneur en éléments minéraux, en auxine, acide abscissique, zéatine et zéatine-riboside. CR Acad sci Paris 303:13–18

Bennett MD (1984) Nuclear architecture and its manipulation. In: Gustafson JP (ed) Gene manipulation in plant improvement. Plenum, New York, pp 469–502

Bernard-Durix F, Kovoor A (1987) Do mitochondria mediate organogenesis? In: Bonga JM, Durzan DJ (eds) Cell and tissue culture in forestry, vol 1. Nijhoff, Dordrecht, pp 372–388

Bolstad PV, Libby WJ (1982) Comparison of radiata pine cuttings of hedge and tree-form origin after seven growing seasons. Silv Genet 31:9–13

Bon MC (1988) Nucleotide status and protein synthesis in vivo in the apices of juvenile and mature *Sequoiadendron giganteum* during budbreak. Physiol Plant 72:796–800

Bonga JM (1980) Plant propagation through tissue culture, emphasizing woody species. In: Sala F, Parisi B, Cella R, Cliferri O, Plant cell cultures: results and perspectives. Elsevier/North Holland Biomedical Press, Amsterdam, pp 253–264

Bonga JM (1982) Vegetative propagation in relation to juvenility, maturity, and rejuvenation. In: Bonga JM, Durzan DJ (eds) Tissue culture in forestry. Nijhoff, Dordrecht, pp 387–412

Bonga JM (1987) Clonal propagation of mature trees: Problems and possible solutions. In: Bonga JM, Durzan DJ (eds) Cell and tissue culture in forestry, vol 1. Nijhoff, Dordrecht pp 249–271
Bonga JM, von Aderkas P (1988) Attempts to micropropagate mature *Larix decidua* Mill. In: Ahuja MR (ed) Somatic cell genetics of woody plants. Kluwer, Dordrecht, pp 155–168
Borchert R (1976a) The concept of juvenility in woody plants. Acta Hortic 56:21–36
Borchert R (1976b) Differences in shoot growth patterns between juvenile and adult trees and their interpretation based on systems analysis of trees. Acta Hortic 56:123–130
Boulay M (1987) In vitro propagation of tree species. In: Green CE, Somers DA, Hackett WP, Biesboer DD (eds) Plant tissue and cell culture. Liss, New York, pp 367–382
Boulay M, Gupta PK, Krogstrup P, Durzan DJ (1988) Development of somatic embryos from cell suspension cultures of Norway spruce (*Picea abies* Karst.). Plant Cell Rep 7:134–137
Bruck DK, Walker DB (1986) Cell determination during embryogenesis in *Citrus jambhiri*. III. Graft formation and non formation in embryonic tissues. Can J Bot 64:2057–2062
Burrows GE (1987) Leaf axil anatomy in the Araucariaceae. Aust J Bot 35:631–640
Cherfas J (1985) When is a tree more than a tree. New Sci 106:42–45
Chwirot WB, Dygdala RS, Chwirot S (1985) Ultraweak photon emission during microsporogenesis in *Larix europaea*. Cytobios 44:95–102
Copes DJ (1983) Effects of annual crown pruning and serial propagation on rooting of stem cuttings from Douglas-fir. Can J For Res 13:419–424
Cullis CA (1986) Unstable genes in plants. In: Jennings DH, Trewavas AJ (eds) Plasticity in plants. Synp Soc Exp Biol 40, Cambridge, pp 77–84
Davies E (1987) Action potentials as multifunctional signals in plants: a unifying hypothesis to explain apparently disparate wound responses. Plant Cell Environ 10:623–631
De Champs E (1986) Étude de l'utilisation possible du rapport K/Ca comme marqueur de juvénilité. Ann, Rech Sylvicoles AFOCEL, pp 133–157
Dickinson HG (1987) Nucleocytoplasmic interaction. Ann Bot London 60:61–73
Dumas E (1986) Micropropagation d'un clone agé de pin maritime en vue de l'obtention de pieds-méres. Annu Rech Sylvic AFOCEL, pp 95–107
Dunstan DI, Mohammed GH, Thorpe TA (1987) Morphogenetic response of vegetative bud explants of adolescent and mature *Picea glauca* (Moench) Voss in vitro. New Phytol 106:225–236
Durzan DJ (1984) Special problems: adult vs. juvenile explants. In: Sharp WR, Evans DA, Ammirato PV, Yamada Y (eds) Handbook of plant cell culture, vol 2: Crop species. Macmillan, New York, pp 471–503
Durzan DJ, Gupta PK (1987) Somatic embryogenesis and polyembryogenesis in Douglas-fir cell suspension cultures. Plant Sci 52:229–235
Economou AS, Read PE (1986) Microcutting production from hardy deciduous azalea shoot tips. HortScience 21:137–139
Edmunds LN, Jr, Adams KJ (1981) Clocked cell cycle clocks. Science 211:1002–1013
Evers PW (1987) Correlations within the tree. In: Bonga JM, Durzan DJ (eds) Cell and tissue culture in forestry, vol 2. Nijhoff, Dordrecht, pp 218–229
Fortanier EJ, Jonkers H (1976) Juvenility and maturity of plants as influenced by their ontogenetical and physiological ageing. Acta Hortic 56:37–44
Fouret Y, Arnaud Y, Larrieu C, Miginiac E (1986a) *Sequoia sempervirens* as an in vitro rejuvenation model. NZ J For Sci 16:319–327
Fouret Y, Arnaud Y, Maldiney R, Sotta B, Miginiac E (1986b) Relation entre rhizogenése et teneur en auxine et acide abscissic chez trois clones de *Sequoia sempervirens* (Endl.) issus d'arbres d'âge-différent. C R Acad Sci Paris 303:135–138
Franclet A, Boulay M, Bekkaoui F, Fouret Y, Verschoore-Martouzet B, Walker N (1987) Rejuvenation. In: Bonga JM, Durzan DJ (eds) Cell and tissue culture in forestry, vol 1. Nijhoff, Dordrecht, pp 232–248
Gill DE, Halverson TG (1984) Fitness variation among branches within trees. In: Shorrocks B (ed) Evolutionary ecology. Blackwell, Oxford, pp 105–116
Gorst J, Overall RL, Wernicke W (1987) Ionic currents traversing cell clusters from carrot suspension cultures reveal perpetuation of morphogenetic potential as distinct from induction of embryogenesis. Cell Differ 21:101–109
Greenwood MS (1984) Phase change in loblolly pine: shoot development as a function of age. Physiol Plant 61:518–522

Greenwood MS (1987) Rejuvenation of forest trees. Plant Growth Regul 6:1–12
Grime JP, Crick JC, Rincon JE (1986) Ecological significance of plasticity. In: Jennings DH, Trewaves AJ (eds) Plasticity in plants. Symp Soc Exp Biol 40, Cambridge, pp 5–30
Guern J (1987) Regulation from within: The hormone dilemma. Ann Bot (London, Suppl 4) 60:75–102
Gupta PK, Durzan DJ, Finkle BJ (1987) Somatic polyembryogenesis in embryogeneic cell masses of *Picea abies* (Norway spruce) and *Pinus taeda* (loblolly pine) after thawing from liquid nitrogen. Can J For Res 17:1130–1134
Hackett WP (1985) Juvenility, maturation, and rejuvenation in woody plants. Hortic Rev 7:109–155
Haines RJ, Nikles DG (1987) Mass propagation of superior families and individual genotypes of some conifers in Queensland. Aust For 50:106–111
Hakman I, Fowke LC (1987) An embryogenic cell suspension culture of *Picea glauca* (white spruce). Plant Cell Rep 6:20–22
Hakman I, Von Arnold S (1988) Somatic embryogenesis and plant regeneration from suspension cultures of *Picea glauca* (white spruce). Physiol Plant 72:579–587
Halpervin W (1986) Attainment and retention of morphogenetic capacity in vitro. In: Vasil IK (ed) Cell culture and somatic cell genetics of plants, vol 3: Plant regeneration and genetic variability. Academic Press, New York London Orlando, pp 3–47
Hartman C, Henry Y, De Buyser J, Aubry C, Rode A (1989) Identification of new mitochondrial genome organizations in wheat plants regenerated from somatic tissue cultures. Theor Appl Genet 77:169–175
Holliday R (1984) The biological significance of meiosis. In: Evans CW, Dickinson HG (eds) Controlling events in meiosis. Symp Soc Exp Biol Univ Reading 38, pp 381–395
Holliday R, Pugh JE (1975) DNA modification mechanisms and gene activity during development. Science 187:226–232
Hood JV, Libby WJ (1978) Continuing effects of maturation state in radiata pine and a general maturation model. In: Hughes KW, Henke R, Constantin M (eds) Propagation of higher plants through tissue culture—bridge between research and application. Tech Inf Cent, US Dep Energ, Conf—7804111, pp 220–232
Ikemori YK (1987) Epicormic shoots from the branches of *Eucalyptus grandis* as an explant source for in vitro culture. Commonw For Rev 66:351–356
John A, Webb KJ (1987) Sitka spruce (*Picea sitchensis* (Bong.) Carr.). In: Bonga JM, Durzan DJ (eds) Cell and tissue culture in forestry, vol 3. Nijhoff, Dordrecht pp 30–41
Johnson JD (1988) Stress physiology of forest trees: The role of plant growth regulators. Plant Growth Regul 6:193–215
Jones HG (1987) Repeat flowering in apple caused by water stress or defoliation. Trees 1:135–138
Kaplan DR (1983) Heteroblastic development and phenotypic plasticity in higher plants. Am J Bot 70 (5 pt 2, Abstr): 9–10
Karki L (1985) Genetically narrow-crowned trees combine high timber quality and high stem wood quality at low cost. In: Tigerstedt PMA, Puttonen P, Koski V (eds) Crop physiology of forest trees. Univ Press, Helsinki, pp 245–256
Kartha KK, Fowke LC, Leung NL, Caswell KL, Hakman I (1988) Induction of somatic embryos and plantlets from cryopreserved cell cultures of white spruce (*Picea glauca*). J Plant Physiol 132:529–539
König C, Ebert I, Greilhuber J (1987) A DNA cytophotometric and chromosome banding study in *Hedera helix* (Araliaceae), with reference to differential DNA replication associated with juvenile-adult phase change. Genome 29:498–503
Krogstrup P (1987) Formation of adventitious buds in needle primordia isolated from embryonic shoots of adult *Picea abies*. Biol Plant 29:437–444
Leakey RRB (1985) The capacity for vegetative propagation in trees. In: Cannell MGR, Jackson JE (eds) Attributes of trees as crop plants. Inst Terr Ecol, Monks Wood, Huntingdon, UK, pp 110–133
Lehnhardt B, Nitzsche W (1988) Loss of apomixis in *Arabis holboellii* HORNEM. when regenerated in tissue cultures. Angew Bot 62:225–232
Lesney MS, Johnson JD, Korhnak T, McCaffery MW (1988) in vitro manipulation of slash pine (*Pinus elliottii*). In: Hanover JW, Keatley DE (eds) Genetic manipulation of woody plants. Plenum, New York, pp 43–55

Lev-Yadun S, Liphschitz N (1987) The ontogeny of gender of *Cupressus sempervirens* L. Bot Gaz 148:407–412
Li FL, Dickinson HG (1987) The structure and function of nuclear invaginations characteristic of microsporogenesis in *Pinus banksiana*. Ann Bot (London) 60:321–330
Lu CY, Thorpe TA (1987) Somatic embryogenesis and plantlet regeneration in cultured immature embryos of *Picea glauca*. J Plant Physiol 128:297–302
McCown BH (1986) Woody ornamentals, shade trees, and conifers. In: Zimmerman RH, Griesbach RJ, Hammerschlag FA, Lawson RK (eds) Tissue culture as a plant production system for horticultural crops. Nijhoff, Dordrecht, pp 333–342
Mission JP, Giot-Witgot P (1984) Rajeunissement d'un clone de Thuja en vue de sa multiplication in vitro. Annu Rech Syvicoles AFOCEL, pp 187–197
Monnier M, Norstog K (1986) Effect of in ovulo period on the differentiation and regulation of immature embryos of *Zamia* cultured in vitro. J Exp Bot 37:1633–1642
Monteuuis O (1984) La multiplication végétative du sequoia géant en vue du clonage. Ann AFOCEL, pp 140–171
Monteuuis O (1986) Microgreffage de points végétatifs de *Sequoiadendron giganteum* Buchhloz séculaires sur de jeunes semis cultivés in vitro. CR Acad Sci Paris 302:223–225
Monteuuis O (1987) Profils méristématiques de sequoias géants (*Sequiadendron giganteum* Buchholz) jeunes et âgés durant les stades de repos végétatif et de débourrement. C R Acad Sci Paris 305:715–720
Monteuuis O, Bon MC (1985) Microbouturage du sequoia géant. Annu Rech Sylvicoles AFOCEL, pp 49–87
Monteuuis O, Pages C, Sarran P (1986) De l'amélioration des conditions de bouturage en cascade du *Sequoia sempervirens*. Annu Rech Sylvicoles AFOCEL, pp 111–131
Mullins MG, Nair Y, Sampet P (1979) Rejuvenation in vitro: Induction of juvenile characters in an adult clone of *Vitis vinifera* L. Ann Bot 44:623–627
Nagl W, Popp FA (1983) A physical (electromagnetic) model of differentiation. 1 Basic considerations. Cytobios 37:45–62
Namkoong G, Barnes RD, Burley J (1980) Screening for yield in forest tree breeding. Commonw For Rev 59:61–68
Pennell RI, Bell PR (1987a) Intracellular RNA during meiosis in microsporangia of *Taxus baccata*. Am J Bot 74:444–450
Pennell RI, Bell PR (1987b) Megasporogenesis and the subsequent cell lineage within the ovule of *Taxus baccata* L. Ann Bot (London) 59:693–704
Philip VJ, Nainar SAZ (1988) In vitro transformation of root meristem to shoot and plantlets in *Vanilla planifolia*. Ann Bot (London) 61:193–199
Raff JW, Hutchinson JF, Knox RB, Clarke AE (1979) Cell recognition:Antigenic determinants of plant organs and their cultured callus cells. Differentiation 12:179–186
Raff RA, Kaufman TC (1983) Embryos, genes and evolution. The developmental-genetic basis of evolutionary change. Macmillan, New York
Rohr R, von Aderkas P, Bonga JM (1989) Ultrastructural changes in haploid embryoids of *Larix decidua* during early embryogenesis. Am J Bot 76:1460–1467
Sachs T (1984) Axiality and polarity in vascular plants. In: Barlow PW, Carr DJ (eds) Positional controls in plant development. Univ Press, Cambridge, pp 193–224
Sangwan RS, Sangwan-Norreel BS (1987) Ultrastructural cytology of plastids in pollen grains of certain androgenic and nonandrogenic plants. Protoplasma 138:11–22
Singh H (1978) Embryology of gymnosperms. Encyclopedia of plant anatomy X, 2. Bornträger, Berlin
Sriskanadarajah S, Mullins MG, Nair Y (1982) Induction of adventitious rooting in vitro in difficult-to-propagate cultivars of apple. Plant Sci Lett 24:1–9
Stange L (1984) Cellular interactions during early differentiation. Encyclopedia of plant physiology, new Ser vol 17. Springer, Berlin Heidelberg New York, pp 424–452
St. Clair JB Sr, Kleinschmit J, Svolba J (1985) Juvenility and serial vegetative propagation of Norway spruce clones (*Picea abies* Karst.). Silv Genet 34:42–48
Summerbell D, Lewis JH, Wolpert L (1973) Positional information in chick limb morphogenesis. Nature (London) 244:492–496
Sussex I (1976) Phase change: physiological and genetic aspects. Acta Hortic 56:275–280

Thompson JE, Legge RL, Barber RF (1987) The role of free radicals in senescence and wounding. New Phytol 105:317–344
Tranvan H, David A (1985) Greffage in vitro du pin maritime (*Pinus pinaster*). Can J Bot 63:1017–1020
Trewavas AJ (1985) Growth substances, calcium and the regulation of cell division. In: Bryant JA, Francis D (eds) The cell division cycle in plants. Soc Exp Biol, Sem Ser 26. Univ Press, Cambridge, pp 133–156
Trewavas AJ (1986) Resource allocation under poor growth conditions. A major role for growth substances in developmental plasticity. In: Jennings DH, Trewavas AJ (eds) Plasticity in plants. Symp Soc Exp Biol 40, Cambridge, pp 31–76
Tufuor K (1973) Comparitive growth performance of seedlings and vegetative propagules of *Pinus radiata* (D. Don) and *Sequoia sempervirens* (D. Don) Endl. PhD Diss, Univ Cal, Berkeley
Ununger J, Ekberg I (1987) Rejuvenation of juvenile growth characters in *Picea abies*. Scand J For Res 2:441–453
Van't Hof J (1987) Extrachromosomal DNA and cell differentiation in cultured pea roots (*pisum sativum*). In: Green CE, Somers DA, Hackett WP, Biesboer DD (eds) Plant tissue and cell culture. Liss, New York, pp 33–44
Von Aderkas P, Bonga JM (1988) Formation of haploid embryoids of *Larix decidua*: early embryogenesis. Am J Bot 75:690–700
Von Arnold S (1984) Importance of genotype on the potential for in vitro adventitious bud production of *Picea abies*. For Sci 30:314–318
Von Arnold S, Roomans GM (1983) Analyses of mineral elements in vegetative buds and needles from young and old trees of *Picea abies*. Can J For Res 13:689–693
Walbot V, Cullis CA (1985) Rapid genomic change in higher plants. Annu Rev Plant Physiol 36:367–396
Walker N (1985) *Sequoia sempervirens*. réjuvénilisation et culture de méristémes en cascade. Annu Rech Sylvicoles AFOCEL, pp 25–47
Walker N, Dumas E, Franclet A, Bekkaoui F (1984) Technique de culture in vitro des méristémes de *Sequoia sempervirens* et *Pinus pinaster*. Annu Rech Sylvicoles AFOCEL, pp 87–109
Wareing PF, Frydman VM (1976) General aspects of phase change with special reference to *Hedera helix* L. Acta Hortic 56:57–68
Williamson JD, Quatrano RS (1988) ABA-regulation of two classes of embryo-specific sequences in mature wheat embryos. Plant Physiol 86:208–215
Wühlisch G (1984) propagation of Norway spruce cuttings free of topophysis and cyclophysis effects. Silv Genet 33:215–219
Yeoman MM (1987) Bypassing the plant. Ann Bot (London, Suppl 4) 60:157–174
Young E, Hanover JW (1976) Accelerating maturity in *Picea* seedlings. Acta Hortic 56:105–114
Zimmerman RH (1973) Juvenility and flowering of fruit trees. Acta Hortic 34:139–142

13 Conifer Somatic Embryogenesis and Clonal Forestry

M.R. Becwar[1]

1 Introduction

Somatic embryogenesis, the formation and development of embryos from somatic (vegetative) tissues under in vitro conditions, was first demonstrated with carrot (Steward et al. 1958). The list of plants that can be regenerated by somatic embryogenesis continues to grow (Tisserat et al. 1979; Ammirato 1983; see references in this chapter). In fact, somatic embryogenesis has been demonstrated in essentially all major groups of higher plants, angiosperms (monocots and dicots) and gymnosperms (coniferous and nonconiferous).

The development of somatic embryogenesis in combination with encapsulated somatic or "artificial" seed technology holds immense potential for clonal forestry. This is especially true for coniferous trees that are currently propagated in seedling nurseries, because the potential exists for artificial seed technologies to easily interface with the existing seedling production systems.

Until recently most advances in somatic embryogenesis of woody plants had been with nonforest trees, e.g., citrus and palms (Litz 1985). Progress made during the 1970s and 1980s on somatic embryogenesis in nonconiferous woody plants has been reviewed (Tulecke 1987; Wann 1989). In addition, there have been recent reviews on progress in tree biotechnology research (Dunstan 1988; Cheliak and Rogers 1990). This chapter describes the process of regenerating plants from conifer somatic embryos; with reference to current literature, in an attempt to synthesize and summarize progress on conifer somatic embryogenesis.

2 Historical Background

Many studies, primarily with herbaceous plants, have shown several factors to be important in the production of somatic embryos (Thorpe 1988; Brown 1988). These include culture medium, explant (starting tissue), and genotype of the source plant. Carrot has served as an experimental model for somatic embryogenesis in plants in many ways (Ammirato 1983; Thorpe 1988). For instance, it

[1]Westvaco Corporation, Forest Science Laboratory, P.O. Box 1950, Summerville, SC 29484, USA

Clonal Forestry I, Genetics and Biotechnology
Ed. by M.R. Ahuja and W.J. Libby

was demonstrated with carrot that an auxin was critical for initiation of embryogenesis in vitro and that its removal induced embryo development (Halperin and Wetherell 1964). The importance of the level and form of nitrogen for somatic embryogenesis was also demonstrated with carrot tissue cultures (Halperin and Wetherell 1965). Much progress has been made on regenerating cereal crops, once considered extremely recalcitrant, via somatic embryogenesis. A key to this success has been the use of explants obtained at precise stages of development (Vasil 1987). The most responsive explants of grasses have been immature embryos, and young inflorescence and leaves. Genotypic effects on somatic embryogenesis and regeneration have been reported in several crop plants (Hodges et al. 1986, Brown 1988).

The application of tissue culture to clonal propagation of trees imposes unique problems that are not encountered with most herbaceous crop plants, because of the long life cycle of trees, in addition to the need to regenerate from mature or genetically tested trees. Tissues from mature trees are often recalcitrant in terms of their in vitro regeneration capacity (Bonga 1987). Even so, several nonconiferous trees have been regenerated by somatic embryogenesis from tissues derived from mature trees (Litz 1985, 1987). Initial progress was made with tropical trees by using the nucellus tissue as an explant. The nucellus is diploid maternal tissue within the seed which has not undergone meiosis and genetic recombination. There has also been success with regeneration of tropical nonconiferous trees from young leaf tissue of mature trees (Sondahl and Sharp 1977; Litz 1988). Regeneration by somatic embryogenesis in temperate-zone trees has primarily been from juvenile explant tissue (Merkle and Sommer 1986). However, somatic embryogenesis was obtained from young leaves derived from a 20-year old *Populus* interspecific hybrid (Park and Son 1988). Reviews by Tulecke (1987) and Wann (1989) contain references to numerous other studies and reviews on somatic embryogenesis in nonconiferous woody plants.

3 Somatic Embryogenesis in Conifers

The first reproducible report of diploid somatic embryogenesis in conifers was from immature embryos of *Picea abies* (Hakman et al. 1985). The embryogenic cultures regenerated plantlets (Hakman and von Arnold 1985). Nagmani and Bonga (1985) were the first to report haploid embryogenesis and plantlet regeneration with conifers; from megagametophytes of *Larix decidua.* Since these initial reports of in vitro embryogenesis in 1985, there have been numerous reports to verify and extend the findings to other coniferous species. With the exception of the recent report on *Sequoia sempervirens.* (Bourgkard and Favre 1988), all studies have been on members of the Pinaceae. The most progress has been made with the genus *Picea.*

Figure 1 is a diagram of the regeneration process via somatic embryogenesis; its use for mass propagation and relationship to genetic modification (transformation). Starting with a piece of plant tissue referred to as the explant, an embryogenic culture is initiated and proliferated on a nutrient medium. The

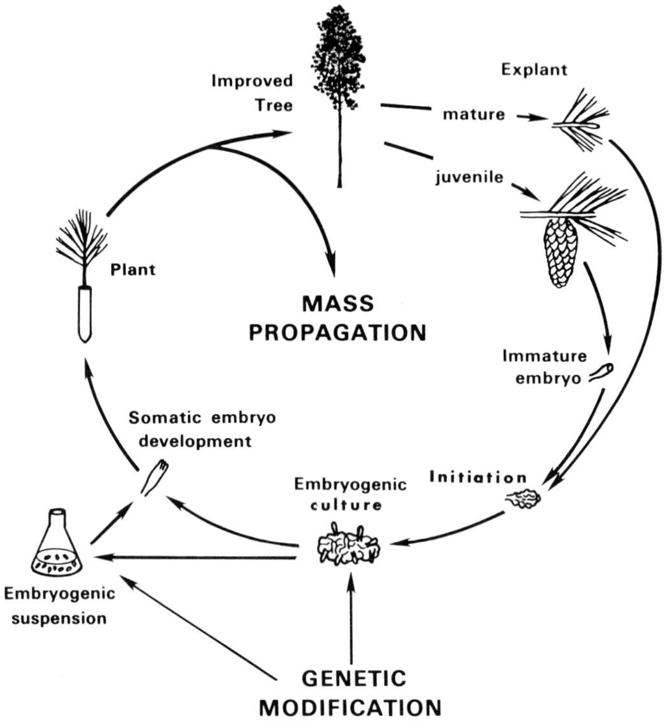

Fig. 1. The conifer somatic embryogenesis process for use in mass propagation and genetic transformation. (Modified from Becwar et al. 1987b).

embryogenic culture can be maintained in a proliferative state on semi-solid medium or as a suspension culture in liquid medium. Somatic embryo development is induced by transferring the embryogenic culture to a modified culture medium. Typically, the in vitro development process ends with the production of a plantlet which can then be transferred to nonaxenic conditions. The terminology used in this chapter to describe the sequential steps of the plant regeneration process via somatic embryogenesis are listed in Table 1. Each of these steps will be considered in more detail in the following sections.

3.1 Explant

Diploid embryogenic cultures of conifers have only been initiated from juvenile explant tissues derived from developing seed, mature seed, or germinated seed. In general, immature zygotic embryos have proven to be the most responsive explants (Figs. 2, 3), although several reports have used fully developed (mature) zygotic embryos from dry seed (Gupta and Durzan 1986a; von Arnold and Hakman 1986; Bourgkard and Favre 1988; Lu and Thorpe 1988; Mo et al. 1989; Jain et al. 1988; von Arnold and Woodward 1988; Verhagen and Wann 1989;

Table 1. Terminology used to describe the steps in the plant regeneration process via somatic embryogenesis in conifers

Regeneration Process	Terminology	Description
Culture establishment	Initiation	In vitro formatin of embryogenic tissue (ET)
↓	↓	
	Maintenance	Proliferation of ET
↓	↓	
Somatic embryo (SE) development	Embryo development	Switch from proliferative to morphological and physiological development mode
↓	↓	
	Embryo maturation	Completion of physiological and biochemical development of SE
↓	↓	
Production of plantlet	Germination	SE root development, hypocotyl elongation, and cotyledon enlargement
↓	↓	
Acclimatization	Conversion	Survival and growth of plant ex vitro

Krogstrup 1990; Tremblay 1990), or tissues excised from germinating embryos (Krogstrup 1986; Lelu et al. 1987, 1990; Attree et al. 1990; Lelu and Bornman 1990). It is difficult to obtain sufficient immature embryos of some conifers for initiation of embryogenic cultures, e.g., *Picea abies*, because of poor seed development. Prolonged storage of the immature seed cones, which contain immature embryos, is difficult due to fungal contamination. Therefore, the ability to initiate from fully developed dry seed is of considerable utility, because it removes the narrow time constraints imposed by using immature embryos.

The developing immature embryo with the intact megagametophyte was first used as an explant for initiation of diploid embryogenic cultures of *Pinus radiata* (Smith et al. 1985). The megagametophyte surrounds the embryo and supplies nutrients to the developing zygotic embryo(s). A higher initiation frequency was obtained with the culture of immature embryos of *Pinus lambertiana* with intact megagametophyte than with isolated immature embryos (Gupta and Durzan 1986b). The specific effect of the megagametophyte on culture initiation has not been determined. It most likely provides nutrients and/or endogenous phytohormones that are suboptimal in the culture medium.

The most significant progress on initiating embryogenic cultures from older (but still juvenile) explants has been made with *Picea* species. Krogstrup (1986) and Lelu et al. (1987, 1990) initiated embryogenic cultures from cotyledons derived from 7-day-old germinated embryos. Attree et al. (1990) initiated embryogenic cultures from 12 to 30 day-old germinants of *Picea glauca* and

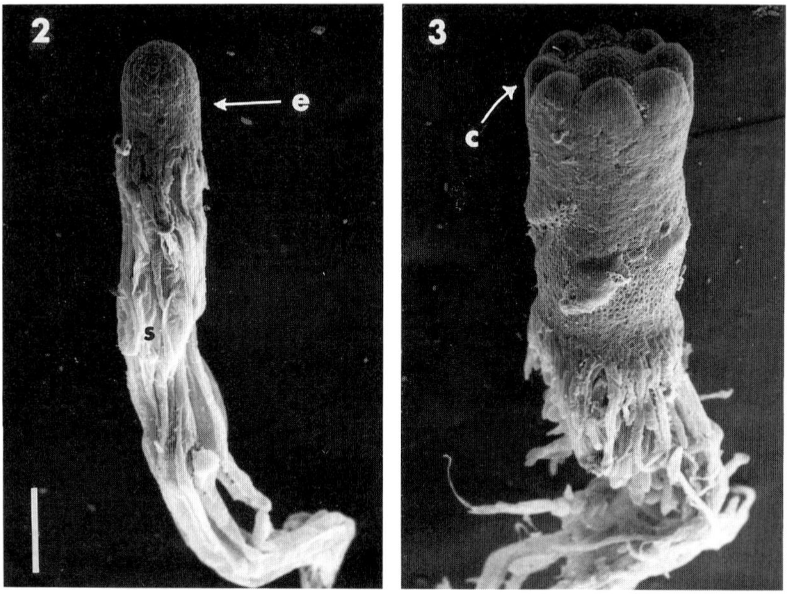

Figs. 2, 3. Immature zygotic embryo explants of *Pinus taeda*. **Fig. 2.** Precotyledonary stage of embryo development. Note elongated suspensor cells (*s*) attached to embryonic region (*e*). **Fig. 3.** Cotyledonary stage of embryo development. Note cotyledon primordia (*c*). Bar = 100 μm. (Becwar, unpubl. 1988).

Picea mariana. As these authors pointed out, it may be possible by continued optimization of conditions for initiating embryogenesis to extend the responsive age to the point where tissues from mature trees are responsive.

One alternative approach to the problem of regenerating directly from tissues of elite genetically proven trees is regeneration from juvenile (embryo) explants derived from controlled crosses of elite parents, cryopreservation of the resulting embryogenic cell lines, and concurrent field testing of regenerates from the individual cell lines in order to select the superior genotype (Cheliak and Rogers 1990). When superior genotypes have been identified, the embryogenic culture can be brought out of cryogenic storage, maintained in culture, and used for mass production and clonal deployment of the desired genotypes. The technology for this approach exists. Embryogenic cultures of conifers have been successfully cryopreserved, and plants have been regenerated from the cryopreserved embryogenic tissue (Gupta et al. 1987, Kartha et al. 1988, Bercetche et al. 1990).

Although this chapter concentrates on the application of diploid embryogenesis to clonal forestry in conifers, haploid embryogenesis offers considerable potential with regards to breeding and tree improvement strategies. Embryogenesis from haploid explants of conifers provides a way to produce homozygous diploids through chemical treatments that induce chromosome doubling. Conifers offer unique opportunities in this regard. Unlike angiosperms, conifers have a large haploid megagametophyte tissue which is quite easy to culture. Embryogenesis and regeneration of haploid plants from conifer megagametophyte explants has been studied in the most detail with *Larix* (Nagmani and Bonga 1985; von Aderkas et al. 1987; 1990; von Aderkas and Bonga 1988; Rohr et al. 1989). Haploid embryogenesis has also been reported from megagametophytes of *Picea abies* (Simola and Santanen 1990)

3.2 Culture Medium and Environment

The nutrient medium for growth of plant tissue cultures is comprised of five groups of ingredients: inorganic nutrients, vitamins, organic supplements, a carbon source, and phytohormones (Gamborg et al. 1976). Hereafter, basal medium refers to culture medium prior to addition of phytohormones.

Studies on initiation of somatic embryogenesis with *Picea* species have most frequently used the basal medium (coded LP) of von Arnold and Eriksson (1981) at full- or half-strength. Although full-strength LP has been effective for initiation of embryogenic tissue from immature embryos of *P. abies*, studies have shown that modifications in the ammonium nitrate levels were necessary to obtain high levels of initiation (50%) from mature embryo explants (von Arnold 1987). Verhagen and Wann (1989) were able to obtain initiation (28%) on a quite dissimilar medium, half-strength BLG (Amerson et al. 1985), which contained no ammonium nitrate, but high levels of glutamine.

Embryogenic cultures in *Pinus* species have been initiated on several basal media, including DCR (Gupta and Durzan 1985) (Gupta and Durzan 1986b, Finer et al. 1989, Becwar et al. 1990), MSG (a modification of BLG) (Becwar et al. 1990), and LP (Laine and David 1990). Similarly, in *Larix* species a variety of basal media have proven effective for initiation, including LM (Litvay et al. 1985) (Nagmani and Bonga 1985, Klimaszewska 1989a), DCR and MSG (Klimaszewska 1989a). Embryogenic cultures in *Abies* were initiated on SH medium (Schenk and Hilderbrandt 1972) (Schuller et al. 1989, Lang and Kohlenbach 1989) and in *Sequoia* on DCR medium (Bourgkard and Favre 1988). The positive results of medium optimization studies in *Picea* mentioned above, suggest that initiation frequencies can be improved in conifers from the low levels reported in *Pinus*, e.g., 3 to 5% (Gupta and Durzan 1986b, Chandler et al. 1989, Becwar et al. 1990, Laine and David 1990).

In terms of phytohormones used to initiate and maintain conifer embryogenic cultures, reports of initiation of *Picea* species have included both auxin and cytokinin; the most commonly used forms are 2,4-dichlorophenoxyacetic acid (2,4-D) and N6-benzyladenine (BA), respectively. With the auxin napthalene-1-acetic acid, there was an absolute requirement for BA for induction of embryogenic tissue from mature embryos of *P. abies* (von Arnold 1987). Initiation of haploid embryogenic tissue from megagametophytes of *Larix decidua* was optimum on basal medium supplemented with 2,4-D only, although low levels of initiation were obtained on medium lacking phytohormones (von Aderkas et al. 1987).

Similar to *Picea*, auxin and cytokinin were used to initiate somatic embryogenesis in *Pinus taeda* (Gupta and Durzan 1987a); but initiation did not require exogenous phytohormones when immature embryos with intact megagametophytes of *P. taeda* were used as explants (Becwar et al. 1990). Somatic embryogenesis was also initiated in *Pinus serotina* (Becwar et al. 1988) and *Pinus radiata* (Chandler et al. 1989) from immature zygotic embryos within the intact megagametophyte on medium with no exogenous phytohormones.

The composition of the gaseous environment, specifically ethylene and CO_2, two of the most common gases found in the culture vessel, have a significant effect on growth and somatic embryo production in embryogenic suspensions of *Picea glauca* (Kumar et al. 1989). The results showed that although there was continued production of ethylene and CO_2 during the process of multiplication of somatic embryos, the accumulation of these gases was inhibitory. The method of closure of the cultures, therefore, may have a significant effect on culture growth and morphogenesis. Closure devices which allow free gas exchange are probably best.

3.3 Initiation of Embryogenic Cultures

Embryogenic cultures of conifers have a distinctly different phenotype than previously reported for conifer callus. All reports have described embryogenic tissue as translucent to white (even when grown in light), and moist or mucilaginous; quite different than opaque (green when grown in light) and friable nonembryogenic callus of conifers. Embryogenic conifer tissue contained proplastids which lacked the internal organization (grana) typical of mature chloroplasts but closely resembled those found in early zygotic embryos (Feirer 1988; Becwar et al. 1988). In contrast, the chloroplasts in nonembryogenic callus were more typical of mature chloroplasts found in vivo in light-grown plants.

Conifer embryogenic tissue has been characterized biochemically (Wann et al. 1987, 1989). The embryogenic tissue evolved less ethylene, contained lower amounts of glutathione and other reducing agents, and synthesized or turned over protein at a faster rate relative to nonembryogenic callus. These results suggest that conifer embryogenic tissue is biochemically and metabolically similar among species, but distinctly different from nonembryogenic callus.

3.3.1 Terminology and Early Embryogeny

The term callus refers to unorganized growth of plant cells on a semi-solid medium. Embryogenic cultures of conifers have been referred to as callus (Hakman et al. 1985; Nagmani and Bonga 1985) even though they contained organized structures; very early stage somatic embryos. The use of the term callus in this case is not unlike its usage in cereal tissue culture, where "type II callus" refers to proliferative embryogenic tissue which contains somatic embryos arrested in the globular stage of embryogeny (Schaeffer 1990). The type II callus of cereals, similar to embryogenic tissue of conifers, can be subcultured and it maintains this morphology. In this chapter embryogenic cultures of conifers maintained on a semi-solid medium are referred to as embryogenic tissue.

The terms "embryonal suspensor mass" and "somatic polyembryogenesis" have been used to describe, respectively, proliferating embryogenic cultures of conifers, and the in vitro embryo formation process in the cultures (Gupta and Durzan 1986b, 1987a). This terminology was based on the observation of multinucleate cells (coenocytes) and the apparent similarity of the in vitro embryo formation process to cleavage polyembryony which occurs in vivo in *Pinus* (Gupta and Durzan 1987a). In addition, Gupta and Durzan (1987a) suggested that there is no true callus or unorganized phase of proliferation in embryogenic cultures of *P. taeda*. Becwar et al. (1990) and Laine and David (1990) reported two types of embryogenic cultures in *Pinus taeda* and *Pinus caribaea*: (1) relatively undifferentiated cultures that contained dense clusters of embryogenically competent cells and unaggregated suspensor-like cells, but did not contain differentiated somatic embryos; and (2) more differentiated cultures which contained early stage somatic embryos. The different interpretations of how somatic embryos form in conifers, and whether there is an "undifferentiated" culture phase, may be explained by the occurrence of several mechanisms of embryo formation in vitro.

Hakman et al. (1987) suggested three possible mechanisms of somatic embryo formation in conifer embryogenic cultures: (1) development from single cells by an initial asymmetric division; (2) development from small meristematic cells within the suspensor or embryonic region; and (3) development of new embryos by cleavage of the embryonic region, similar to in vivo cleavage polyembryony. Although these deductions were based on observations of early stages of embryogeny in established embryogenic cultures, they may also apply to how the process of somatic embryogenesis actually initiates from explant tissue. In vitro embryo formation in *Picea abies*, *Picea glauca* and *Pinus elliottii* occurred via an asymmetric division of a single cell (Hakman and Fowke 1987a, Nagmani et al. 1987, Jain and Newton 1989). Free nuclear divisions, a characteristic of conifer zygotic embryogenesis, were not observed. Furthermore, there was evidence for an unorganized (callus) phase preceding embryo formation (Nagmani et al.1987, Becwar et al. 1991).

Free nuclear division of cells and the occurrence of coenocytes has been reported with several conifer embryogenic cultures (Gupta and Durzan 1987a, Durzan and Gupta 1987, von Aderkas and Bonga 1988). A study using conifer protoplasts derived from embryogenic cultures of *Larix × eurolepis* provided indirect evidence which suggests that coenocytes are not involved in the formation of conifer somatic embryos (Klimaszewska 1989b). A percol density gradient technique was used to fractionate the protoplasts by centrifugation according to their density and number of nuclei. Only the fraction with mostly small, densely cytoplasmic, and uninucleate protoplasts formed somatic embryos. In contrast, the fraction of multinucleate protoplasts, which may have been derived from coenocytes, did not form somatic embryos. Ultimately, only the tracking of regeneration from single isolated coenocytes will provide definitive evidence as to the importance of coenocytes in embryogenic cultures of conifers.

3.3.2 Site of Origin on Explant

Hakman et al. (1985) reported that embryogenic tissue originated from just beneath the cotyledons (near the interface of the cotyledon and hypocotyl tissues) of *Picea abies* zygotic embryos (Fig. 4). They also indicated that suspensor cells of the most immature embryos may have given rise to embryogenic tissue. Subsequent studies verified the hypocotyl and/or the interface of the hypocotyl and cotyledons was the site of embryogenic tissue formation from both immature and mature embryos of several *Picea* species (Krogstrup 1986; Nagmani et al. 1987; Lu and Thorpe 1987; von Arnold and Woodward 1988; Verhagen and Wann 1989; Webb et al. 1989). Embryogenic tissue also originated from cotyledons excised from germinated *P. abies* zygotic embryos (Krogstrup 1986; Lelu et al. 1987; 1990; Lelu and Bornman 1990; Attree et al. 1990) and from shoots excised from 12 to 30 day old germinants of *Picea mariana* and *P. glauca* (Attree et al. 1990). Histological studies showed that the embryogenic tissue resulted from mitotic activity of the outer cell layers of the hypocotyl of both *P. abies* and *Picea glauca* (Nagmani et al. 1987) and of the excised cotyledons of *P. abies* (Lelu et al 1987, 1990). Nagmani et al. (1987) reported that the tissue which formed at the radicle of the immature zygotic embryos was nonembryogenic. In contrast, Gupta and Durzan (1986a) reported that embryogenic tissue originated from the radicle of mature *P. abies* embryos. Embryogenic tissue was also reported to originate form the radicle end of immature *Picea sitchensis* zygotic embryos (Krogstrup et al. 1988).

Whereas most evidence in *Picea* species indicated that embryogenic tissue originated from the outer cell layers of the hypocotyl and cotyledons of zygotic embryos, there have not been similar reports for *Pinus* species. Embryogenic tissue from immature and mature zygotic embryos of *Pinus lambertiana* (Gupta and Durzan 1986b) and immature embryos of *Pinus taeda* (Becwar et al. 1990, 1991) originated from the suspensor region (Fig. 5). The embryogenic tissue extruded from *P. taeda* megagametophytes containing developing zygotic embryos also originated from the suspensor region of the zygotic embryos (Fig. 6). It should be emphasized that the origin of embryogenic tissue from the suspensor region of the zygotic embryos does not necessarily imply its origin from suspensor cells, per se. The embryogenic tissue may initiate from small cells of the embryo at the interface of the suspensor and embryo (the second mechanism cited in previous section, suggested by Hakman et al. 1987).

Isozyme analysis has been used to genotype embryogenic tissue initiated from multiple zygotic embryos contained within seeds of *Pinus taeda* (Becwar et al. 1991). Simple polyembryony, the formation of multiple zygotic embryos within an individual seed due to different fertilization events, occurs in most conifers. The study provided evidence that somatic embryogenesis can be initiated from more than one of the immature zygotic embryo of an individual seed of *P. taeda*, and that the resulting multiple embryogenic cell lines can be genetically different. The results show that: (1) somatic embryogenesis can be initiated from zygotic embryos other than the dominant embryo which develops to maturity and (2) one can not rule out the possibility of genetic mosaic cultures when initiating from seeds containing multiple zygotic embryos.

It has been suggested that the differences in mechanisms of initiation between *Pinus* and *Picea* species reflects the in vivo occurrence (in the case of *Pinus*) or the absence (in the case of *Picea*) of cleavage polyembryony (Becwar et al. 1988). In other words, in *Pinus* species it appears that somatic embryogenesis initiates de novo from cells in the suspensor region and/or results from continual

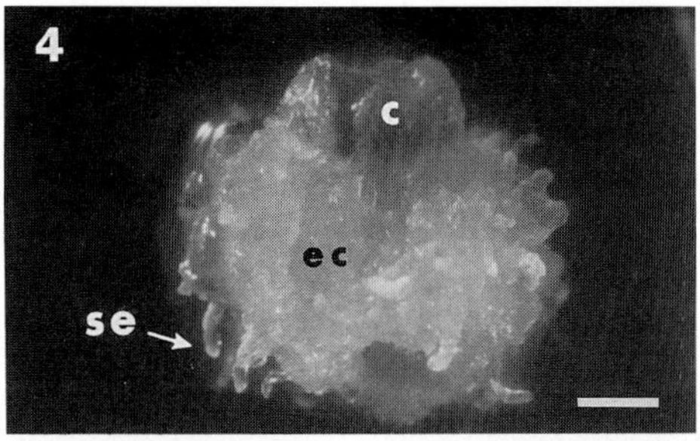

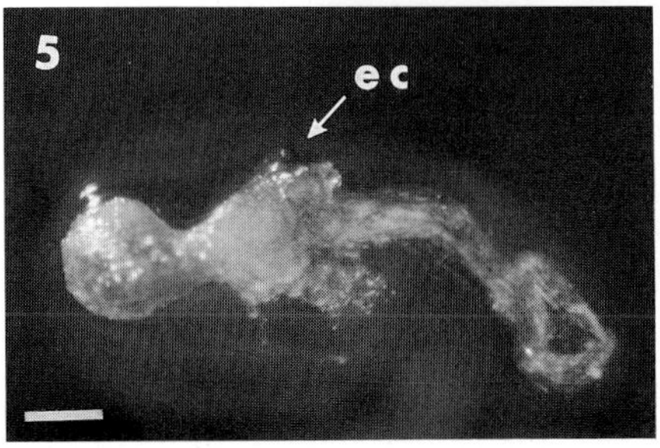

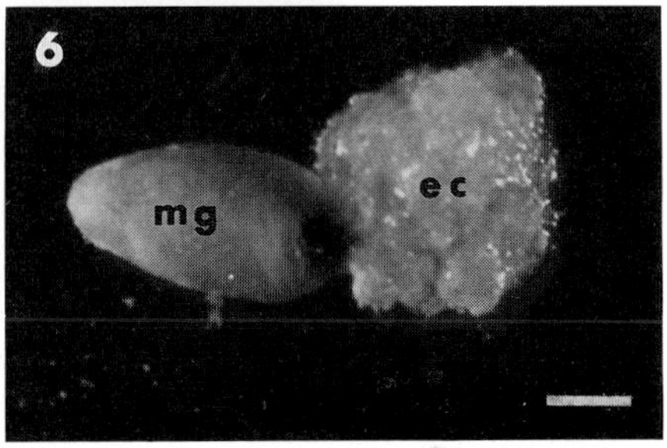

Figs. 4-6. Initiation of embryogenic tissue in conifers. **Fig. 4.** Embryogenic tissue initiated from hypocotyl region of an immature *Picea abies* zygotic embryo. Note numerous somatic embryos (*se*) on surface of embryogenic tissue (*ec*) and swollen cotyledons (*c*) of explant. **Fig. 5.** Embryogenic tissue initiated from a precotyledonary immature embryo of *Pinus taeda*. **Fig. 6.** Embryogenic tissue extruded from the micropylar end of a megagametophyte (mg) explant of *P. taeda*. The embryogenic tissue originates from the immature embryos contained within the megagametophyte. Bars = 2 mm Figs. 4, 6; 1 mm Fig. 5. (Becwar, unpubl. 1988)

cleavage of the embryonic region. By contrast, in *Picea* species (which do not undergo cleavage in vivo) the suspensor region is less responsive and the initiation of embryogenesis from the hypocotyl and cotyledons can not be due to cleavage, because it occurs via an unorganized (callus) phase (Nagmani et al. 1987). This hypothesis relates only to how embryogenesis is initiated from the zygotic embryo explant. Formation of somatic embryos by cleavage has been reported in established cultures of species which do not undergo cleavage in vivo, e.g., *Picea glauca* and *Pseudotsuga menziesii* (Hakman et al. 1987, Durzan and Gupta 1987).

3.3.3 Optimum Stage of Explant Development

In *Pinus taeda* the optimum stage for initiation of embryogenic tissue occurred at the precotyledonary stage of zygotic embryo development (Fig. 2) (Becwar et al. 1990). As the embryos developed and formed cotyledon primordia (Fig. 3), the frequency of embryogenic tissue decreased. Other studies with *Pinus caribaea* (Laine and David 1990), *Pinus strobus* (Becwar et al. 1988, Finer et al. 1989), and *Pinus radiata* (Chandler et al. 1989) also suggested that in *Pinus*

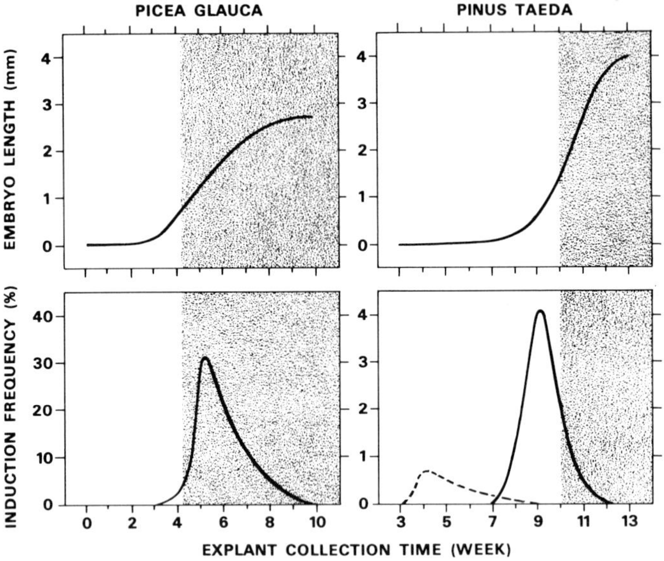

Fig. 7. The relationship of the stage of development of immature zygotic embryos, as measured by embryo length (top graphs), to the frequency of initiation of embryogenic tissue (bottom graphs) in *Picea glauca* and *Pinus taeda*. The shaded region corresponds to when at least 50% of the zygotic embryos formed cotyledons. The solid lines in the bottom graphs represent initiation from isolated immature zygotic embryo explants and the dashed line initiation from immature zygotic embryos with intact megagametophytes. (Becwar et al. 1988, 1990).

species the optimum stage of immature zygotic embryo development for initiation was precotyledonary. By contrast, in *Picea* species the optimum stage of zygotic embryo development occurred just when cotyledon primordia formed (Hakman and Fowke 1987b, Lu and Thorpe 1987, Becwar et al. 1988, Webb et al. 1989). A comparison of initiation of embryogenic tissue in *Picea glauca* and *P. taeda* is shown in Fig. 7.

One report has examined the biochemical characteristics of explants of conifers in relation to their morphogenic competence (Roberts et al. 1989). Electrophoretic analysis of total proteins from cotyledonary zygotic embryos of interior spruce revealed that competence to form embryogenic tissue was limited to the cotyledonary stage of development prior to accumulation of storage proteins. The phytohormone abscisic acid (ABA) may play an important role in the accumulation of storage proteins during zygotic embryogenesis (Finkelstein et al. 1985). Changes in embryo maturation regulated by ABA may be important in determining the ability of explants to form embryogenic tissue (Roberts et al. 1989).

3.3.4 Effect of Genotype on Initiation

Frequently, reports have failed to indicate if differences were observed in somatic embryogenesis either among source trees from which the zygotic

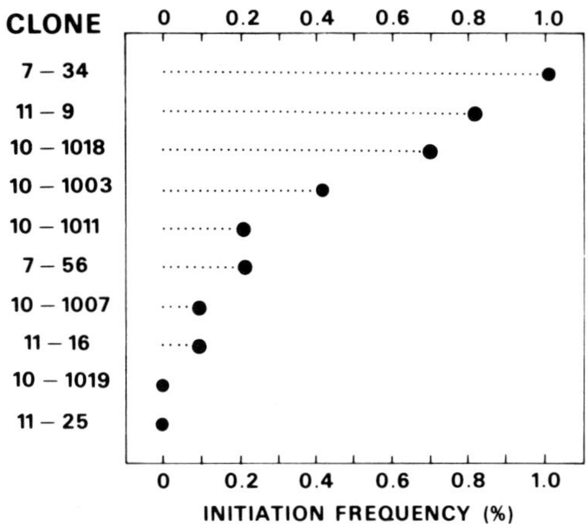

Fig. 8. Comparison of initiation frequency of embryogenic tissue from immature embryos derived from ten clones of *Pinus taeda*. (Becwar et al. 1990)

embryos were collected, or among the resulting embryogenic cell lines. Significant differences in the level of somatic embryogenesis have been reported among cultures derived from different immature zygotic embryo explants (different genotypes) collected from a *Picea abies* tree (Becwar et al. 1987b). Genotypic differences have also been observed among embryogenic cultures derived from different megagametophyte explants of *Larix decidua* (von Aderkas and Bonga 1988). In a survey of ten genetically different trees of *Pinus taeda*, embryogenic tissue was initiated from immature zygotic embryos of eight of the ten trees (Becwar et al. 1990) (Fig. 8). There were differences in the initiation frequency from the explants derived from different trees. Over 70% of the embryogenic cultures were derived from explants of three trees. In a survey of five *Picea glauca* trees and one *Picea engelmannii* tree, initiation of embryogenic tissue ranged from 6 to 15% among different trees surveyed (Webb et al. 1989). Collectively, these results provide evidence for genotype specificity in establishing an efficient somatic embryo regeneration system in conifers.

3.4 Maintenance of Embryogenic Cultures

Embryogenic cultures can be maintained on the same medium and phytohormones on which they are initiated. This semi-solid medium contains agar or a similar gelling agent. Alternatively, they can be maintained as a liquid suspension on the same medium without the gelling agent.

Liquid culture systems are amenable to automation in terms of culture maintenance, embryo development, and embryo harvest. It is therefore likely that suspension cultures will play an important role in the application of conifer somatic embryogenesis to clonal forestry. Embryogenic suspensions capable of plantlet regeneration have been reported for several conifer species (Durzan and Gupta 1987, Boulay et al. 1988, Hakman and von Arnold 1988, Dunstan et al.

1988, Attree et al. 1989a, Tautorus et al. 1990, Laine and David 1990, Krogstrup 1990). A detailed study of factors affecting conifer suspension culture growth has been done with embryogenic cultures of *Picea sitchensis* (Krogstrup 1990). Rapid proliferation of highly embryogenic suspensions was dependent on culture density and an interaction between the volume of medium and the volume of air space above the medium. In addition, suspension density affected the regeneration capacity when the cultures were plated out for embryo development. The *P. sitchensis* regeneration system made use of several novel modifications, including plating on a filter paper supported by polyester fleece to allow efficient exchange of medium during the embryo development period.

There is evidence for long-term stability of conifer embryogenic cultures maintained on either gelled or liquid medium. Embryogenic cultures of *Picea glauca* have been maintained on gelled medium for 16 months (Hakman and von Arnold 1988). Subsequently, this embryogenic tissue was used to initiate liquid suspension cultures that maintained plantlet regeneration capacity for over 6 months. Embryogenic cultures of *Picea abies* retained both the capacity to regenerate plantlets after 2 years in culture and the diploid amount (32 pg/nucleus) of DNA in the regenerated plantlets (Mo et al. 1989).

3.5 Somatic Embryo Development and Plant Regeneration

The process of somatic embryo development and plant regeneration can be divided into the following steps according to the stage of morphogenesis and the culture conditions used; embryo development, embryo maturation, germination, and conversion (Table 1). Hakman and von Arnold (1988) proposed the following system for identification of the stage of conifer somatic embryo development: stage 1, small embryos consisting of an embryonic region of small, densely cytoplasmic cells subtended by a suspensor comprised of long and highly vacuolated cells. Stage 2, embryos with a prominent embryonic region that is more opaque and with a more smooth and glossy surface than stage 1 embryos. Stage 3, embryos with small cotyledons. This system of identification of stages of embryo development will be used in this chapter.

3.5.1 Embryo Development and Maturation

Frequently, conifer embryogenic cultures contain stage 1 embryos when grown under proliferative conditions on a medium containing auxin and cytokinin (Figs. 9, 10). However, as previously indicated, some embryogenic cultures of *Pinus* were maintained in a less differentiated state under similar conditions (Becwar et al. 1990, Laine and David 1990). Such cultures contained long suspensor-like cells and small dense clusters of embryogenically competent cells while on maintenance medium. Well formed somatic embryos were only observed after the culture was transferred to medium devoid of auxin. Regardless of the level of differentiation while on maintenance medium, auxin or auxin and cytokinin in combination promote proliferation of the culture and arrest embryo development. Further development of somatic embryos to stage 2 and

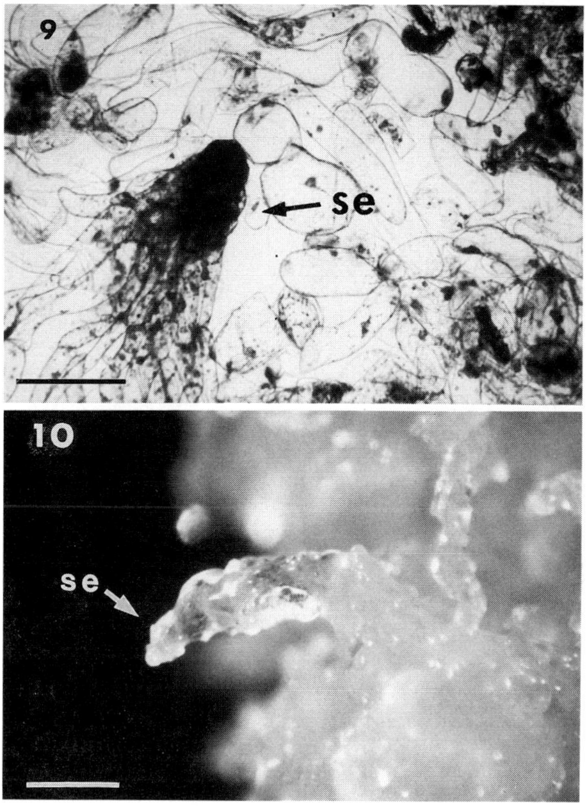

Figs. 9, 10. Stage 1 somatic embryos (se) of *Pinus taeda*. **Fig. 9.** Micrograph of embryogenic tissue with somatic embryo visible (Bar = 250 μm). **Fig. 10.** Somatic embryo on surface of embryogenic tissue (Bar = 500 μm) (Becwar and Nagmani, unpubl. 1988)

beyond is induced by transferring the embryogenic culture to a modified medium (Figs. 11, 12).

Activated charcoal, which acts as an absorbent of phytohormones (Ebert and Taylor 1990) has been used in media during induction of embryo development (Hakman et al. 1985; Becwar et al. 1987a, 1987b; Verhagen and Wann 1989; Roberts et al. 1990a, 1990b; Webster et al. 1990).

The most significant difference between culture media used for proliferation of embryogenic tissue and for development is the level and form of phytohormones. Typically, auxin and cytokinin are absent from embryo development medium, or their concentrations are greatly reduced. In addition, the more recent embryo development protocols include ABA in the medium for induction of embryo development.

Endogenous levels of ABA increase during embryo development in vivo in angiosperms (Walton 1980). Exogenous ABA was first used at relatively low concentrations (1–2 μM) to induce conifer somatic embryo development (Becwar et al. 1987a, 1987b; Gupta and Durzan 1987b). Numerous subsequent

Figs. 11–18. Regeneration of *Picea abies* plants from embryogenic tissue and comparison of plants grown from somatic embryos and seedlings grown from zygotic embryos. **Fig. 11.** Immature somatic embryos (mostly stage 2) after 21 days on development medium. **Fig. 12.** Cotyledonary (stage 3) somatic embryos. **Fig. 13.** Somatic embryo germination, primary root (pr) growth. **Fig. 14.** Conversion, survival and growth of somatic plantlet in soil mix. **Fig. 15.** Somatic embryo plant with dormant vegetative bud (db) prior to overwintering. **Fig. 16.** Renewed vegetative growth (arrow) of control seedling (ze) after overwintering. **Fig. 17.** Renewed vegetative growth of somatic plant (se). Figs. 16 and 17 photos taken on same date. **Fig. 18.** Seedling plant (left) and somatic plant (right). Bars = 1 mm Figs. 11–13; 1 cm Figs. 14–17; 5 cm Fig. 18. (Becwar et al. 1989)

studies have verified the effectiveness of ABA for improving conifer somatic embryo development and maturation. In addition, higher concentrations of ABA ($\approx$ 40–60 μM) have proven much more effective that the low concentrations originally reported (Roberts et al. 1990a, 1990b; Webster et al. 1990). In conifer somatic embryogenesis, ABA increases: (1) the frequency of calli which form stage 2 somatic embryos (von Arnold and Hakman 1988); (2) the number of plantlets regenerated per ml of embryogenic suspension (Hakman and von Arnold 1988); (3) both the density of immature somatic embryos (Dunstan et al. 1988; Becwar et al. 1989; Finer et al. 1989; Feirer et al. 1989) and (4) the number of morphologically normal stage 3 somatic embryos (Roberts et al. 1990a, 1990b).

It appears that one of the important physiological effects of exogenous ABA on conifer somatic embryo development is to increase nutrient storage reserves. A histological study provided evidence that lipids and storage proteins accumulated during treatment of somatic embryos with ABA (Hakman and von Arnold 1988). Feirer et al. (1989) provided quantitative evidence that ABA promoted accumulation of triacylglycerides (storage lipids). Storage lipids comprised 30 to 50% of the dry weight of conifer zygotic embryos; their deficiency in conifer somatic embryos may hinder subsequent embryo germination. Increasing storage reserves of somatic embryos, therefore, may greatly improve germination. Four major storage proteins have been identified in *Picea* zygotic and somatic embryos, ranging in molecular mass from 22 to 42 kDa (Roberts et al. 1990a; Hakman et al. 1990).

The manner in which the term "maturation" has been used in the literature on conifer somatic embryogenesis may be misleading. Somatic embryo maturation has frequently been used to refer to the attainment of full morphological development; e.g., development of embryos to the cotyledonary stage (stage 3), similar in appearance to zygotic embryos removed from fully developed seed (Becwar et al. 1989). Upon closer examination, it appears that the complex process of embryo maturation has not been fully completed in many cotyledonary stage conifer somatic embryos. Embryo maturation includes the completion of biochemical and physiological development. For instance, the biochemical buildup of nutrient storage reserves, and the physiological changes that occur in order for the embryo to become tolerant of desiccation. Without the completion of the maturation phases of somatic embryo development, germination may be blocked or may be precocious and result in poor quality plants.

3.5.2 Germination

A germinating *P. abies* somatic embryo is shown in Fig. 13. Reports of regeneration from conifer somatic embryos from 1985 to 1988 recorded relatively low germination frequencies; typically 10 to 20% of the cotyledonary stage somatic embryos developed roots. Von Arnold and Hakman (1988) obtained the highest level of germination (35%) with *P. abies* somatic embryos which had developed to the cotyledonary stage in dark culture conditions. There have been very few reports of germination of *Pinus* somatic embryos. One to five plantlets were obtained per gram of embryogenic tissue with *P. taeda* (Gupta and Durzan 1987a). Nearly 50% of 69 somatic embryos produced from embryogenic tissue of *Pinus caribaea* germinated (Laine and David 1990).

Significant improvements have been made in germination of conifer somatic embryos. The partial drying of cotyledonary stage somatic embryo at high relative humidity increased the germination frequency of interior spruce somatic embryos to 90%, as compared to 3% for untreated controls (Roberts et al. 1990b). This treatment also increased the germination vigor and synchronized germination. The high relative humidity treatment did not induce desiccation tolerance, but it was suggested that the treatment substitutes for desiccation to switch gene expression from a mode of maturation to germination.

High germination frequencies have also been obtained with somatic embryos of *Picea glauca* and *P. mariana*, not by a partial drying treatment, but by transfer to ABA free medium for 14 days (subsequent to removal from development medium containing ABA), prior to harvesting embryos for germination (Attree et al. 1991). Over 1800 plantlets were recovered and germination frequencies of 64 to 74% were obtained with this method. It was suggested that the maturation treatment allowed the endogenous levels of ABA in the cotyledonary somatic embryos to decline prior to germination.

3.5.3 Conversion

The process of embryo-to-plant development has been referred to as "embryo conversion" to distinguish such development from somatic embryo germination (Redenbaugh et al. 1986). Conversion, therefore, refers to development of "seedling-quality" plants (Fig. 14) capable of growth and development beyond the initial stages of germination. The conversion process includes: (1) the acclimatization process that in vitro derived plantlets undergo in order to survive in nonaxenic conditions (ex vitro), and (2) subsequent continued growth and development under nonaxenic conditions (Becwar et al. 1989).

Plants derived from somatic embryos of *Picea abies* responded to environmental changes in a manner similar to seedlings grown and overwintered under identical conditions (Becwar et al. 1989). They set dormant terminal buds (Fig. 15), survived overwintering to −5 °C, and renewed vegetative growth synchronously with control seedlings (Figs. 16, 17). The somatic plants continued to grow and appeared phenotypically normal relative to seedlings (Fig. 18). Subsequently, the somatic plants and seedlings set dormant buds, overwintered, and produced new vegetative growth during their second growing season (Figs. 19, 20).

The first large-scale evaluation of conifer somatic embryo plants under ex vitro conditions was with interior spruce, a mixture of two closely related species, *Picea glauca* and *P. engelmannii* (Webster et al. 1990). Survival of most of the somatic embryo derived plants, following transfer to soil, acclimatization, and the first season's growth in the nursery was 80% or greater for most of the 71 genotypes tested. Growth rates, final height, shoot and root morphology, and frost hardiness of the somatic embryo plants were similar to seedlings following

Figs. 19, 20. Plants derived from zygotic embryo (**Fig. 19**) and from somatic embryo (**Fig. 20**) after resumption of vegetative growth during the second growing season. Bar = 10 cm (Becwar, unpubl. 1988)

the first growing season. This study clearly demonstrated the potential for somatic embryogenesis to be used for the production of planting stock for a range of conifer genotypes.

3.6 Regeneration from Protoplasts

The ability to induce protoplasts (single cells with cell walls removed) of conifers to form cell colonies capable of somatic embryogenesis and regeneration could be of value to: (1) facilitate the study of the earliest biochemical events associated with morphogenic cell division, and (2) utilize gene transfer techniques that are only effective with protoplasts; e.g., electroporation and passive DNA uptake (Sederoff and Stomp, Chap. 15, Vol. 1).

Regeneration of cell colonies from conifer protoplasts capable of forming somatic embryos was first reported with *Pinus taeda* (Gupta and Durzan 1987b). This and several other studies to follow have used the approach of isolating protoplasts from rapidly growing embryogenic suspension cultures of conifers (Bekkaoui et al. 1987; Attree et al. 1987, 1987a; 1989b; Bekkaoui and Dunstan 1989; Tautorus et al. 1990; Laine and David 1990). Attree et al. (1989a) and Klimaszewska (1989b) were the first to verify that somatic embryos derived from protoplasts of conifers were capable of continued development and plantlet regeneration.

The utility of protoplasts derived from embryogenic cultures for transient expression of foreign genes has been demonstrated with several conifer species (Gupta et al. 1988; Bekkaoui et al. 1988, 1990; Tautorus et al. 1989). In these reports, electroporation was used as the means of introducing the foreign gene. Regardless of the rapid advancements made with transient expression of foreign genes in conifer protoplasts, it appears that other gene insertion techniques, that do not require protoplasts; e.g., microprojectile-mediated transfer, are preferred for stable integration of foreign DNA into conifers (Sederoff and Stomp, Chap. 15, Vol. 1).

4 Utility of Somatic Embryogenesis for Clonal Forestry

One of the most promising applications of somatic embryogenesis, and of most importance to clonal forestry, is the potential to efficiently and economically mass produce clonal planting stock. For example, a one liter liquid conifer embryogenic culture contains about 10^6 somatic embryos (Becwar et al. 1988). The fact that the cultures can be grown in liquid makes possible the development of large scale bioreactors, similar to those used for microbial culture. Thus, many of the manipulations required to produce somatic embryos can be automated. Furthermore, somatic embryos can be encapsulated (Redenbaugh et al. 1986) to form artificial seeds, providing an efficient propagule delivery system.

The potential also exists for combining readily available gene transfer techniques with somatic embryogenesis systems to mass produce forest trees with economically useful traits, e.g. increased pest, disease, and stress resistance, that are difficult or impossible to obtain by conventional breeding.

5 Summary and Conclusions

In terms of improving the efficiency of regeneration, the use of the phytohormone ABA played a key role in improving conifer somatic embryo development and maturation. These encouraging empirical results serve to emphasize the need for more basic and mechanistically oriented research on the physiology and biochemistry of zygotic and somatic embryo development and maturation. There are numerous research laboratories worldwide concentrating on developing in vitro embryogenesis protocols for a particular species, but there are relatively few scientists specifically studying the basic process of somatic embryogenesis (Thorpe 1988). Indeed, somatic embryogenic systems in conifers offer unprecedented opportunities with regards to supplying embryos and plant material for basic research. With the advent of highly embryogenic cultures of several conifers it is feasible to routinely obtain large quantities of somatic embryos at various stages of development. This will make it easier to do detailed biochemical and physiological characterization of the embryogenesis process, and for the first time to apply techniques of molecular biology to study mechanisms of genetic control and regulation of embryogenesis and regeneration. Fundamental knowledge gained in such research will have direct application to somatic embryogenesis in relation to clonal forestry.

Because of the considerable amount of research and development that needs to be completed in order to apply somatic embryogenesis to clonal forestry, it is easy to lose sight of the enormous amount of progress that has been made since 1985 toward achieving this goal. It is worthwhile to emphasize that nearly all integral parts of the envisioned protocol outlined in Fig. 1 have been achieved with conifers. In summary therefore, there is reason to believe that the problems of developing an efficient somatic embryo regeneration system in conifers can be surmounted.

Verification of genetic fidelity of plants regenerated via somatic embryogenesis from individual explants is of paramount importance for the application of somatic embryogenesis to clonal forestry. For the first time, it is now possible to do so with plants regenerated from conifer embryogenic cultures. (Literature review completed February 1991)

Acknowledgments. The support of Westvaco Corporation and the other member companies of the Institute of Paper Science and Technology (formerly The Institute of Paper Chemistry) is gratefully acknowledged. The figures and several of the author's citations in this chapter were based on research done at IPST from 1985 to 1988.

References

Amerson HV, Frampton LJ, McKeand SE, Mott RL, Weir RJ (1985) Loblolly pine tissue culture: laboratory, greenhouse, and field studies. In: Henke RR, Hughes KW, Constantin MJ, Hollaender A (eds) Tissue culture in forestry and agriculture. Plenum, New York, pp 271–287

Ammirato PV (1983) Embryogenesis. In: Evans DA, Sharp WR, Ammirato PV, Yamada Y (eds) Handbook of plant cell culture, vol 1. Macmillan, New York, pp 82–123

Attree SM, Bekkaoui F, Dunstan DI, Fowke LC (1987) Regeneration of somatic embryos from protoplasts isolated from an embryogenic suspension of white spruce (*Picea glauca*). Plant Cell Rep 6:480–483

Attree SM, Dunstan DI, Fowke LC (1989a) Initiation of embryogenic callus and suspension cultures, and improved embryo regeneration from protoplasts of white spruce (*Picea glauca*). Can J Bot 67:1790–1795

Attree SM, Dunstan DI, Fowke LC (1989b) Plantlet regeneration from embryogenic protoplasts of white spruce (*Picea glauca*). Bio/Technology 7:1060–1062

Attree SM, Budimir S, Fowke LC (1990) Somatic embryogenesis and plantlet regeneration from cultured shoots and cotyledons of seedlings from stored seeds of black and white spruce (*Picea mariana* and P. glauca) Can J Bot 68:30–34

Attree SM, Tautorus TE, Dunstan DI, Fowke LC (1991) Somatic embryo maturation, germination, and soil establishment of plants of black and white spruce (*Picea mariana* and *Picea glauca*). Can J Bot 68:2583–2589

Becwar MR, Wann SR, Noland TL (1987a) A method for quantification of the level of somatic embryogenesis among Norway spruce callus lines. Plant Cell Rep 6:35–38

Becwar MR, Wann SR, Noland TL (1987b) Somatic embryo development and plant regeneration from embryogenic Norway spruce callus. Tappi J 70:155–160

Becwar, MR, Wann SR, Johnson MA, Verhagen VA, Feirer RP, Nagmani R (1988) Development and characterization of in vitro embryogenic systems in conifers. In: Ahuja MR (ed) Somatic cell genetics of woody plants. Kluwer, Dordrecht, pp 1–18

Becwar MR, Noland TL, Wyckoff JL (1989) Maturation, germination, and conversion of Norway spruce (*Picea abies* L.) somatic embryos to plants. In Vitro Cell Dev Biol 25:575–580

Becwar MR, Nagmani R, Wann SR (1990) Initiation of embryogenic cultures and somatic embryo development in loblolly pine (*Pinus taeda*). Can J For Res 20:810–817

Becwar MR, Blush TD, Brown DW, Chesick EE (1991) Multiple paternal genotypes in embryogenic tissue derived from individual immature loblolly pine seeds. Plant Cell Tissue Org Cult 26:37–44

Bekkaoui F, Dunstan DI (1989) Permeabilization of *Picea glauca* protoplasts to macromolecules. Can J For Res 19:1316–1321

Bekkaoui F, Saxena P, Attree SM, Fowke LC, Dunstan DI (1987) The isolation and culture of protoplasts from an embryogenic cell suspension culture of *Picea glauca* (Moench) Voss. Plant Cell Rep 6:476–479

Bekkaoui F, Pilon M, Laine E, Raju DSS, Crosby WL, Dunstan DI (1988) Transient gene expression in electroporated *Picea glauca* protoplasts. Plant Cell Rep 7:481–484

Bekkaoui F, Datla RSS, Pilon M, Tautorus TE, Crosby WL, Dunstan DI (1990) The effects of promoter on transient expression in conifer cell lines. Theor Appl Genet 79:353–359

Bercetche J, Galerne M, Dereuddre J (1990) Efficient regeneration of plantlets from embryogenic callus of *Picea abies* (L.) Karst after freezing in liquid nitrogen. C R Acad Sci Paris 310 Ser 3:357–363

Bonga JM (1987) Clonal propagation of mature trees: Problems and possible solutions. In: Bonga JM, Durzan DJ (eds) Cell and tissue culture in forestry. Nijhoff, Dordrecht, pp 249–271

Boulay MP, Gupta PK, Krogstrup P, Durzan DJ (1988) Development of somatic embryos from cell suspension cultures of Norway spruce (*Picea abies* Karst.). Plant Cell Rep 7:134–137

Bourgkard F, Favre JM (1988) Somatic embryos from callus of *Sequoia sempervirens*. Plant Cell Rep 7:445–448

Brown DC (1988) Germplasm determination of *in vitro* somatic embryogenesis in alfalfa. HortScience 23:526–531

Chandler SF, Bateman C, Blomstedt C, Willyams D, Young R (1989) Forestry biotechnology at calgene pacific. Aust J Biotechnol 3:281–284

Cheliak WM, Rogers DL (1990) Integrating biotechnology into tree improvement programs, Can J For Res 20:452–463

Dunstan DI (1988) Prospects and progress in conifer biotechnology. Can J For Res 18:1497–1506

Dunstan DI, Bekkaoui F, Pilon M, Fowke LC, Abrams SR (1988) Effects of abscisic acid and analogues on the maturation of white spruce (*Piecea glauca*) somatic embryos. Plant Sci 58:77–84

Durzan DJ, Gupta PK (1987) Somatic embryogenesis and polyembryogenesis in Douglas-fir cell suspension cultures. Plant Sci 52:229–235

Ebert A, Taylor HF (1990) Assessment of the changes of 2,4-dichlorophenoxyacetic acid concentrations in plant tissue culture media in the presence of activated charcoal. Plant Cell Tissue Org Cult 20:165–172

Feirer RP (1988) Chloroplast ultrastructure and gene expression in embryogenic conifer callus. In: Cheliak WM, Yapa AC (eds) Molecular genetics of forest trees. Proceedings, 2nd IUFRO Working party Molecular Genetics, 15–18 June, 1987. Petawawa Natl For Inst, Chalk River, Ontario, Can, pp 89–95

Feirer RP, Conkey JH, Verhagen SA (1989) Triglycerides in embryogenic conifer calli: a comparison with zygotic embryos. Plant Cell Rep 8:207–209

Finer JJ, Kriebel H, Becwar MR (1989) Initiation of embryogenic callus and suspension cultures of white pine (*Pinus strobus* L.). Plant Cell Rep 8:203–206

Finkelstein RR, Tenbarge KM, Shumway JE, Crouch ML (1985) Role of ABA in maturation of rapeseed embryos. Plant Physiol 78:630–636

Gamborg OL, Murashige T, Thorpe TA, Vasil IK (1976) Plant tissue culture media. In Vitro Cell Dev Biol 12:473–478

Gupta PK, Durzan DJ (1985) Shoot multiplication from mature trees of Douglas-fir (*Pseudotsuga menziesii*) and sugar pine (*Pinus lambertiana*). Plant Cell Rep 4:177–179

Gupta PK, Durzan DJ (1986a) Plantlet regeneration via somatic embryogenesis from subcultured callus of mature embryos of *Picea abies* (Norway spruce). In Vitro Cell Dev Biol 22:685–688

Gupta PK, Durzan DJ (1986b) Somatic polyembryogenesis from callus of mature sugar pine embryos. Bio/Technology 4:643–645

Gupta PK, Durzan DJ (1987a) Biotechnology of somatic polyembryogenesis and plantlet regeneration in loblolly pine. Bio/Technology 5:147–151

Gupta PK, Durzan DJ (1987b) Somatic embryos from protoplasts of loblolly pine proembryonal cells. Bio/Technology 5:710–712

Gupta PK, Durzan DJ, Finkle BJ (1987) Somatic polyembryogenesis in embryogenic cell masses of *Picea abies* (Norway spruce) and *Pinus taeda* (loblolly pine) after thawing from liquid nitrogen. Can J For Res 17:1130–1133

Gupta PK, Dandekar AM, Durzan DJ (1988) Somatic proembryo formation and transient expression of a luciferase gene in Douglas fir and loblolly pine protoplasts. Plant Sci 58:85–92

Hakman I, Fowke LC (1987a) An embryogenic cell suspension culture of *Picea glauca* (white spruce). Plant Cell Rep 6:20–22

Hakman I, Fowke LC (1987b) Somatic embryogenesis in *Picea glauca* (white spruce) and *P. mariana* (black spruce). Can J Bot 65:656–659

Hakman I, von Arnold S (1985) Plantlet regeneration through somatic embryogenesis in *Picea abies* (Norway spruce). J Plant Physiol 121:149–158

Hakman I, Fowke LC, von Arnold S, Eriksson T (1985) The development of somatic embryos in tissue cultures initiated from immature embryos of *Picea abies* (Norway spruce). Plant Sci 38:53–59

Hakman I, von Arnold S (1988) Somatic embryogenesis and plant regeneration from suspension cultures of *Picea glauca* (white spruce). Physiol Plant 72:579–587

Hakman I, Rennie P, Fowke L (1987) A light and electron microscope study of *Picea glauca* (white spruce) somatic embryos. Protoplasma 140:100–109

Hakman I, Stabel P, Engstrom P , Eriksson T (1990) Storage protein accumulation during zygotic and somatic embryo development in *Picea abies* (Norway spruce). Physiol Plant 80:441–445

Halperin W, Wetherell DR (1964) Adventive embryony in tissue cultures of the wild carrot, Daucus carota. Am J Bot 51:274–283

Halperin W, Wetherell DR (1965) Ammonium requirement for embryogenesis in vitro. Nature (London) 205:519–520

Hodges TK, Kamo KK, Imbrie CW, Becwar MR (1986) Genotype specificity of somatic embryogenesis and regeneration in maize. Bio/Technology 4:219–223

Jain SM, Newton RJ (1989) Somatic embryogenesis in slash pine (*Pinus elliottii*) form immature embryos cultured in vitro. Plant Sci 65:233–241

Jain SM, Newton RJ, Soltes EJ (1988) Enhancement of somatic embryogenesis in Norway spruce (*Picea abies* L.). Theor Appl Genet 76:501–506

Kartha KK, Fowke LC, Leung NL, Caswell KL, Hakman I (1988) Induction of somatic embryos and plantlets from cryopreserved cell cultures of white spruce (*Picea glauca*) J Plant Physiol 132:529–539

Klimaszewska K (1989a) Plantlet development from immature zygotic embryos of hybrid larch through somatic embryogenesis. Plant Sci 63:95–103

Klimaszewska K (1989b) Recovery of somatic embryos and plantlets from protoplast cultures of *Larix × eurolepis*. Plant Cell Rep 8:440–444

Krogstrup P (1986) Embryolike structures from cotyledons and ripe embryos of Norway spruce (*Picea abies*). Can J For Res 16:664–668

Krogstrup P (1990) Effect of culture densities on cell proliferation and regeneration from embryogenic cell suspensions of *Picea sitchensis*. Plant Sci 2:115–123

Krogstrup P, Eriksen EN, Moller JD, Roulund H (1988) Somatic embryogenesis in Sitka spruce [*Picea sitchensis* (Bong.) Carr.]. Plant Cell Rep 72:594–597

Kumar PP, Joy RW, Thorpe TA (1989) Ethylene and carbon dioxide accumulation and growth of cell suspension cultures of *Picea glauca* (white spruce). J Plant Physiol 135:592–596

Laine E, David A (1990) Somatic embryogenesis in immature embryos and protoplasts of *Pinus caribaea*. Plant Sci 69:215–224

Lang H, Kohlenbach HW (1989) Cell differentiation in protoplast cultures from embryogenic callus of *Abies alba* L. Plant Cell Rep 8:120–123

Lelu MAP, Bornman CH (1990) Somatic embryogenesis in excised cotyledons of *Piceae glauca* and *Picea mariana*. Plant Physiol Biochem 28:785

Lelu MA, Boulay M, Arnaud Y (1987) Formation of embryogenic calli from cotyledons of *Picea abies* (L.) Karst. collected from 3 to 7 day old seedlings. C R Acad Sci Paris 305:105–107

Lelu MAP, Boulay M, Bornman CH (1990) Somatic embryogenesis in cotyledons of *Picea abies* is enhanced by an adventitious bud-inducing treatment. New For 4:125–135

Litvay JD, Verma DC, Johnson MA (1985) Influence of loblolly pine (*Pinus taeda* L.) culture medium and its components on growth and somatic embryogenesis of wild carrot (*Daucus carota* L.). Plant Cell Rep 4:325–328

Litz RE (1985) Somatic embryogenesis in tropical fruit trees. In: Henke RR, Hughes KW, Constantin MP, Hollaender A (eds) Tissue culture in forestry and agriculture. Plenum, New York, pp 179–193

Litz RE (1987) Application of tissue culture to tropical fruits. In: Green CE, Sommers DA, Hackett WP, Biesboer DD (eds) Plant tissue and cell culture. Liss, New York, pp 407–418

Lu CY, Thorpe TA (1987) Somatic embryogenesis and plantlet regeneration in cultured immature embryos of *Picea glauca*. J Plant Physiol 128:297–302

Lu CY, Thorpe TA (1988) Shoot-bud regeneration in subcultured callus of Engelmann spruce. In Vitro Cell Dev Biol 24:239–242

Merkle SA, Sommer HE (1986) Somatic embryogenesis in tissue cultures of *Liriodendron tulipifera*. Can J Res 16:420–422

Mo LH, von Arnold S, Lagercrantz U (1989) Morphogenic and genetic stability in longterm embryogenic cultures and somatic embryos of Norway spruce [*Picea abies* (L.) Karst.]. Plant Cell Rep 8:375–378

Nagmani R, Bonga JM (1985) Embryogenesis in subcultured callus of *Larix decidua*. Can J For Res 15:1088–1091

Nagmani R, Becwar MR, Wann SR (1987) Single-cell origin and development of somatic embryos in *Picea abies* (L.) Karst. (Norway spruce) and *P. glauca* (Moench) Voss (white spruce). Plant Cell Rep 6:157–159

Park YG, Son SH (1988) *In vitro* organogenesis and somatic embryogenesis from punctured leaf of *Populus nigra* × *P. maximowiczii*. Plant Cell Tissue Org Cult 15:95–105

Redenbaugh K, Paasch BD, Nichol JW, Kossler ME, Viss PR, Walker KA (1986) Somatic seeds: encapsulation of asexual plant embryos. Bio/Technology 4:797–801

Roberts DR, Flinn BS, Webb DT, Webster FB, Sutton BCS (1989) Characterization of immature embryos of interior spruce by SDS-PAGE and microscopy in relation to their competence for somatic embryogenesis. Plant Cell Rep 8:285–288

Roberts DR, Flinn BS, Webb DT, Webster FB, Sutton BCS (1990a) Abscisic acid and indole-3-butyric acid regulation of maturation and accumulation of storage proteins in somatic embryos of interior spruce. Physiol Plant 78:355–360

Roberts DR, Sutton BCS, Flinn BS (1990b) Synchronous and high frequency germination of interior spruce somatic embryos following partial drying at high relative humidity. Can J Bot 68:1086–1090

Rohr R, von Aderkas P, Bonga JM (1989) Ultrastructural changes in haploid embryoids of *Larix decidua* during early embryogenesis. Am J Bot 76:1460–1467

Schaeffer WI (1990) Terminology associated with cell, tissue and organ culture, molecular biology and molecular genetics. In Vitro Cell Dev Biol 26:97–101

Schenk RU, Hilderbrandt AL (1972) Medium and techniques for induction and growth of monocotyledonous and dicotyledonous plant cell cultures. Can J Bot 50:199–204

Schuller A, Reuther G, Geier T (1989) Somatic embryogenesis from seed explants of *Abies alba*. Plant Cell Tissue Org Cult 17:53–58

Simola LK, Santanen A (1990) Improvement of nutrient medium for growth and embryogenesis of megagametophyte and embryo callus lines of *Picea abies*. Physiol Plant 80:27–35

Smith DR, Singh AP, Wilton L (1985) Zygotic embryos of *Pinus radiata* in vivo and in vitro. In: Smith DR (ed) Abstr Int Conifer tissue culture working group, 12–16 Aug 1985. For Res Inst, N Z For Serv, Rotorua, New Zealand, p 21

Steward FC, Mapes MO, Mears K (1958) Growth and organized development of cultured cells. II. Organization in cultures grown from freely suspended cells. Am J Bot 45:705–708

Sondahl MR, Sharp WR (1977) High frequency induction of somatic embryos in cultured leaf explants of *Coffea arabica* L. Z Pflanzenphysiol 81:395–408

Tautorus TE, Bekkaoui F, Pilon M, Datla RSS, Crosby WL, Fowke LC, Dunstan DI (1989) Factors affecting transient gene expression in electroporated black spruce (*Picea mariana*) and jack pine (*Pinus banksiana*) protoplasts. Theor Appl Genet 78:531–536

Tautorus TE, Attree SM, Fowke LC, Dunstan DI (1990) Somatic embryogenesis from immature and mature zygotic embryos, and embryo regeneration from protoplasts in black spruce (*Picea mariana* Mill) Plant Sci 67:115–124

Thorpe TA (1988) In vitro somatic embryogenesis. ISI Atlas Sci Anim Plant Sci 1:81–88

Tisserat B, Esan EB, Murashige (1979) Somatic embryogenesis in angiosperms, In: Janick J (ed) Horticultural reviews, vol 1. AVI, Westport, Conn, pp 1–78

Tulecke W (1987) Somatic embryogenesis in woody perennials. In: Bonga JM, Durzan DJ (eds) Cell and tissue culture in forestry, vol 2, Nijhoff, Dordrecht, pp 61–91

Tremblay FM (1990) Somatic embryogenesis and plantlet regeneration from embryos isolated from stored seeds of *Picea glauca*. Can J Bot 68:236–242

Vasil IK (1987) Developing cell and tissue culture systems for the improvement of cereal and grass crops. J Plant Physiol 128:193–218

Verhagen SA, Wann SR (1989) Norway spruce somatic embryogenesis: high frequency initiation from light cultured mature embryos. Plant Cell Tissue Org Cult 16:103–111

von Aderkas P, Bonga JM (1988) Formation of haploid embryoids of *Larix decidua*: early embryogenesis. Am J Bot 75:690–700

von Aderkas P, Bonga JM, Nagmani R (1987) Promotion of embryogenesis in cultured megagametophytes of *Larix decidua*. Can J For Res 17:1293–1296

von Aderkas P, Klimaszewska K, Bonga JM (1990) Diploid and haploid embryogenesis in *Larix leptolepis*, *L. decidua*, and their reciprocal hybrids. Can J For Res 20:9–14

von Arnold S (1987) Improved efficiency of somatic embryogenesis in mature embryos of *Picea abies*. J Plant Physiol 128:233–244

von Arnold S, Eriksson T (1981) In vitro studies of adventitious shoot formation in *Pinus contorta*. Can J Bot 59:870–874

von Arnold S, Hakman I (1986) Effect of sucrose on initiation of embryogenic callus cultures from mature zygotic embryos of *Picea abies* (L) Karst (Norway spruce). J Plant Physiol 122:261–265

von Arnold S, Hakman I (1988) Regulation of somatic embryo development in *Picea abies* by abscisic acid (ABA). J Plant Physiol 132:164–169

von Arnold S, Woodward S (1988) Organogenesis and embryogenesis in mature zygotic embryos of *Picea sitchensis*. Tree Physiol 4:291–300

Walton DC (1980) Biochemistry and physiology of abscisic acid. Annu Rev Plant Physiol 31:453–489

Wann SR (1989) Somatic embryogenesis in woody species. In: Janick J (ed) Horticultural reviews, vol 10. Timber, Portland, Or, pp 153–181

Wann SR, Johnson MA, Noland TL, Carlson JA (1987) Biochemical differences between embryogenic and non-embryogenic callus of *Picea abies* (L.) Karst. Plant Cell Rep 6:39–42

Wann SR, Becwar MR, Nagmani R, Feirer RP, Johnson MA (1989) Biochemical differences between embryogenic and nonembryogenic calli of conifers. Trees 3:173–178

Webb DT, Webster F, Flinn BS, Roberts DR, Ellis DE (1989) Factors influencing the induction of embryogenic and caulogenic callus from embryos of *Picea glauca* and *Picea engelmannii*. Can J For Res 19:1303–1308

Webster FB, Roberts DR, McInnis SM, Sutton BCS (1990) Propagation of interior spruce by somatic embryogenesis. Can J Res 20:1759–1765

14 Molecular Structure of Forest Trees

H.B. Kriebel[1]

1 Introduction

Knowledge of genome structure is a prerequisite to the genetic engineering of trees and their subsequent clonal propagation, because some characterization of nuclear and organellar DNA must precede its manipulation. We need information on nuclear genome size and how much of the DNA is composed of highly repeated, moderately repeated, and near-unique or single-copy genes. We also need to locate and characterize genes of interest with regard to their molecular structure, repetition frequency, and base sequence. It is useful to know what genes have a highly conserved structure, making them similar in DNA sequence to genes in other organisms. Some genes are steady-state, i.e., expressed in all tissues and at all stages of development, and some are stage- or tissue-specific. Information is also required on the mechanisms regulating expression of genes in forest trees if we are to be successful in eliciting their expression in recombinant cells.

Even allowing for the expected rapid increase in research publication between the time of writing and the time of publication of this book, our current knowledge about the molecular structure of forest trees is very limited. Until the recent application of molecular techniques, knowledge of forest tree genetics was based on the results of conventional breeding and variation experiments. With at most two or three generations of breeding experience, we do not have the information on the genetics and cytogenetics of trees available to breeders of maize, soybean, or tobacco. Fortunately, however, it now appears that significant levels of base sequence homology often exist between many forest tree genes, even genes of gymnosperms, and those of other plants with better-known genetic structure. When a foreign gene sequence is hybridized to tree DNA, binding of the two will often locate the homologous tree gene in its DNA. Control sequences from other organisms will often function in tree cells, leading to expression of the transferred gene. Thus knowledge of the molecular structure of the DNA of other plants in relation to gene expression has accelerated the acquisition of information on the gene structure of trees and the application of biotechnology to forestry.

[1] Division of Forestry, The Ohio State University, Wooster, OH 44691, USA

Clonal Forestry I, Genetics and Biotechnology
Ed. by M.R. Ahuja and W.J. Libby

2 Nuclear Genome Structure and Organization

2.1 Genome Size

The first research on the molecular genetics of forest trees began in the middle 1960s, was concerned with the size of the nuclear genome, and was some of the earliest research on the DNA characteristics of higher plants. Measurements made on root tip squashes by Feulgen cytophotometry showed that gymnosperms have a very large amount of DNA in the nucleus, exceeding that of most other plants and animals, including humans (Miksche 1967). Intraspecific variation was also reported in *Picea* (Miksche 1968, 1971) and Douglas-fir [*Pseudotsuga menziesii* (Mirb.) Franco] (El Lakany and Sziklai 1971, 1973). DNA estimates made from nuclear volumes of 236 gymnosperms were reported by Price et al. (1974). All of these inter- and intraspecific measurements, although meaningful as rough approximations of relative genome size, were later shown to be imprecise because of the absence of a uniform calibration standard such as red blood cells (erythrocytes), which are very constant in size (Dhillon et al. 1977). Revised estimates of haploid DNA amount were based on microspectrophotometric measurement using chicken erythrocytes, or in one case dormant pine embryos, as an internal standard. The estimates for 17 species of conifers varied from 17 picograms (pg) per nucleus in *Picea glauca* (Moench [Voss]) to 41 pg per nucleus in *Pinus resinosa* Ait. (Dhillon 1980, 1987, 1988;

Table 1. Haploid genome size of forest trees in base (nucleotide) pairs[a]

Species	Genome size (bp)[b]
Gymnosperms	
Tsuga canadensis	12×10^9
Picea abies	14×10^9
Picea mariana	10×10^9
Picea glauca	8×10^9
Larix decidua	9×10^9
Larix laricina	8×10^9
Pinus strobus	18×10^9
Pinus resinosa	19×10^9
Pinus sylvestris	13×10^9
Pinus taeda	10×10^9
Pinus rigida	8×10^9
Pinus contorta	11×10^9
Pinus radiata	10×10^9
Thuja occidentalis	6×10^9
Angiosperms	
Populus deltoides	1×10^9
Liriodendron tulipifera	3×10^9
Fraxinus americana	1×10^9

[a]Values converted from picograms (pg) in Dhillon (1987), except *Liriodendron tulipifera* from Merkle et al. (1988).
[b]Conversion basis: 1 base pair (bp) = 660 daltons and 1 dalton = 1.66024×10^{-12} pg; therefore 1 pg = 0.9126×10^9 bp.

Rake et al. 1980; Miksche and Dhillon 1981; Berlyn et al. 1987). The reported haploid values for angiospermous forest trees are much lower (Schaefer and Miksche 1977; Dhillon 1987). Table 1 summarizes published reports of haploid nuclear genome size of forest trees.

2.2 Intraspecific Variation in Genome Size

A positive correlation between DNA content and provenance latitude was reported in two spruces (Mergen and Thielges 1967: Miksche 1971). However, there were no internal standards. No such correlation was found in a later study of one of the spruces, *P. glauca*, by Teoh and Reese (1976), who also found no significant differences in DNA content between or within provenances of *Picea engelmannii* Parry or of *Pinus contorta* Dougl. Dhillon et al. (1978) used refined photometric technique and found the haploid DNA value of *Pinus rigida* Mill, to be uniform over a wide range of latitude, but examination of three widely dispersed varieties of *Pinus caribaea* Mor. showed that genome size increased with increase in northern latitude of the variety (Berlyn et al. 1987). Thus there are reports of both presence and absence of a correlation between DNA content and latitude, even within the genus *Pinus*.

2.3 Genetic Stability of the Genome

The stability of nuclear DNA, in vivo during the life of a tree and in vitro over a period of repeated subculturing, is a major concern in applications of biotechnology to tree improvement where tissue culture is involved. There is very little current information on the nature of DNA changes during aging of trees, although qualitative and quantitative changes, including increased methylation, have been reported in other organisms (Kriebel and Whitmore unpubl. review 1983,). Eighteen- to 35-year-old Norway spruce trees have been found to have a higher level of methylation of cytosine nucleotides in extracted DNA than seeds and 2- and 6-year-old trees (Westcott 1988).

Nuclear DNA content may change over time in vitro (Renfroe and Berlyn 1985). Conversely, high levels of DNA stability of protoplasts, cell cultures, and regenerants have been reported in several hardwoods as well as conifers. Stability was measured by ploidy level in conifers (David et al. 1986; Franklin et al. 1985, 1986;), DNA content per cell in *Pinus* (Berlyn et al. 1987), and DNA restriction patterns in hardwoods (Giasson and Lalonde 1987; Merkle et al. 1988). Giasson and Lalonde restricted genomic DNA of three species of Betulaceae with ten endonucleases, and compared banding patterns in cell cultures by densitometric scanning, after staining with ethidium bromide. Results suggested genetic stability. Merkle et al. used a cloned repeated sequence of yellow poplar (*Liriodendron tulipifera* L.), to study restriction fragment length polymorphisms in nuclear DNA. In cultures, embryogenic sublines and regenerated plantlets from a given line were stable, although polymorphisms could be detected among lines and among plantlets from different lines.

2.4 Whole Genome Characterization

The discovery of repeated sequences in the DNA of higher organisms led to the development of useful technology for the analysis of proportions of single-copy and multiple-copy sequences and the abundance of sequences in major DNA fractions identified by the analysis. The technique, known as DNA reassociation kinetics or "Cot analysis" (Britten and Kohne 1968), was first applied to trees by Miksche and Hotta (1973). They found that *Pinus resinosa*, *Pinus banksiana* Lamb., and *Picea glauca*, all previously found to have a high nuclear DNA content relative to angiospermous plants, also contained a very high proportion of repeated sequences. Subsequent analysis indicated that 20–30% of the DNA of these conifers and of *Pinus lambertiana* Dougl. consisted of unique or near-unique sequences (Rake et al. 1980; Miksche and Dhillon 1981).

Rake et al. (1980) carried analysis of the three pines and *P. glauca* farther by attempting to determine the proportion of total DNA in each of what appeared to be five fractions of the Cot curve. They fitted straight lines to these fractions and estimated that 4–12% of the genome was composed of highly repeated sequences, 61–75% of intermediate repetition frequency, and 21–32% "unique", i.e., single-copy. Kriebel (1985) analyzed the *Pinus strobus* L. genome using the least-squares computer program of Pearson et al. (1977). About 24% of the DNA was in the "single-copy" fraction. However, the *P. strobus* analysis led to the conclusion that a large conifer genome can best be described as consisting of a continuity of repetition frequencies ranging from highly repetitious to unique; description in terms of major frequency classes is a simplification of a complex pattern of sequence organization.

As in previous studies of conifer DNA reassociation, the kinetic estimate of genome size in *P. strobus* was very different from the cytophotometric estimate. The estimate from the Cot analysis was only 30% of Miksche and Dhillon's genome size estimate of 18×10^9 base pairs (Table 1). This discrepancy, which has also been reported in other plants, might be explained by the hypothesis of Thompson and Murray (1981). They speculated that most of the DNA that in large plant and animal genomes behaves as single-copy DNA under generally used kinetic conditions is actually repeated-sequence DNA that has become highly diverged in the course of evolution. These repeated sequences continue to diverge until they cease to react with each other and reassociate with the kinetics of single-copy sequences. In gymnosperms, the proportion of genes that are truly unique may be very small, possibly less than 1% of total DNA (Kriebel 1985). It seems probable that "single-copy" sequences are, in fact, present in two to several copies. We still know little about the structure and function of this fraction of tree DNA.

At the time of writing, only one analysis had been published of the organization of DNA in an angiospermous forest tree. From reassociation kinetics, Merkle et al. (1988) estimated that 22% of the DNA sequences in the genome of yellowpoplar (*Liriodendron tulipifera* L.) were highly repeated and 68% were single-copy sequences.

3 Nuclear genes

In forest trees, the initial focus in gene identification and characterization has been on repeated genes, which can be grouped into those exhibiting "dosage repetition" and those with "variant repetition" (Long and Dawid 1980). Genes with dosage repetition are those with RNAs present in every cell in high concentrations, for which repetition is necessary to produce the required amounts of protein products. They are principally genes synthesizing ribosomal RNAs, although they also include histone genes. Genes exhibiting variant repetition can produce large amounts of protein product from a single cell through the accumulation of mRNA. They are not identical but related. In conifers, many of them may be in the previously described low-frequency fraction of the DNA.

3.1 Ribosomal RNA Genes

Gymnosperms, in addition to having large genomes, also have large numbers of ribosomal RNA genes (Hotta and Miksche 1974). Estimates of 25S-18S rDNA haploid genome copy number in eight gymnosperm species of six genera varied from 1250 to 13400 (Ingle et al. 1975). Copy number can be variable within a species. In a densitometric analysis of Douglas-fir filter hybridization, copy number tended to increase with latitude, elevation, and longitude, i.e., to be higher in cold climates. There was more polymorphism in ribosomal RNA gene number within than among populations (Strauss and Tsai 1988). In pines (Hotta and Miksche 1974), as in other eukaryotes (Long and Dawid 1980), the rRNA molecules are encoded on a single transcription unit. In plant genomes, the 25S-18S repeat unit includes a highly conserved coding sequence and a non-transcribed spacer unit, organized in tandem arrays at one or a few chromosomal loci (Flavell 1986). DNA spacers, unlike coding sequences, diverge rapidly (Federoff 1979) and would be expected to show little homology between angiosperms and pines.

The high level of ribosomal DNA coding sequence homology in higher plants has made it possible to locate pine ribosomal DNA sequences with flax, soybean, and maize probes. In *Pinus radiata* D. Don, the entire repeat, at 27 kilobases (Fig. 1), is more than twice as long as the longest angiosperm rDNA

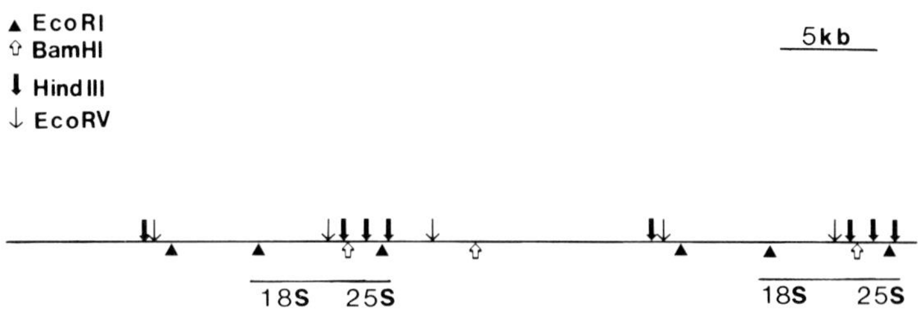

Fig. 1. Restriction map of a major class of ribosomal genes in *Pinus radiata*. A number of polymorphisms exist within this class of genes but the variable sites have not been indicated. (After Cullis et al. 1988)

unit so far characterized (Cullis et al. 1988). Similarly, the repeat unit in *Pinus taeda* L. appears to be at least 22 to 24 kb long; 20 kb or more of this is estimated to be spacer (Sederoff et al. 1987). The coding region of the 5S DNA of *P. radiata* had an 83 to 91% homology with the ten angiosperm species with which it was compared.

The length of the 5S repeat unit of *P. strobus* and of nine species of southern pines is about 500 base pairs (Sederoff et al. 1987). In *P. radiata*, the 5S RNA gene was sequenced and found to have a 120 base pair coding region and a total length of 524 base pairs. Both the 18S-25S and 5S ribosomal RNA genes appear to be located at a larger number of sites within the genome than in angiosperms (Cullis et al. 1988).

3.2 Repeated Genes of Unknown Function

Only a minority of the repeated DNA sequences in eukaryotic genomes can be associated with defined genetic functions (see Thompson and Murray 1981 for a discussion of the significance of repetitive DNA). Research has suggested that some classes of ubiquitous, dispersed repeats in higher organisms may be involved in DNA amplification, in the initiation of DNA replication, in the regulation of transcription, and in the transposition of larger fragments leading to genome changes associated with species evolution (Georgiev et al. 1983; Hastie 1985). We have very little information at this point on the structure and function of the large amount of middle repetitive DNA in trees, especially conifers.

Gustafsson and Sitbon (1986) cloned a fragment of *Pinus sylvestris* L. DNA that hybridization by Southern blot showed to be part of a repeated gene. The cloned probe hybridized identically to four other species of pine, including a soft pine, and to *Larix sibirica* Ledeb. but not to pea DNA. The results showed that the cloned DNA was highly conserved within conifers with respect to both fragment length and overall base composition. Sederoff et al. (1987) cloned and screened *Pinus taeda* DNA by a procedure that selects for highly repeated nuclear sequences. They also isolated two clones of *Pinus lambertiana* repeated sequence DNA by a different procedure. These two clones were partially sequenced. Only 7.6% of the nucleotides differed between the two clones. C. Cullis (pers. commun.) isolated about 20 clones of dispersed highly repetitive sequences of *Pinus radiata* from a cDNA library. Each clone has a unique pattern when hybridized to repeated sequence DNA; the pattern is conserved across the gene pool.

Highly repeated sequences have been isolated from at least one angiospermous forest tree as well as from gymnosperms. Merkle et al. (1988) used DNA reassociation kinetics to locate and isolate the sequences from the nuclear genome before restricting and cloning the fragments. The genetic stability of these repeated sequences was discussed in Section 2.3.

3.3 Genes of Variant Repetition and Unique Genes

Only a few nuclear genes of forest trees have been isolated, cloned, and characterized, although the number is expanding rapidly with the increasing availability of hybridizable cloned probes from higher plants and other organisms. Multigene families have received most attention because they are easier to isolate than single- or low-copy genes. Among them are the genes coding for actin, alcohol dehydrogenase, the small subunit of ribulose-1,5-bisphosphate carboxylase, the light-harvesting chlorophyll a/b binding protein, lectins, ubiquitin, and genes transcribing messengers for stage-specific proteins of as yet unknown function.

Plant actins are located in subcortical filament bundles. They appear to play a diverse role in cell functioning through their effects on force generation affecting cytoplasmic streaming and on the movement of organelles (Williamson 1980). Actin genes are ubiquitous and highly conserved. An actin gene of *Pinus contorta* was the first forest tree gene to be isolated and cloned. The carboxy-terminal portion of one gene was isolated with a soybean actin probe, cloned and sequenced, and subsequently the partial nucleotide sequence of a second clone was determined. Both pine actin amino acid sequences were highly homologous to the soybean actin sequence SAc3 (Kenny 1985, 1988).

Alcohol dehydrogenase (*adh*) and aldolase (*ald*) genes of *Pinus radiata* were isolated by hybridization of *Zea mays* L. probes to genomic pine DNA under low stringency conditions. The hybridization results suggested that the radiata pine *ald* gene(s) shares approximately 70% homology with the maize *ald* coding sequence, that *adh* coding sequences are more highly conserved between maize and radiata pine than *adh* noncoding sequences, and that *ald* genomic sequences are more highly conserved than *adh* genomic sequences (Kinlaw et al. 1988; Harry et al. 1988).

The small subunit of ribulose-1,5-bisphosphate carboxylase (ssu-*rbc*S) and the chlorophyll a/b (*cab*) binding protein are prominent light-induced gene products controlled by the photoreceptor phytochrome (Kuhlemeier et al. 1987). Two cDNA clones encoding these two proteins were isolated from *Pinus thunbergii* Parl. and used to estimate the transcription of each gene in seedlings grown in dark and light conditions. Northern blotting showed that in pine seedlings these two genes are expressed in the dark as well as in the light (Yamamoto et al. 1988). In an initial study of the structure of light-induced gene expression in *Larix laricina* (DuRoi) K. Koch, clones of ssu-*rbc*S were isolated from cDNA libraries with a maize probe. A portion of the larch gene that was sequenced showed strong conservation with the homologous sequence from other plants. The transcript produced was expressed in foliage but not in roots (Hutchison et al. 1988). Jansson and Gustafsson (1988) isolated several clones of a *cab* gene from Scots pine and confirmed the identity of one clone by sequencing. The DNA sequence had a homology of 83% to 97% with sequenced *cab* genes of several angiosperms.

Plant lectins are multigene families of sugar-binding proteins that are found mainly in seeds but also in other organs. They appear to function in intercellular and intracellular recognition phenomena and may have other functions (Lis and Sharon 1981). F.W. Whitmore (pers. commun.) found a lectin in ovules of *Pinus strobus* that is at its highest activity around the time of fertilization and early embryo development, decreasing rapidly as the ovule matures. Although most lectins are not stage-specific, an endogenous lectin has been found in *Xenopus laevis* that may play a role in early embryogenesis (Outenreath et al. 1988). The pine lectin has a molecular weight of around 20 kDa and coincides in periodicity with an mRNA for a 23 kDa protein seen in translation experiments. It is planned to test this coincidence by immunoassay, determine the amino-terminal sequence of the lectin, and construct an oligonucleotide probe for experimental hybridization.

Ubiquitin may be the most highly conserved of all proteins (Ecker et al. 1987). It is stress-induced, e.g., by heat, is involved in protein degradation, and, in yeast, has a role in DNA repair (Ferguson et al. 1988; Vierstra 1987; Jentsch et al. 1987). Ubiquitin genes are ogranized as tandem repeats (Sharp and Li 1987). Ubiquitin is present in gymnosperms. Several genomic clones have been isolated in *Pinus radiata* with an *Arabidopsis* probe (C Cullis pers. commun.).

Stage-specific or developmentally regulated genes are little known in forest trees. Their role in plant embryogenesis is, however, well established (Dure 1985). For this reason, knowledge of these genes will be particularly important in clonal forestry when somatic embryogenesis is the propagation method. In *Pinus strobus*, in vitro translation of mRNA has shown that a messenger coding for a protein of about 23 kDa appears as a prominent component about the time of fertilization, remains present for several days, then disappears (Fig. 2), suggesting that it may be the result of gene expression activated by fertilization.

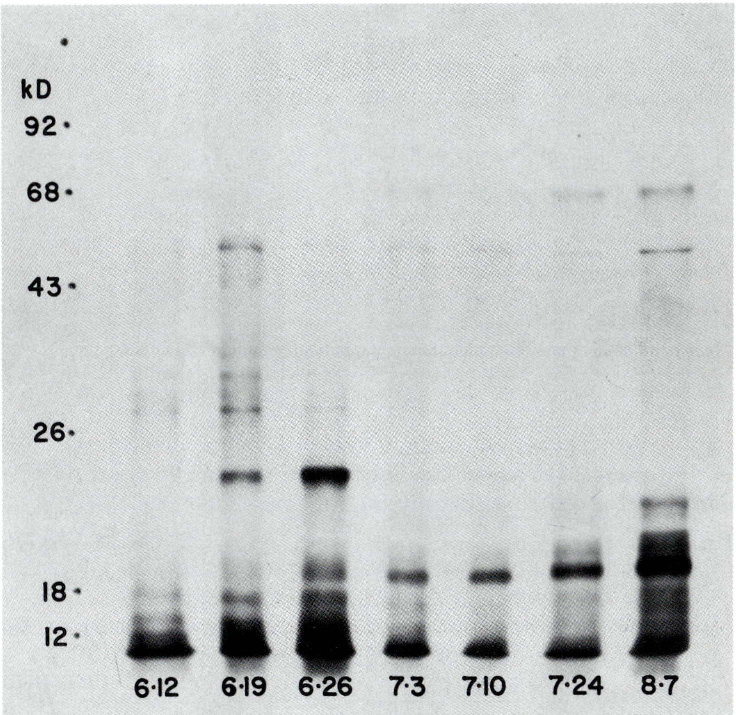

Fig. 2. Example of developmentally regulated gene expression. Fluorograph of ^{35}S-labeled translation products from mRNA extracted from *Pinus strobus* ovules collected at weekly intervals during embryogenesis (1984 dates indicated), showing the presence of a prominent 23 kDa mRNA during a short period after fertilization. (Whitmore and Kriebel 1987)

The translation product is located in the integument and nucellar tissue and does not appear to be a storage protein because of rapid turnover (Whitmore and Kriebel 1987). A cDNA library that was constructed from the transient mRNA is being screened with differential probes for presence of the sequences expressing the 23 kDa protein (Rice et al. 1988).

There is experimental evidence of the expression of a developmentally regulated gene in pines during seed germination as well as during embryogenesis. The gene expression is a response to light-induced translation of a performed mRNA. The presence of the mRNA was demonstrated in dry seeds and dark-imbibed seeds by in vitro translation, but translation in vivo occurred only after exposure of imbibed seeds to light (Yamamoto 1982).

4 Organelle Genomes

4.1 The Chloroplast Genome

The chloroplast DNA (cpDNA) molecule of most of the higher plants is about 150 kilobases (kb) in length. It is circular, highly conserved within a species, and in a majority of species studied contains a large (20–28 kb) repeat, the segments of which are separated by a large and a small single-copy DNA region. At least 30% of the cpDNA sequences are shared by numerous unrelated plant species (Whitfield and Bottomley 1983). The identification of plastid genes in numerous plants and the complete sequencing of the cpDNA of the liverwort *Marchantia polymorpha* (Ohyama et al. 1986) and tobacco (Shinozaki et al. 1986) made it possible to estimate that the genome codes for about 123 genes (Mullet 1988).

4.2 Inheritance and Polymorphism of Conifer cpDNA

In angiosperms, chloroplasts of most species have maternal inheritance, although a few species have been found to have biparental inheritance. None have been found to have strictly paternal inheritance (see short review by Neale and Sederoff 1988). However, a number of studies have shown that this does not hold for conifers.

Analysis of restriction fragment length polymorphisms in Douglas-fir and coast redwood (*Sequoia sempervirens* D. Don Endl.) showed that inheritance is paternal. Most restriction enzyme-probe combinations were invariant, but small percentages of insertion/deletion variants were found in both species (Neale et al. 1986; Neale and Sederoff 1988).

Wagner et al. (1987) took sample trees of lodgepole pine and jack pine distributed throughout their natural ranges and assayed them for Sal I and Sst I restriction fragment banding patterns. Allopatric populations of the two species had distinct banding patterns. At least two distinct Sal I and 13 distinct Sst restriction fragment banding patterns occur in these closely related species. Chloroplast inheritance was paternal in controlled matings between lodgepole pine (♀) and jack pine (♂). A study was also made of chloroplast DNA polymorphism in the sympatric region between the two species by sampling several hundred trees in each of two populations. In approximately 4.5% of the individuals, cpDNA variants were found that did not occur in trees in either of the allopatric regions (Wagner et al. 1988).

Paternal inheritance appears to be general in *Larix*, although hybrids from controlled crosses between *L. decidua* Mill. and *L. leptolepis* Sieb. & Zucc. were variable in restriction fragment length patterns. One hybrid had maternal chloroplast DNA patterns, but two other hybrids had mixed Bam-HI patterns, suggesting recombination between maternal and paternal chloroplast DNA (Szmidt et al. 1987).

There is also evidence of paternal inheritance of cpDNA in *Picea*. In an analysis of reciprocal crosses of *P. pungens* and *P. glauca*, all progeny exhibited the restriction pattern of the pollen parent (Stine and Keathley 1988).

4.3 Characterization of Forest Tree Chloroplast DNA

Sitbon and Gustafsson (1988) cloned a 1600 base pair fragment of Scots pine cpDNA that banded prominently on gels. The cloned DNA hybridized specifically to cpDNA of members of the Pinaceae and was highly conserved within that family when tested against both angiosperms and gymnosperms. It was, in fact, more highly conserved than a heterologous chloroplast psbA gene or a plant nuclear 18S rRNA gene hybridized to the same filters. Although the cloned sequence shows a homology to tobacco chloroplast DNA, its biological function, if any, is presently unknown.

The chloroplast genomes of angiosperms characteristically include a pair of large inverted repeats containing genes for rRNA, tRNA and ribosomal proteins (Whitfield and Bottomley 1983). In the gymnosperms, the repeat pair occurs in *Ginkgo biloba* L. (Palmer and Stein 1986), but Douglas-fir, radiata pine, and lodgepole pine lack one of the repeats. Restriction site maps were constructed in Douglas-fir and radiata pine by overlap hybridization (Fig. 3). With the exception of the single 40 to 50 kb inversion, the linear arrangement of 32 genes appeared to be identical in the two species and nucleotide sequence differentiation was estimated to be only 3.8% (Strauss et al. 1988). Lodgepole pine cpDNA, in addition to having a single inversion, also differed from other higher plants in having a duplication of the psbA gene (Fig. 4). The psbA gene encodes the D-1 protein, which is thought to be an integral part of the photosystem II (PS II) reaction center. The pine psbA genes were partially sequenced; homology to the spinach psbA gene probe was 87% at the nucleotide level (Lidholm et al. 1988).

In three pines (Scots, radiata, and lodgepole) and Douglas-fir, chloroplast genome size has been estimated to be approximately 120 kb, i.e., about 30 kb less than that of plant genomes containing a pair of inverted repeats (Szmidt et al. 1986; Strauss et al. 1988).

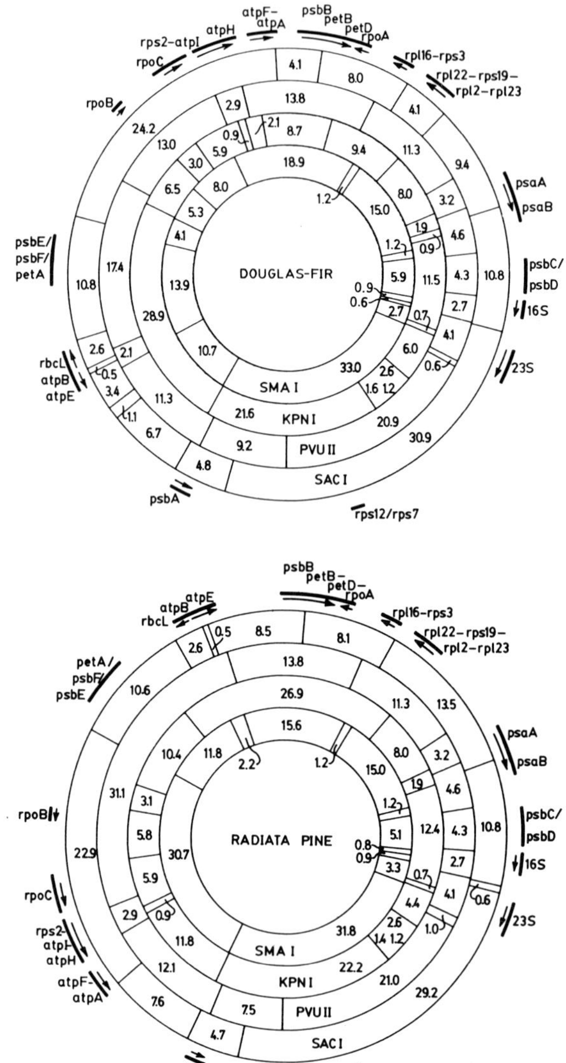

Fig. 3. Circular restriction site and gene maps of Douglas-fir and radiata pine chloroplast genomes. The order of genes separated by *dashes* has been inferred from their arrangement in tobacco. The linear order of genes separated by *slashes* was also determined by analogy to tobacco, but the orientation of the gene cluster with respect to the remainder of the genome could not be determined. Transcriptional directions of other genes and the location and transcriptional directions of 16S and 23S rRNA have been assigned by analogy to tobacco. *Dark bands* represent the lengths of gene probes used in hybridizations, rather than the lengths of the genes themselves, except for 16S and 23S rRNA. (Strauss et al. 1988)

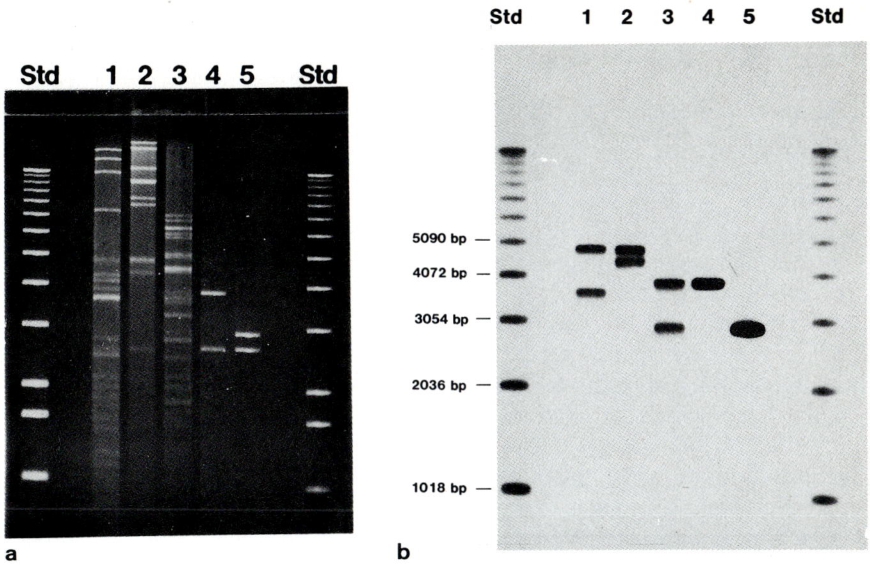

Fig. 4a, b. Duplication of the *psbA* gene in the chloroplast genome of *Pinus radiata*. Agarose gel (**a**) and southern blot (**b**), probed with a spinach psbA probe. The cpDNA restriction digests are: *Eco*RI (*lane 1*), *Sac*I (*Lane 2*), and *Bam*HI (*lane 3*). *Lanes 4* and *5* are *psb*A-positive clones of *P. radiata* cpDNA, pPC121 and pPC932 respectively, cut with *Bam*HI. (Lidholm et al. 1988)

4.4 The Mitochondrial Genome

The structure and organization of the mitochondrial genome of higher plants was reviewed by Lonsdale (1984). Surveys have only recently begun to determine the extent and nature of mitochondrial DNA (mtDNA) variation in conifers, using cloned gene probes from maize. Restriction fragment length polymorphisms have been detected for loblolly pine, Douglas-fir, and redwood. Maternal inheritance, typical of angiosperms, was found in loblolly pine progenies of full-sib crosses cut with EcoRI and hybridized with the 18-5S rRNA mtDNA clone from maize (Neale and Sederoff 1988). Maternal inheritance is not a consistent pattern for gymnosperm mitochondrial genomes, however. Ultrastructure analysis suggests that inheritance may be paternal in most or all of the Cupressaceae (Chesnoy and Thomas 1971).

5 Conclusions

Except for early investigations of conifer genomes by DNA reassociation, some of it before the advent of restriction enzyme technology, there was no research on the molecular genetics of forest trees until work began in Ohio in the early 1980s. In the 3 years prior to the time of writing (August 1988), many new projects have been started. Initial study of the nuclear genome has mostly focused on highly repeated genes, especially the ribosomal RNA genes and multigene families coding for abundant proteins. This has been possible because of the increasing availability of heterologous cloned probes constructed from the complementary DNA of other higher organisms, especially, but not entirely, angiospermous plants. It is now apparent that many of these cloned gene sequences are moderately to highly conserved and can, therefore, be used to locate genes in trees. As more tree genes are isolated and mapped, including low-frequency and single-copy genes, we will rapidly expand our knowledge of the structure of tree genomes.

In 1983, Rivin et al. described techniques for evaluation of genomic variability at the nucleic acid level, including discrimination between inbred lines of maize on the basis of restriction fragment length polymorphisms. The technique is now being widely applied to the study of variation in forest trees. Its application to chloroplast DNA has attracted more initial interest than probing the more complex nuclear genome, especially in conifers. We can expect a rapid expansion of our knowledge of the nuclear genome, however, because – if for no other reason – we are very nearly starting from a zero baseline. Genome mapping will be a slow process, facilitated by the accumulation of marker genes. A very important and intriguing aspect of research on the nuclear genome will be the investigation of the structure and function of developmentally regulated genes controlling life processes vital to tree embryogenesis, growth and survival and, therefore, of great importance to forestry. Our ability to manipulate and transfer these genes would be a major contribution to clonal forestry.

6 Summary

In the most intensive application of clonal forestry, nuclear and organellar DNA will be restructured by genetic engineering. This will require more knowledge of tree genomes than we have at this time. Most of our current knowledge of the molecular structure of forest trees is based on conifers, which are unique in having a large nuclear genome with a correspondingly large complement of repeated sequence DNA, much with unknown function. There are changes in the quantity and structure of nuclear DNA over space and time, although there is also evidence of the stability of coniferous DNA in vitro. The highly repeated ribosomal RNA genes are similar to those in other plants, although the rDNA repeat in at least two pines is twice as long as that found in any angiosperm to date. A few repetitive sequences of unknown function have been cloned and found to be highly conserved. Several important multigene families have been isolated, cloned and to some extent characterized. In conifers, the chloroplast genome has some distinct characteristics and is paternally inherited. Analysis of chloroplast DNA restriction fragment length polymorphism has become a standard procedure for the study of variation in forest trees.

References

Berlyn, GP, Anuruo AO, Beck RC, Cheng J (1987) DNA content polymorphism and tissue culture regeneration in Caribbean pine. Can J Bot 65: 954–961

Britten RJ, Kohne DE (1968) Repeated sequences in DNA. Science 161: 529–540

Chesnoy L, Thomas MJ (1971) Electron microscopy studies on gametogenesis and fertilization in gymnosperms. Phytomorphology 21: 50–63

Cullis CA, Creissen GP, Gorman SW, Teasdale RD (1988) The 25S, 18S, and 5S ribosomal RNA genes from *Pinus radiata* D. Don. In: Cheliak WM, Yapa AC (eds) Molecular genetics of forest trees. Proc 2nd Worksh IUFRO Working party S2. 04-06. Can For Serv PNFI Inf Rep PI-X-80, pp 34–40

David H, de Boucaud, M-T, Gaultier, J-M, David A (1986) Sustained division of protoplast-derived cells from primary leaves of *Pinus pinaster*, factors affecting growth and change in nuclear DNA content. Tree Physiol 1: 21–30

Dhillon SS (1980) Nuclear volume, chromosome size, and DNA content in three species of *Pinus*. Cytologia (Tokyo) 45: 555–560

Dhillon SS (1987) DNA in tree species. In: Bonga JM, Durzan DJ (eds) Cell and tissue culture in forestry, vol 1. Nijhoff, Dordrecht, pp 298–313

Dhillon SS (1988) DNA analysis during growth and development. In: Hanover JW, Keathley DE (eds) Genetic manipulation of woody plants. Plenum, New York, pp 265–274

Dhillon SS, Berlyn GP, Miksche JP (1977) Requirement of an internal standard for microspectrophotometric measurements of DNA. Am J Bot 64: 117–121

Dhillon SS, Berlyn GP, Miksche JP (1978) Nuclear DNA content in populations of *Pinus rigida*. Am J Bot 65: 192–196

Dhillon SS, Miksche JP, Cecich, RA (1984) DNA changes in senescing leaves of *Populus deltoides*. Suppl Plant Physiol (Bethesda) 75: 120

Dure L III (1985) Embryogenesis and gene expression during seed formation. In: Miflin BJ (ed) Oxford surveys of plant molecular and cell biology, vol 2. Oxford Univ. Press New York, pp 179–197

Ecker DJ, Butt TR, Marsh J, Khan MI, Sternberg EJ, Margolis N, Monia BP, Jonnalgadda S, Weber PL (1987) Ubiquitin nature's most conserved protein as a model for protein engineering. In: Oxender DL (ed) Protein structure, folding and design. Liss, New York, pp 119–130 (DuPont-UCLA Symp Molecular and cellular biology, New Ser, vol 69)

El Lakany MH, Sziklai O (1971) Intraspecific variation in nuclear characteristics of Douglas-fir. In: Proc NW Sci Assoc Univ Idaho, Moscow, pp 363–378

El Lakany MH, Sziklai O (1973) Further investigations of intraspecific variation in DNA contents of Douglas-fir, *Pseudotsuga menziesii* (Mirb.) Franco. Egypt J. Genet Cytol 2: 345–354

Federoff NV (1979) On spacers. Cell 16: 697–710

Ferguson D, Guikema J, Paulsen G (1988) Ubiquitin-protein conjugation in heat-stressed roots of wheat. Suppl Plant Physiol (Bethesda) 86: 49

Flavell (1986) The structure and control of expression of ribosomal RNA genes. In: Miflin BJ (ed) Oxford surveys of plant molecular and cell biology, vol 3. Oxford Univ. Press, New York, pp 251–274

Franklin CI, Mott RL, Miksche JP (1985) Stability of nuclear DNA content in long-term callus cultures of loblolly pine. In: Henke RR, Hughes KW, Constantin MJ, Hollaender, A (eds) Tissue culture in forestry and agriculture. Plenum., New York, 319 pp

Franklin CI, Mott RL, Vuke TM (1986) Stable ploidy levels in different parts of loblolly pine seedlings and callus cultures. Suppl Plant Physiol (Bethesda) 80: 37

Georgiev GP, Kramerov DA, Ryskov AP, Skryabin KG, Lukanidin EM (1983) Dispersed repetitive sequences in eukaryotic genomes and their possible biological significance. Cold Spring Harbor Symp Quant Biol 47: 1109–1121

Giasson L, Lalonde M (1987) Restriction pattern analysis of deoxyribonucleic acid isolated from callus and cell suspension of actinorhizal and non-actinorhyzal Betulaceae. Physiol Plant 70: 304–310

Gustafsson P, Sitbon F (1986) Studies on genetic differences between pine species by modern DNA technology: the nuclear genome. In: Lindgren D (ed) Provenances and forest tree breeding for high latitudes. Proc Frans Kempe Symp, Swed Univ Agric Sci Dep For Genet Plant Physiol Rep 6, Umea, pp 253–268

Harry DE, Kinlaw CS, Sederoff RR (1988) The anaerobic stress response and its use for studying gene expression in conifers. In: Hanover, JW, Keathley DE (eds) Genetic manipulation of woody plants. Plenum, New York, pp 275–290

Hastie N (1985) Middle repetitive DNA: amplification, homogeneity and mobility. Trends Genet 1:37

Hotta Y, Miksche JP (1974) Ribosomal genes in four coniferous species. Cell Differ 2:299–305

Hutchison KW, Singer PB, Greenwood MS (1988) Molecular genetic analysis of development and maturation in larch. In: Cheliak WM, Yapa AC (1988) Molecular genetics of forest trees. Proc 2nd Worksh IUFRO Working Party S2. 04-06. Can For Serv PNFI Inf Rep PI-X-80, pp 26–33

Ingle J, Timmis JN, Sinclair J (1975) The relationship between satellite DNA, ribosomal RNA redundancy, and genome size in plants. Plant Physiol (Bethesda) 55:496–501

Jansson S, Gustafsson P (1988) Cloning of a cDNA encoding the light-harvesting chlorophyll a/b binding protein (CAB) from Scots pine (*Pinus sylvestris* L.). In: Cheliak WM, Yapa AC (1988) Molecular genetics of forest trees. Proc 2nd Worksh IUFRO Working Party S2. 04–06. Can For Serv PNFI Inf Rep PI-X—80, pp 19–25

Jentsch S, McGrath JP, Varshavsky A (1987) The yeast DNA repair gene *rad*-6 encodes a ubiquitin-conjugating enzyme. Nature (London) 329:131–134

Kenny JR (1985) Isolation, cloning and sequencing of the actin gene(s) in *Pinus contorta* var. *latifolia* (Engelm.). In: Kriebel HB (ed) Molecular genetics of forest trees-research techniques and strategies. Program Abstr 1st Meet IUFRO Working Party S2.04–06, Avon Lake, Ohio, p 5

Kenny J (1988) Partial nucleotide sequence of an entire pine actin gene. In: Hanover JW, Keathley DE (eds) Genetic manipulation of woody plants. Plenum, New York, p 473

Kinlaw CS, Harry DE, Sleeter DD, Sederoff RR (1988) Use of heterologous probes to isolate and characterize conifer genes. In: Cheliak WM, Yapa AC (eds) Molecular genetics of forest trees. Proc 2nd Worksh IUFRO Working Party S2. 04–06. Can For Serv PNFI Inf Rep PI-X-80, pp 9–18

Kriebel HB (1985) DNA sequence components of the *Pinus strobus* nuclear genome. Can J For Res 15:1–4

Kuhlemeier C, Green P, Chua N-H (1987) Regulation of gene expression in higher plants. Annu Rev Plant Physiol 38: 221–257

Lidholm J, Szmidt A, Gustafsson P (1988) Lack of one rDNA repeat and duplication of psbA in the chloroplast genome of lodgepole pine (*Pinus contorta*) Dougl.). In: Cheliak WM, Yapa AC (eds) Molecular genetics of forest trees. Proc. 2nd Worksh IUFRO Working Party S2. 04–06. Can For Serv PNFI Inf Rep PI-X-80, pp 67–74

Lis H, Sharon N (1981) Lectins in higher plants. Biochem Plants 6:371–447

Long, EO, Dawid IB (1980) Repeated genes in eukaryotes. Annu Rev Plant Biochem 49:727–764

Lonsdale DM (1984) A review of the structure and organization of the mitochondrial genome of higher plants. Plant Mol Biol 3:201–206

Mergen F, Thielges BA (1967) Intraspecific variation in nuclear volume in four conifers. Evolution 21:720–724

Merkle SA, Chou PL, Sommer HE (1988) Stability of highly repeated sequences in the DNA of embryogenic cultures of yellow poplar. In: Cheliak WM, Yapa AC (eds) Molecular genetics of forest trees. Proc 2nd Worksh IUFRO Working Party S2. 04–06. Can For Serv PNFI Inf Rep PI-X-80, pp 85–88

Miksche JP (1967) Variation in DNA content of several gymnosperms. Can J Genet Cytol 9:717–722

Miksche JP (1968) Quantitative study of intraspecific variation of DNA per cell in *Picea glauca* and *Pinus banksiana*. Can J Genet Cytol 10:590–600

Miksche JP (1971) Intraspecific variation of DNA per cell between *Picea sitchensis* (Bong.) Carr. provenances. Chromosoma 32:343–352

Miksche JP, Dhillon SS (1981) Nucleic acid research techniques in forest genetics. In: Khosla PK (ed) Advances in forest genetics. Ambika, New Delhi, pp 329–371

Miksche JP, Hotta Y (1973) DNA base composition and repetitious DNA in several conifers. Chromosoma 41:29–36

Mullet JE (1988) Chloroplast development and gene expression. Annu Rev Plant Physiol Plant Mol Biol 39:475–502

Neale DB, Sederoff RR (1988) Inheritance and evolution of conifer organelle genomes. In: Hanover JW, Keathley DE (eds) Genetic manipulation of woody plants. Plenum, New York, pp 251–264

Neale DB, Wheeler NC, Allard RW (1986) Paternal inheritance of chloroplast DNA in Douglas-fir. Can J For Res 16:1152–1154

Ohyama K, Fukuzawa H, Kohchi T, Shirai H, Sano T, Sano S, Umesono K, Shiki Y, Takeuchi M, Chang Z, Aota S-I, Inokuchi H, Ozeki H (1986) Chloroplast gene organization deduced from complete sequence of liverwort *Marchantia polymorpha* chloroplast DNA. Nature (London) 322:572–574

Outenreath RL, Roberson MM, Barondes SH (1988) Endogenous lectin secretion into the extracellular matrix of early embryos of *Xenopus laevis*. Dev Biol 125:187–194

Palmer JD, Stein DB (1986) Conservation of chloroplast genome structure among vascular plants. Curr Genet 10:823–833

Pearson WR, Davidson EH, Britten RJ (1977) A program for least squares analysis of reassociation and hybridization data. Nucleic Acids Res 4:1727–1737

Price HJ, Sparrow AH, Nauman AN (1974) Evolutionary and developmental considerations of the variability of nuclear parameters in higher plants. I. Genome volume, interphase chromosome volume, and estimated DNA content of 236 gymnosperms. Brookhaven Symp Biol 25:390–421

Rake AV, Miksche JP, Hall RB, Hansen KM (1980) DNA reassociation kinetics of four conifers. Can J Genet Cytol 22:69–79

Renfroe MH, Berlyn GP (1985) Variation in nuclear DNA content in *Pinus taeda* L. tissue cultures of diploid origin. J Plant Physiol 121:131–139

Rice JB, Baker S, Whitmore FW, Kamalay JC (1988) Isolation of early embryo specific gene sequences from white pine. In: Abstr Present Ohio State Univ Biotechnol Conf, Columbus Abstr 34

Rivin CJ, Zimmer EA, Cullis CA, Walbot V, Huynh, T, Davis RW (1983) Evaluation of genomic variability at the nucleic acid level. Plant Mol Biol Rep 1:9–16

Schaefer VG, Miksche JP (1977) Microspectrophotometric determination of DNA per cell and polyploidy in *Fraxinus americana*. Silv Genet 26:184–192

Sederoff R, Stomp A-M, Gwynn B, Ford E, Loopstra C, Hodgskiss P, Chilton WS (1987) Application of recombinant DNA techniques to pines: a molecular approach to genetic engineering in forestry. In: Bonga JM, Durzan DJ (eds) Cell and tissue culture in forestry, vol 1. Nijhoff, Dordrecht, pp 314–329

Sharp PM, Li W-H (1987) Ubiquitin genes as a paradigm of concerted evolution of tandem repeats. J Mol Evol 25:58–64

Shinozaki K, Ohme M, Tanaka M, Wakasugi T, Hayashida N, Matsubayashi T, Zaita N, Chunwongse J, Obokata J, Yamaguchi-Shinozaki K, Ohto C, Torazawa K, Meng BY, Sugita M, Deno H, Kamogashira T, Yamada K, Kusuda J, Takaiwa F, Kato A, Tohdoh N, Shimada H, Sugura M (1986) The complete nucleotide sequence of tobacco chloroplast genome: its organization and expression. EMBO J 5:2043–2049

Sitbon F, Gustafsson P (1988) Cloning and characterization of a highly conserved, Pinaceae-specific chloroplast DNA component from Scots pine (*Pinus sylvestris* L.). In: Cheliak WM, Yapa AC (eds) Molecular genetics of forest trees. Proc 2nd Worksh IUFRO Working Party S2. 04-06. Can for Serv PNFI Inf Rep PI-X-80, pp 41–48

Stine M, Keathley DE (1988) Evidence for paternal inheritance of plastids in interspecific hybrids of *Picea*. In: Hanover JW, Keathley DE (eds) Genetic manipulation of woody plants. Plenum, New York, p 486

Strauss SH, Tsai C-H (1988) Genetic polymorphism for nuclear ribosomal gene number in Douglas-fir (*Pseudotsuga menziesii* [Mirb.]Franco). In: Cheliak WM, Yapa AC (1988) Molecular genetics of forest trees. Proc 2nd Worksh IUFRO Working Party S2. 04-06. Can For Serv PNFI Inf Rep PI-X- = 80, pp 79–84

Strauss SH, Palmer JD, Howe GT, Doerksen AH (1988) Chloroplast genomes of two conifers lack a large inverted repeat and are extensively rearranged. Proc Natl Acad Sci USA 85:3892–3902

Szmidt AE, Lidholm J, Hallgren J-E (1986) DNA extraction and preliminary characterization of chloroplast DNA from *Pinus sylvestris* and *Pinus contorta*. In: Lindgren D (ed) Provenances and forest tree breeding for high latitudes. Proc Frans Kempe Symp, Swed Univ Agric Sci Dep For Genet Plant Physiol Rep 6, Umea, pp 269–280

Szmidt AE, Alden T, Hallgren J-E (1987) Paternal inheritance of chloroplast DNA in *Larix*. Plant Mol Biol 9:59–64

Teoh SB, Rees H (1976) Nuclear DNA amounts in populations of *Picea* and *Pinus* species. Heredity 36:123–137

Thompson WF, Murray MG (1981) The nuclear genome: structure and function. In: Stumpf PK, Conn EE (eds) The biochemistry of plants, vol 6. Academic Press, New York London, pp 1–81

Vierstra RD (1987) Ubiquitin a key component in the degradation of plant proteins. Physiol Plant 70:103–106

Wagner DB, Furnier GR, Saghai-Maroof MA, Williams SM, Dancik BP, Allard RW (1987) Chloroplast DNA polymorphisms in lodgepole and jack pines and their hybrids. Proc Natl Acad Sci USA 84:2097–2100

Wagner DB, Govindaraju DR, Dancik BP (1988) Chloroplast DNA polymorphism in a sympatric region. In: Cheliak WM, Yapa AC (eds) Molecular genetics of forest trees. Proc 2nd Worksh IUFRO Working Party S2. 04.06. Can For Serv PNFI Inf Rep PI-X-80, pp 75–78

Westcott R (1988) Difference in the methylation of DNA extracted from trees of Norway spruce (*Picea abies*) of different ages. In: Hanover JW, Keathley DE (eds) Genetic manipulation of woody plants. Plenum, New York, 493 pp

Whitfield PR, Bottomley W (1983) Organization and structure of chloroplast genes. Annu Rev Plant Physiol 34:279–310

Whitmore FW, Kriebel HB (1987) Expression of a gene in *Pinus strobus* ovules associated with fertilization and early embryo development. Can J For Res 17:408–412

Williamson RE (1980) Actin in motile and other processes in plant cells. Can J Bot 58:766–772

Yamamoto N (1982) Preformed mRNA and light-induced germination of pine (*Pinus thunbergii*) seed. Plant Cell Physiol 23:865–869

Yamamoto N, Mukai Y, Matsuoka M, Ohashi Y, Kano-Murakami Y, Tanaka Y, Ozeki Y (1988) Expression of *rbc*S and *cab* genes in both dark and light conditions in pine (*Pinus thunbergii*) seedlings. In: Hanover, JW, Keathley DE (eds) Genetic manipulation of woody plants. Plenum, New York, 494 pp

15 DNA Transfer in Conifers

R.R. Sederoff and A.M. Stomp[1]

1 Introduction

Within the past two decades, it has become possible to isolate, identify, and manipulate specific genes to modify the inherited characteristics of a wide variety of plants and animals. This technology has been widely applied to biomedical sciences and agriculture. It is now apparent that this technology can also be applied to forest trees.

It is often argued that genetic engineering will produce great benefits for agriculture because it can bypass the long time periods required to reach sexual maturity, and produce specific changes without many cycles of selection or backcrossing. If this concept is correct for annual crops, where generations are measured in months, it only serves to emphasize the greater potential that exists for improvement of forest trees through genetic engineering, where generation times can be decades.

At the present time, this potential of genetic engineering is far from being realized. In woody plants, more so than in annual crop plants, very little is known about specific genes that affect processes important for productivity. Insufficient effort has been directed toward learning about the underlying basic processes of growth, development, metabolism, and responses to environmental and biological stress.

Recently, studies of molecular biology in forest tree species have provided a beginning for the biochemical and genetic dissection of these basic processes. The first DNA sequences from conifers have been isolated and characterized (Sederoff et al. 1987; Kenny et al. 1988; Hutchison et al. 1988; Kinlaw et al. 1990). Methods for the transfer and expression of foreign genes into cells and tissues, and in some cases regeneration of transgenic plants, have been developed for a few important tree species. It is already clear that these methods can provide new and powerful approaches to many traditional problems of forest biology and, in addition, extend the potential of genetic engineering from agriculture to forest trees.

The central methodology of molecular genetics has been the ability to take DNA that has been characterized or modified in vitro, to transfer it into a particular experimental organism, and to observe the resulting phenotype. Plant molecular genetics has required the development of new methods for DNA

[1]Department of Forestry, North Carolina State University, Raleigh, NC 27695-8008, USA

Clonal Forestry I, Genetics and Biotechnology
Ed. by M.R. Ahuja and W.J. Libby

transformation. To prove that a specific DNA segment controls or codes for a specific phenotype, it is now possible to demonstrate that directed DNA sequence changes will alter the phenotype.

The ability to create transgenic plants is a pivotal technology for advancing our understanding of the genetic regulation of important economic traits in forest tree species. The probability of obtaining a transgenic plant can be viewed as the product of two probabilities: (1) the probability of successful gene transfer into any particular cell, and (2) the probability that the newly transformed cell will divide and participate in the organization of a plant. In addition, method development requires reliable marker genes with which to "label" transformed cells to trace their further growth and development.

In this chapter, our discussion will focus primarily on DNA transfer and expression in coniferous tree species. In angiosperm tree species, transgenic poplars (Fillatti et al. 1987) have been produced, as well as transgenic walnuts (McGranahan et al. 1988); however, the major impact on forestry of genetic engineering awaits the production of transgenic conifers. Here, we review recent developments in conifer transformation, referring to promoters, marker genes, and strategies related to conifer studies. The reader is also referred to recent reviews on gene transfer in woody plants and forest tree species (Ahuja 1988a, b).

2 Systems for DNA Transfer

The most successful and most widely used method for introducing foreign DNA sequences into plants has been that involving *Agrobacterium* sp. (Matzke and Chilton 1981; Horsch et al. 1985). Host plant genotypes need to be identified which are highly susceptible to the infecting *Agrobacterium* strain (Hinchee et al. 1988). Alternatively, an *Agrobacterium* strain, which is highly infective over the broad range of genotypes within the species of interest, would be best. Strain Bo542 and its derivatives have relatively broad infectivity for conifers (Stomp et al. 1990a). Also, wounded plant tissue facilitates transformation because the production of acetosyringone and other compounds that activate the mobilization and transfer process (Stachel and Zambryski 1986). Exogenously applied acetosyringone can activate the virulence genes, thus substituting for wounded plant tissue. In addition, transformation efficiency is also controlled by the developmental state of the plant cells (An 1985).

DNA transformation experiments, using nonbiological vector systems, show transient gene expression within hours after transfer. Transient expression systems are useful to test for activity of specific gene constructs. Two kinds of systems have been developed for transient expression, electroporation of protoplasts and microprojectile bombardment. Both DNA transfer systems can give rise at low frequency to stable transformed cells which can then be cultured to regenerate plants.

Plant cells can be induced to take up DNA from a surrounding solution (Saul et al. 1988). Uptake of DNA usually involves cells bathed in a solution of salts, protective agents, and DNA. This method of DNA transfer most frequently uses protoplasts as recipient cells; however, several reports have used intact cells or tissues. Electroporation is a variation of direct DNA uptake developed by Fromm et al. (1986). DNA uptake is increased by passing a short pulse of high voltage direct current through a salt solution containing DNA and protoplasts, causing temporary pores in the cell membrane, thus increasing DNA uptake. Electroporation has been used successfully in several species, including monocots which are not susceptible to *Agrobacterium* infection. As with microinjection, the major limitation of this method lies in the difficulty of regeneration from protoplasts in many species of plants.

Microprojectile-mediated DNA transfer is the newest method for DNA transfer in plants. DNA is first precipitated on the surface of metallic particles which are then accelerated to velocities sufficient to penetrate plant cells placed in the particles' path. Stable transformation has been demonstrated in tobacco (Tomes et al. 1990), soybean (McCabe et al. 1988), and maize (Klein et al. 1988). Microprojectile-mediated DNA transfer can avoid the time, expense, and increased somaclonal variation of prolonged tissue culture, as virtually any tissue can serve as a source of recipient cells.

3 Marker Genes

The success of DNA transfer continues to be limited by the availability of marker genes that are easily assayed, are reliable indicators of DNA transfer, and allow selection of the transformed genotype. For a more detailed discussion and protocols for markers and reporters for DNA transfer, the reader is referred to Herrera-Estrella et al. (1988) and Reynaerts et al. (1988). Our discussion will review specific features of those markers used on conifers and briefly mention a new marker that also should be useful.

One of the most widely used marker genes, derived from the *E. coli* transposon Tn5, is one that codes for neomycin phosphotransferase (*npt*II) and confers resistance to the antibiotic, kanamycin (Reiss et al. 1984). It is used most widely as a selectible marker under the control of plant promoters (Horsch et al. 1985). The gene for kanamycin resistance confers upon transformed cells the ability to grow on levels of kanamycin or related aminoglycoside antibiotics that kill untransformed cells.

Phytohormone biosynthetic genes transferred by *Agrobacterium* into recipient plant cells can be used as selectable markers. The expression of these genes in recipient cells allows these cells to grow without exogenously supplied phytohormones in vitro (hormone autotrophy). These genes are not useful selectable markers for the production of transgenic plants because the hormone imbalance created by the genes often prevents plant regeneration. In addition, untransformed cells in culture can habituate and grow in the absence of exogenous phytohormone (Binns 1983).

The most widely used nonselectable marker used in conifers and other higher plants is beta-glucuronidase (GUS) from *E. coli* (Jefferson et al. 1986, 1987). The enzyme expressed by this gene can be detected at very low levels in a fluorescence assay and in histochemical assays. In histochemical staining reactions, activity can be detected in single cells (Stomp et al. 1990b). The histochemical assay has been widely used to demonstrate tissue specificity in transgenic plants containing GUS fused to new promoters. One disadvantage of GUS is that its histochemical assay is destructive. Although endogenous GUS activity is widely believed to be absent from plant tissue, endogenous beta-glucuronidase activity has been reported for several plants including conifers (Fishman 1955; Gahan et al. 1967; Bekkaoui et al. 1988).

Two types of markers based on bioluminescence are becoming more widely used for plant transformation, firefly luciferase (Ow et al. 1986) and those coded by the *lux* genes from *Vibrio harveyi* (Olsson et al. 1989). The light assay can be extremely efficient because small quantities of light can be detected. Luciferase has an advantage over GUS in that the assay for luciferase activity does not kill cells.

Opines are unusual metabolic products that have been useful markers for transformation in conifers (Stomp et al. 1990a). Opines are not usually found in bacteria or in untransformed plant cells under normal conditions. Opines are synthesized at high levels in plant cells transformed by *Agrobacterium.* Opine biosynthetic genes are transferred with the T-DNA and are expressed under control of promoters active in plant cells. There are at least two dozen different opines that have been identified from gall tissue induced by different *Agrobacterium* strains (Reynaerts et al. 1988).

The gene coding for chloramphenicol acetyl transferase (CAT) provides resistance to chloramphenicol in bacteria by acetylation and can be expressed in conifers and other plants (Herrera-Estrella et al. 1988). Although CAT has an advantage over opine synthetases because of its quantitative assay, it suffers the same disadvantage as the opine genes in that there is no visible, histochemical assay.

Recently a marker called *bar* that confers resistance to the herbicide bialaphos, a tripeptide antibiotic, produced by *Streptomyces hygroscopicus*, has proven to be a powerful selectible marker (de Block et al. 1987; Thompson et al. 1987). It has not yet been tested on conifers, but it has been used in several other plants and is potentially superior to kanamycin for selection in transformation experiments.

4 DNA Vectors

Agrobacterium-mediated DNA transfer depends on cis-acting border sequences that span the regions to be transferred and upon trans-acting *vir* functions encoded by the virulence region of the Ti plasmid and the *Agrobacterium* chromosome (Lundquist et al. 1984; Stachel and Nester 1986). In binary strains, now widely used, the T-DNA is located on a mini-plasmid, separate from the DNA of the helper plasmid containing the mobilizing and transfer genes of the virulence region.

One widely used set of vectors is the pBI series (Bevan 1984; Jefferson et al. 1987) pBI-121 contains the selectable neomycin phosphotransferase II (NPTII) marker inserted between the nopaline synthase (*nos*) promoter and terminator, and contains the bacterial beta-glucuronidase gene inserted between the cauliflower mosaic virus 35S promoter (CaMV 35S) and the *nos* terminator. A polylinker near the translation start site for GUS allows insertion or substitution of many different promoter elements. For direct DNA uptake or microprojectile bombardment, a smaller derivative pBI-221 is useful which contains the CaMV-polylinker-GUS-nos cassette inserted into the bacterial plasmid pUC-18. Another versatile plasmid, pRT99-GUS has both NPTII and GUS under the control of CaMV promoters (Topfer et al. 1988)

5 Promoters

Very little information is available on the activity of specific promoters in forest trees, particularly in conifers and other gymnosperms. There are no reports on the isolation of any promoter regions from any tree species. The work on DNA transfer, however, does provide information on the activity of *Agrobacterium*-derived promoters and the 35S promoter from CaMV. To obtain growth of transformed cells in culture in the absence of added phytohormones, the *Agrobacterium* genes for the production of indole acetic acid must be active. Promoters from opine synthase genes are used to derive other marker genes. The promoter for nopaline synthase (*nos*) is known to be expressed in many different plant tissues; it is inducible by wounding and responds to auxin (An et al. 1988; An et al. 1990). The promoters for the genes that code for synthesis of many other opine synthase enzymes also function in gymnosperms including agropine and mannopine synthase. The nopaline synthase promoter (*nos*) and the mannopine synthase (*mas*) are used in several vector constructs.

The CaMV 35S promoter is one of the most widely used in plant transformation studies including work in many conifers. The 35S promoter has been considered a constitutive promoter in many cells and tissues. Benfey et al. (1989) have shown that the constitutive activity is due to the combined effects of individual promoter elements that show considerable specificity in development and in different tissues, but appear to be constitutive in the combined promoter.

6 Evidence for Physical Transfer

Evidence for transfer and integration of DNA is usually determined by molecular hybridization using Southern blots. This method is widely used to determine if a new DNA sequence is present and if it has been inserted into the genome. In addition to the Southern blotting method for the detection of transferred DNA sequences, a new method has been developed that extends the resolution of the technique by amplification of specific sequences. This method, called polymerase chain reaction (PCR), uses oligonucleotides to initiate disproportionate replication at specific chromosome sites defined by the oligonucleotides that serve as primers (Mullis et al. 1986). The amplification procedure allows the detection of rare sequences with increased sensitivity by many orders of magnitude compared to previous methods. A Southern blot analysis can readily detect a sequence of DNA present in one copy per cell, but the PCR method can potentially detect one copy of a sequence if it is present in one cell out of at least 10,000 cells. Variations of the method allow detection of specific sequences in mRNA, and to examine the unknown flanking sequences adjacent to those defined by the oligonucleotides, using "inverse" PCR (Ochman et al. 1990). This method is currently being applied to DNA transfer studies in conifers (D. Robertson pers. commun.). The presence of new flanking sequences demonstrate integration of transferred DNA into chromosomal sites.

7 Transformation in Conifers

7.1 Overview

As recently as 1985, the evidence for DNA transfer in conifers was limited to a few reports of gall formation in response to *Agrobacterium* inoculation (Smith 1939, 1942; de Cleene and de Ley 1976). De Cleene and de Ley (1976) summarized the status of the *Agrobacterium* host range in conifers. In contrast to the common belief that *Agrobacterium* is specific to angiosperms, in fact, many gymnosperms are good hosts for crown gall disease. Forty-two species are known to be susceptible, representing 16 genera and 6 families. By contrast, angiosperms contain 596 known-sensitive species representing 310 genera and 84 families. These species represent 56% susceptibility for gymnosperms and 58% for angiosperms (de Cleene and de Ley 1976).

Research in the last 5 years has clearly established transfer and expression of marker genes in coniferous species. The majority of this evidence has come from studies using *Agrobacterium*. Expression of genes for opine synthases, phytohormone synthesis, and/or kanamycin resistance has been reported in nine pine species (Sederoff et al. 1986; Stomp et al. 1988, 1990a; Morris et al. 1988; Loopstra et al. 1990), Douglas-fir (Gupta et al. 1988; Morris et al. 1989; Ellis et al. 1989; Stomp et al. 1990a), four spruce species (Clapham and Ekberg 1986; Hood et al. 1989; Ellis et al. 1989), two true fir species (Clapham and Ekberg 1988; Morris et al. 1988), and western hemlock (Morris et al. 1989). Transient gene expression using either beta-glucuronidase, luciferase, or neomycin phosphotransferase has been demonstrated in loblolly pine and Douglas-fir (Gupta et al. 1988), and in jack pine, white spruce, and black spruce (Tautorus et al. 1989; Bekkaoui et al. 1988). Transient expression of beta-glucuronidase using microprojectile bombardment has been demonstrated in loblolly, Monterey, and Scots pine (Stomp et al. 1990b). A summary of the coniferous species in which DNA transfer has been reported is presented in Table 1.

Table 1. DNA transfer in conifers

	Common name	Reference
1. *Agrobacterium*-mediated DNA transfer in conifers		
Class Coniferopsida		
Order Coniferae		
Fam. Pinaceae		
Pinus eldarica	Afghan pine	1
elliotii	Slash pine	1
jeffreyi	Jeffrey pine	1
lambertiana	Sugar pine	1, 2
ponderosa	Ponderosa pine	1, 3
radiata	Monterey pine	1
sylvestris	Scots pine	1, 4
taeda	Loblolly pine	5, 1
virginiana	Virginia pine	1
Pseudotsuga menziesii	Douglas-fir	6, 7, 1, 3, 8
Picea abies	Norway spruce	9, 10, 4
glauca	White spruce	8
engelmanni	Engelmann spruce	8
sitchensis	Sitka spruce	8
Larix decidua	European larch	11
Abies alba	Silver fir	12
cephalonica	Greek fir	13
concolor	White fir	13
firma	Momi fir	13
holophyla	Needle fir	13
normanniana	Nordman fir	12, 9
procera	Noble fir	3
Tsuga heterophylla	Western hemlock	3
Fam. Taxodiaceae		
Cunninghamia lanceolata	China fir	13
Sciadopitys verticillata	Umbrella pine	13
Sequoiadendron gigantea	Giant redwood	6, 14, 13
Sequoia sempervirens	Coast redwood	6, 14, 13, 4
Fam. Cupressaceae		
Libocedrus decurrens	Incense cedar	1, 15
Chamaecyparis lawsoniana	Lawson cypress	6
Cupressus arizonica	Arizona cypress	16
*benthamii**	Mexican cypress	16
bakeri	Modoc cypress	16
duttoni		16
forbesii	Tecate cypress	16
goveniana		16
*knightiana**	Mexican cypress	16
*macnabiana**	Piute cypress	16
macrocarpa	Monterey cypress	16
nevadensis	Piute cypress	16
pygmaea	Mendicino cypress	16
sargentii		16
sempervirens	Italian cypress	16
thurifera		16
Juniperus communis	var. Hibernica	16
phoenecia	Phoenician juniper	16
procera	African pencil cedar	16
virginiana	Red cedar	16

Table 1. (continued)

	Common name	Reference
Thuja occidentalis	American arborvitae	16, 12
orientalis	Oriental arborvitae	16
plicata	Western red cedar	16
dolobrata	Hiba arborvitae	16
*Thujopsis dolobrata**	False arborvitae	16
Fam. Podocarpaceae		
Podocarpus elongatus		13
Fam. Araucariaceae		
Araucariaceae bidwillii	Bunya-bunya	17
Class Taxopsida		
Order Taxales		
Fam. Taxaceae		
Taxus baccata	English yew	18, 13
brevifolia	California yew	13
baccata × cuspidata	Taxus × media	13
Torreya californica	Calfornia nutmeg	13
2. Electroporation DNA transfer		
Picea mariana	Black spruce	19
glauca	White spruce	19
Pinus banksiana	Jack pine	19
taeda	Loblolly pine	20
Pseudotsuga menziesii	Douglas fir	20
3. Microprojectile-mediated DNA transfer		
Pinus radiata	Monterey pine	21
taeda	Loblolly pine	21
sylvestris	Scots pine	21

1. Stomp et al. 1990a; 2. Loopstra et al. 1990; 3. Morris et al. 1989; 4. Ahuja 1988a, b; 5. Sederoff et al. 1986; 6. Smith 1935a; 7. Dandekar et al. 1987; 8. Ellis et al. 1989; 9. Clapham and Ekberg 1986; 10. Hood et al. 1989; 11. Diner and Karnosky 1987; 12. de Cleene and de Ley 1976; 13. Smith 1942; 14. Smith 1935b; 15. Smith 1937; 16. Smith 1939; 17. Smith 1936; 18. Smith 1938; 19. Tautorus et al. 1984; 20. Gupta et al. 1988; 21. Stomp et al. 1990b. Some of the classifications are marked with an *asterisk* when different systems of classification give different names. The reader is referred to Krussmann (1985) for more information. Note, in the *Cupressus*,, *benthamii* and *knightiana* are probably varieties of *lusitiana*; *macnabiana* and *nevadensis* may be the same species. Also *Thuja dolobrata* may be the same as *Thujopsis dolobrata*.

7.2 *Agrobacterium*-Mediated DNA Transfer in Pines

In conifers, DNA transfer has been most extensively studied in the pines. Host range studies have shown that pines are generally susceptible to *Agrobacterium* (Stomp et al. 1988, 1990a; Morris et al. 1989) in contrast with earlier studies showing pines resistant to gall formation (Smith 1939, 1942; de Cleene and de Ley 1976).

In transformation experiments with sugar pine (Stomp et al. 1988; Loopstra et al. 1990), shoots derived from cytokinin-treated cotyledons were inoculated with a modified *Agrobacterium* strain. This binary strain was created by moving a second plasmid, containing a selectable marker (NPTII) for kanamycin resistance (Klee et al. 1985), into wild type Bo542. Bo542 was used because of its general hypervirulence (Hood et al. 1984) and its high rates of gall formation in loblolly pine (Stomp et al. 1988). Callus proliferated from galls was phytohormone autotrophic and grew in the presence of kanamycin (Stomp et al. 1988; Loopstra et al. 1990). Enzyme assays verified the expression of NPTII and the *npt*II gene could be detected by Southern blot analysis of total callus DNA. Detailed mapping of the transferred DNA demonstrated that the borders of the T-DNA transferred into sugar pine were similar to those mapped after transfer to alfalfa or soybean (Loopstra et al. 1990; Hood et al. 1986).

7.3 *Agrobacterium*-Mediated Transfer in Other Conifers

Crown gall formation was first reported in Douglas-fir by Smith (1942). More recently, phytohormone autotrophic callus has been proliferated from tumors induced on micropropagated shoots of Douglas-fir. This callus was shown to produce opines and could be grown in the presence of kanamycin (Dandekar et al. 1987). Other workers have confirmed and extended the evidence for expression and physical transfer of foreign DNA in this species (Ellis et al. 1989; Morris et al. 1989).

Several spruce species have also been studied. Ellis et al. (1989) showed that white (*Picea glauca*), Sitka (*Picea sitchensis*), and Engelmann (*Picea engelmanni*) spruce could be transformed by *Agrobacterium.* Spruce seedlings were inoculated with a number of uncharacterized, wild-type strains which gave variable rates of gall formation. The only characterized strain which gave efficient transformation was the hypervirulent strain A281, a derivative of Bo542. Stable integration of T-DNA was inferred by opine synthesis and hormone autotrophic growth. Transfer and expression of kanamycin resistance and firefly luciferase activity were observed using binary vectors.

Both *Agrobacterium tumefaciens* and *Agrobacterium rhizogenes*, carrying the neomycin phosphotransferase gene, have been used to infect *Larix decidua* seedlings and calli (Diner and Karnosky 1987, 1988). Callus cultures were established from either galls or hairy roots formed at the inoculation sites. Callus growth was obtained on medium containing kanamycin.

Although the data on *Agrobacterium*-mediated DNA transfer are still limited to a few strains and species, it is clear that infection, DNA transfer and integration, and the initial events of gall formation occur in a manner similar to that in angiosperms. The most detailed study of *Agrobacterium* transferred DNA (Loopstra et al. 1990) indicates that the process of bacterial T-DNA integration is very similar to that characterized in alfalfa and soybean (Hood et al. 1984). Mapping studies of the DNA transferred into pine cells show similar patterns to those seen in angiosperms, indicating that the signals used to transfer specific segments of DNA into angiosperms are utilized in a similar manner.

Recent studies (Table 1) have extended the host range of *Agrobacterium* to coniferous species such as pines, earlier thought to be totally resistant to infection. This suggests that the range of susceptibility of conifers to *Agrobacterium* is wider than currently documented. Although the *Agrobacterium* vector system has good potential as a gene transfer system for conifers, several problems must be overcome before this vector system can be truly useful.

Infectivity rates, measured by gall formation, are typically low. In model plant systems, like tobacco, the frequency of gall formation approaches 100% of inoculated plants. In pines (Stomp et al. 1990a) rates of gall formation ranged from 86 to 0%, depending on the *Agrobacterium* strain/species combination. The broad host range strain, C58, from which many of the modified strains are derived, and that has been used to produce transgenic tobacco and tomato plants, does not work particularly well in conifers. Further work to identify *Agrobacterium* strains with much higher infectivity in conifers may be required before efficient DNA transfer methods can be developed. A broad host range, generally infectious strain is Bo542 and its derivatives (Ellis et al. 1989; Stomp et al. 1990a).

Susceptibility to *Agrobacterium* also depends on the genotype of the host plant. Hinchee et al. (1988) working with soybean, argue that a critical component of successful transformation using *Agrobacterium* is the identification of specific plant genotypes that are highly susceptible to infection. This genetic specificity poses potential problems for conifer transformation as the genetic variation among individual trees is known to be quite high, i.e., no cultivars or inbred lines exist. However, some clonal material is available for a few conifers, and methods for production of clonal material are available for many conifer species. To identify specific genotypes which give high rates of transformation will mean finding individual trees or clones which are highly susceptible. However, for transformation to be a useful tool for crop improvement, the efficiency of producing transgenic plants must be high across a broad range of genotypes. To overcome this potential problem, a nuclear population of genotypes selected for their ability to give high rates of both plant regeneration and transgenic plant formation in vitro could be created. The starting material for this population would come from a larger population of trees selected for their superior economic phenotypes found in an established breeding program. This nuclear population would serve both as the source of plant material for genetic engineering and the vehicle through which new genes would be introduced into operational breeding programs (Stomp 1989). This population would have variation that could be a source for breeding, but the population would be preselected for high rates of regeneration and transformability. Further research is needed to determine the magnitude of genotypic effects on both the efficiency of *Agrobacterium* transformation and in vitro plant regeneration.

7.4 Transient Gene Expression System in Conifers

7.4.1 Electroporation in Conifers

Transient expression of reporter genes has been observed in several conifer systems following electroporation. Transient expression in protoplasts of black spruce (*Picea mariana*) and jack pine (*Pinus banksiana*) by electroporation have been reported using the CAT marker gene (Tautorus et al. 1989). Expression of CAT was sensitive to a variety of environmental conditions. Linear plasmid was reported to be 2.5 times more efficient than circular. Ballas et al. (1988) also reported a higher frequency of transient expression with linearized plasmid in protoplasts from petunia and tobacco. Bekkaoui et al. (1988) reported transient gene expression in *Picea glauca* protoplasts using the GUS gene and the CAT gene. GUS proved to be an unsatisfactory marker in *P. glauca*, because of an endogenous activity which produced fluorescence in a substrate-dependent manner (methylumbelliferone-beta-D-glucuronide used as substrate). Gel filtration of protoplast extracts showed that this activity eluted later than the main protein peak. These data raise the possibility of endogenous GUS activity in spruce, which may limit the usefulness of this marker gene in transformation studies. Expression of the luciferase gene has been obtained from electroporated protoplasts derived from somatic embryogenic suspension cultures of Douglas-fir and loblolly pine (Gupta et al. 1988).

7.4.2 Microprojectile-Mediated DNA Transfer in Conifers

Several laboratories are exploring the use of microprojectile bombardment in conifers; however, research results are not yet available in published reports. We report here preliminary findings of our research in loblolly, Monterey, and Scots pine (Stomp et al. 1990b). Transient expression using this method has been achieved using GUS driven by the CaMV 35S promoter. Embryonic tissue (for all three species) and callus (for loblolly pine) were used as the target tissues. All tissues, including cotyledons, hypocotyls, radicles, callus, and megagametophytes, gave expression. GUS expression was stabilized in cotyledons cultured on a shoot induction medium and was seen in multicellular patches in newly organizing shoot primordia.

References

Ahuja MR (1988a) Gene transfer in forest trees. In: Hanover JW, Keathley DW (eds) Genetic manipulation of woody plants. Plenum, New York, pp 25–41

Ahuja MR (1988b) Gene transfer in woody plants: perspectives and limitations. In: Ahuja MR (ed) Somatic cell genetics of woody plants. Kluwer, Dordrecht, pp 83–101

An GM (1985) High efficiency transformation of cultured tobacco cells. Plant Physiol 79: 568–570

An GM, Costa MA, Mitra A, Ha SB, and Marton L (1988) Organ-specific and developmental regulation of nopaline synthase promoter in transgenic tobacco plants. Plant Physiol 88: 547–552

An GM, Costa MA, Ha SB (1990) Nopaline synthase promoter is wound-inducible and auxin-inducible. Plant Cell 2(3): 225–233

Ballas N, Zakai N, Friedberg D, Loyter A (1988) Linear forms are superior to supercoiled structures as active templates for gene expression in plant protoplasts. Plant Mol Biol 11: 517–528

Bekkaoui F, Pilon M, Laine E, Raju DSS, Crosby WL, Dunstan DI (1988) Transient gene expression in electroporated *Picea glauca* protoplasts. Plant Cell Rep 7: 481–484

Benfey PN, Ren L, Chua N-H (1989) The CaMV 35S enhancer contains at least two domains which can confer different developmental and tissue specific patterns. EMBO J 8: 2195–2202

Bevan MW (1984) Binary *Agrobacterium* vectors for plant transformation. Nucleic Acids Res 12:8711–8721

Binns AN (1983) Host and T-DNA determinants of cytokinin autonomy in tabacco cells transformed by *Agrobacterium tumefaciens*. Planta 158:272–279

Clapham DH, Ekberg I (1986) Induction of tumours by various strains of *Agrobacterium tumefaciens* on *Abies nordmanniana* and *Picea abies*. Scand J For Res 1:435–437

Clapham DH, Ekberg I (1988) Induction of tumors by various strains of *Agrobacterium tumefaciens* on *Abies nordmanniana* and *Picea abies*. In: Hanover JW, Keathley DE (eds) Genetic manipulation of woody plants. Plenum, New York 463 pp

Dandekar AM, Gupta PK, Durzan DJ, Knauf V (1987) Transformation and foreign gene expression in micropropagated Douglas-fir (*Pseudotsuga menziesii*). Bio/Technology 5:587–590

De Block M, Botterman J, Vandewiele M, Dockx J, Thoen C, Gossele V, Rao Movva N, Thompson C, van Montague M, Leemans J (1987) Engineering herbicide resistance in plants by expression of a detoxifying enzyme. EMBO J 6:2513–2518

De Cleene M, de Ley J (1976) The host range of crown gall. Bot Rev 42:389–466

De Cleene M, Ley J (1981) The host range of infectious hairy-root. Bot Rev 47:147–194

Diner AM, Karnosky DF (1987) Differential responses of two conifers to in vitro inoculation with *Agrobacterium rhizogenes*. Eur J For Pathol 17:211–216

Diner AM, Karnosky DF (1988) *Agrobacterium*-mediated gene transfer in European larch. In: Hanover JW, Keathley DE (eds) Genetic manipulation of woody plants. Plenum, New York, 465 pp

Ellis D, Roberts D, Sutton B, Lazaroff W, Webb D, Flinn B (1989) Transformation of white spruce and other conifer species by *Agrobacterium tumefaciens*. Plant Cell Rep 8:16–20

Fillatti JJ, Sellmer J, McCown B, Hassig B, Comai L (1987) *Agrobacterium* mediated transformation and regeneration of *Populus*. Mol Gen Genet 206:192–199

Fishman WH (1955) Beta-Glucuronidase. Adv in Enzymol 16:361–409

Fromm ME, Taylor LP, Walbot V (1986) Stable transformation of maize after gene transfer by electroporation. Nature (London) 319:719–793

Gahan PB, McLean J, Kalina JM, Sharma W (1967) Freeze sectioning of plant tissue: the technique and its use in histochemistry. J Exp Bot 18:151–159

Gupta PK, Dandekar AM, Durzan DJ (1988) Somatic proembryo formation and transient expression of a luciferase gene in Douglas fir and loblolly pine protoplasts. Plant Sci 58:85–92

Herrera-Estrella L, Teeri TH, Simpson J (1988) Use of reporter genes to study gene expression in plant cells. In: Gelvin SB, Schilperoort RA (eds) Plant molecular biology manual. Kluwer, Dordrecht, pp B1, 1–22

Hinchee MAW, Connor-Ward DV, Newell CA, McDonnell RE, Sata SJ, Gasser CS, Fischhoff DA, Re DB, Fraley RT, Horsch RB (1988) Production of transgenic soybean plants using *Agrobacterium*-mediated DNA transfer. Bio/Technology 6:915–922

Hood EE, Jen G, Kayes L, Kramer J, Fraley RT, Chilton M-D (1984) Restriction map of pTiBo542, a potential Ti plasmid vector for genetic engineering of plants. Bio/Technology 2:702–708

Hood EE, Chilton WS, Chilton M-D, Fraley RT (1986) T-DNA and opine synthetic loci in tumors incited by *Agrobacterium tumefaciens* on soybean and alfalfa plants. J Bacteriol 168:1283–1290

Hood EE, Ekberg I, Johannson T, Clapham DH (1989) T-DNA presence and opine production in *Agrobacterium-tumefaciens* A281 induced tumors on Norway spruce. J Cell Biochem Suppl 0 (13 Part D):260

Horsch RB, Fry JE, Hoffman NL, Eichholz D, Rogers SG, Fraley RT (1985) A simple and general method for transferring genes into plants. Science 227:1229–1231

Hutchison KW, Singer PB, Greenwood MS (1988) Gene expression during matutation in eastern larch. In: Hallgren J-E (ed) Molecular genetics of forest trees. Proc Frans Kempe Symp, Umea, Sw, June 1988, pp 101–114

Jefferson RA, Burgess SM, Hirsh D (1986) Beta-glucuronidase from *Escherichia coli* as a gene-fusion marker. Proc Natl Acad Sci USA 83:8447–8451

Jefferson RA, Kavanagh TA, Bevan MW (1987) GUS fusions: β-glucuronidase as a sensitive and versatile gene fusion marker in higher plants. EMBO J 6:3901–3907

Kenny JR, Dancik BP, Florence LZ (1988) Nucleotide sequence of the carboxy-terminal portion of a lodgepole pine actin gene. Can J For Res 18:1595–1602

Kinlaw CS, Harry DE, Sederoff RR (1990) Isolation and characterization of alcohol dehydrogenase cDNAs from *Pinus radiata*. Can J For Res 20:343–1350

Klee HJ, Yanosky MF, Nester EW (1985) Vectors for transformation of higher plants. Bio/Technology 3:637–642

Klein TM, Fromm M, Weissinger A, Tomes D, Schaaf S, Sletten M, Sanford JC (1988) Transfer of foreign genes into intact maize cells with high-velocity microprojectiles. Proc Natl Acad Sci USA 85:4305–4309

Krussman G (1985) Manual of cultivated conifers. Timber, Portland, Or

Loopstra CA, Stomp A-M, Sederoff RR (1990) *Agrobacterium* mediated DNA transfer in sugar pine. Plant Mol Biol 15:1–9

Lundquist RC, Close TJ, Kado CI (1984) Genetic complementation of *Agrobacterium tumefaciens* Ti plasmid mutants in the virulence region. Mol Gen Genet 193:1–7

Matzke AJ, Chilton M-D (1981) Site-specific insertion of genes into T-DNA of the *Agrobacterium* tumor-inducing plasmid: An approach to genetic engineering of higher plant cells. J Mol Appl Gen 1:39–49

McCabe D, Swain W, Martinell B, Christou P (1988) Stable transformation of soybean (*Glycine max*) by particle acceleration. Bio/Technology 6:923–926

McGranahan GH, Leslie CA, Uratsu SL, Martin LA, Dandekar AM (1988) *Agrobacterium*-mediated transformation of walnut somatic embryos and regeneration of transgenic plants. Bio/Technology 6:800–804

Morris JW, Castle LA, Morris RO (1988) Transformation of Pinaceous gymnosperms by *Agrobacterium*. In: Hanover JW, Keathley DE (eds) Genetic manipulation of woody plants. Plenum, New York, 481 pp

Morris JW, Castle LA, Morris RO (1989) Efficacy of different *Agrobacterium tumefaciens* strains in transformation of pinaceous gymnosperms. Physiol Mol Plant Pathol 34:451–462

Mullis K, Faloona F, Scharf S, Saiki R, Horn G, Erlich H (1986) Specific enzymatic amplification of DNA in vitro: the polymerase chain reaction. Cold Spring Harbor Symp Quant Biol 51:263–273

Ochman H, Medhora MM, Garza D, Hartl DL (1990) Amplification of flanking sequences by Inverse PCR. In: Innis MA, Gelfand DH, Sninsky JJ, White TJ (eds) PCR protocols: a guide to methods and applications. Academic Press, New York London San Diego, pp 219–227

Olsson O, Escher A, Sandberg G, Schell J, Koncz C, Szalay AA (1989) Engineering of monomeric bacterial luciferases by fusion of lux-a and lux-b genes in *Vibrio harveyi*. Gene 81(2):335–348

Ow DW, Wood KV, Deluca M, De Wet JR, Helinski DR, Howell SH (1986) Transient and stable expression of the firefly luciferase gene in plant cells and transgenic plants. Science 234(4778):856–859

Reiss B, Sprengel R, Will H, Schaller H (1984) A new sensitive method for quantitative and qualitative assay of neomycine phosphotransferase in crude cell extract. Gene 30:211–218

Reynaerts A, De Block M, Hernalsteens J-P, van Montagu M (1988) Selectable and screenable markers. In: Gelvin SB, Schilperoort RA (eds) Plant molecular biology manual. Kluwer, Dordrecht, pp A9, 1–16

Sederoff RR, Stomp A-M, Chilton WS, Moore LW (1986) Gene transfer into loblolly pine by *Agrobacterium tumefaciens*. Bio/Technology 4:647–649

Sederoff RR, Stomp A-M, Gwynn G, Ford E, Loopstra C, Hodgskiss P, Chilton WS (1987) Application of recombinant DNA techniques to pines: a molecular approach to genetic engineering in forestry. In: Bonga JM, Durzan DJ (eds) Cell and tissue culture in forestry. Forestry sciences, vol 1. Nijhoff, Dordrecht pp 314–329

Smith CO (1935a) Crown gall on conifers. Phytopathology 25:894

Smith CO (1935b) Crown gall on the *Sequoia*. Phytopathology 25:439–440

Smith CO (1936) Crown gall on *Araucaria bidwillii*. Phytopathology 26:400–401

Smith CO (1937) Crown gall on incense cedar, *Libocedrus decurrens*. Phytopathology 27:844–849

Smith CO (1938) Crown gall on *Taxus baccata*. Phytopathology 28:153–155

Smith CO (1939) Susceptibility of species of Cupressaceae to crown gall as determined by artificial inoculation. J Agric Res 59:919–925

Smith CO (1942) Crown gall on species of Taxaceae, Taxodiaceae, and Pinaceae, as determined by artificial inoculations. Phytopathology 32:1005–1009

Stachel S, Nester EW (1986) The genetic and transcriptional organization of the vir region of the A6 Ti plasmid of *A. tumefaciens*. EMBO J 5:1445–1454

Stachel S, Zambryski PC (1986) Agrobacterium tumifaciens and the susceptible plant cell. a novel adaptation of extracellular recognition and DNA conjugation. Cell 47:155–157

Stomp A-M (1989) Genetic engineering and traditional breeding: where is the interface? In: Proc 20th Southern forest tree improvement Conf, Charleston, pp 204–210

Stomp A-M, Loopstra C, Sederoff RR, Chiltron S, Fillatti J, Dupper G, Tedeschi P, Kinlaw C (1988) Development of a DNA transfer system for pines. In: Hanover JW, Keathley DE (eds) The genetic manipulation of woody plants. Basic life sciences, vol 44. Plenum, New York, pp 231–241

Stomp A-M, Loopstra C, Chilton WS, Sederoff RR, Moore LW (1990a) Extended host range of *Agrobacterium tumefaciens* in the genus *Pinus*. Plant Physiol 92: 1226–1232

Stomp A-M, Weissinger AK, Sederoff RR (1990b) Characterization and developmental regualtion of microprojectile-mediated gene expression in loblolly pine. Plant Cell Rep 1991 10: 187–190

Tautorus TE, Bekkaoui F, Pilon M, Dalta RSS, Crosby WL, Fowke LC, Dunstan D (1989) Factors affecting transient gene expression in electroporated black spruce, (*Picea mariana*) and jack pine (*Pinus banksiana*) protoplasts. Theor Appl Genet 78: 531–536

Thompson CJ, Rao Movva M, Tizard R, Crameri R, Davies JE, Lauwereys M, Botterman J (1987) Characterization of the herbucide resistant gene *bar* from *Streptomyces hygroscopicus*. EMBO J 6: 2519–2523

Tomes DT, Weissinger AK, Ross M, Higgins R, Drummond B, Schaaf S, Malone-Schoneberg J, Staebell M, Flynn P, Anderson J, Howard J (1990) Transgenic tobacco plants and their progeny derived by microprojectile bombardment of tobacco leaves. Plant Mol Biol 14: 261–269

Topfer R, Schell J, Steinbiss H-H (1988) Versatile cloning vectors for transient gene expression and direct gene transfer in plant cells. Nucleic Acids Res 16: 8725

Glossary

The bold-face numbers following the words and phrases in this Glossary refer to chapters in which the word or phrase is used in context, or in which it is more fully defined. This Glossary was prepared by the Editors, with consultation and help by the Chapter Authors, and by Dr. P. Mathias and Dr. S. Jelaska, Visiting Scholars in Berkeley during 1988–89 and 1990–91. We have drawn on Glossaries (Snyder 1959; Futuyma 1986; Pierik 1987; Curtis and Barnes 1989; Lackie and Dow 1989; Rieger et al. 1991; Donnelly and Vidaver 1988), descriptions in context (Grierson and Covey 1988), and on a dictionary (Guralnik 1976), frequently modifying definitions to better fit the context of this volume.

Abscisic acid (ABA): **13.** A multifunctional phytohormone used to induce development and maturation of some somatic embryos.

Acclimation: **11.** The process of increasing a plant's hardiness to outdoor conditions. See hardening off.

Acicular needles: **3.** Scale (or primary) needles typically borne mostly on the main stem of conifer seedlings during the period of indeterminate growth between germination and first budset.

Activated charcoal: **13.** Powdered carbon used in culture media for adsorption of aromatic molecules.

Additive gene action: **5, 10, 11.** Gene products combine to produce variation in a trait in a simple linear way. See additive genetic variation.

Additive genetic variation: **1, 5, 11.** The proportion of genetic variation that responds to natural selection, mass selection, or pick-the-winner selection. This is the basis of a parent's breeding value, or general-combining-ability (GCA).

Agrobacterium rhizogenes: **15.** A bacterium that, like *A. tumefaciens*, is used as a vector for genetic manipulation of plants.

Agrobacterium tumefaciens: **15.** A bacterium capable of inducing crown-gall disease in a wide variety of higher plants. The ability to disarm its pathogenic functions and substitute other genes gave rise to plant genetic engineering.

Agropine: **15.** An opine formed from mannopine by lactonization (cyclization).

Algorithm: **4.** A formalized series of steps for performing a specific type of calculation.

Alleles: **5, 11.** Alternative forms of a gene (at a given locus), differing in DNA sequence and usually recognized by phenotypic effects.

Allopatric: **14.** Separated. Allopatric populations do not freely exchange genes.

Androgenesis: **7.** Generated from a male organ, as in halpoid tissue-culture originating from pollen.

Aneuploid: **7.** An individual having one or a few chromosomes more or less than the normal diploid (or even-ploid polyploid) number of chromosomes, producing imbalance with respect to normal chromosome complement.

Anova: **12.** An analysis of variance.

Apical meristem: **3.** A group of undifferentiated cells at the tips of roots or shoots that, by cell division, give rise to all primary tissues and thence to all plant organs of the root or shoot.

Artificial seed: **13.** Analog to true (zygotic) seed; composed of a somatic embryo encapsulated in a matrix that protects, feeds and regulates it.

Assortative mating: **5, 11.** Individuals similar (positive assortative mating) or dissimilar (negative) to each other are more likely to mate (or to be mated) than if mates were chosen at random.

Autotrophy: **15.** The ability of cells to grow on or in minimal medium without need for additional nutrients, hormones, etc. Cells transformed by wildtype *Agrobacterium tumefaciens* are often hormone autotrophic.

Auxins: **11, 13.** A class of plant growth regulators involved in apical dominance and cell elongation, and often effective in root organogenesis.

Axenic: **13.** Free of foreign or undesirable organisms.

Bar: **15.** A gene that encodes phosphinothricin acetyltransferase (PAT), leading to resistance to the herbicide bialaphos.

Base population: **11.** The set of trees from which selections will be taken to form advanced-generation breeding and/or production populations.

Beta-glucuronidase: **15.** A bacterial enzyme active on spectrophotometric, fluorometric and histochemical substrates, and thus widely used as a reporter gene in DNA-transfer studies.

Bialaphos: **15.** A naturally occurring tripeptide antibiotic used as an herbicide, inhibiting L-glutamic acid. See BAR.

Binary: **15.** Used to describe a two-plasmid system for gene transfer. For example, in *Agrobacterium*, the virulence functions are on one plasmid and the genes to be transferred are on a second plasmid, flanked by transfer-defining border sequences.

Bioluminescence: **15.** The biological or enzymatically-based emission of light.

Bioreactor: **13.** A vessel for modest-to-large-scale liquid suspension culture of cells, in which conditions are optimized for high production levels.

Breeding value: **5, 11.** See narrow-sense heritability.

Breeding population: **11.** A group of selected parents that are intercrossed to form a population for the next cycle of selection. See production population and multilines.

Breeding orchard: **1.** An orchard where (usually) control-pollinations are made to produce the next generation. Many more parents are usually maintained in a breeding-orchard than in a seed-orchard.

Broad-sense heritability: **5, 9.** The ratio of total genetic variance to phenotypic variance. Used to estimate the degree of genetic control of a trait in a population. Useful for predicting response to clonal selection.

Bulking up: **1, 11.** See vegetative multiplication.

Callus: **11, 13.** Unorganized tissue, often initiating at the cut end of a cutting or explant.

Carbon partitioning: **6.** The pattern of allocation of acquired carbon to various structures and activities.

Cat: **15.** Abbreviation for chloramphenicol acetyl-transferase.

Cauliflower mosaic virus 35S promoter: **15.** This promoter is active at a high level in a wide variety of plants and tissues, and is thus widely used in vectors for DNA transfer.

C-Effects: **5.** Effects in common that are not due to genetic covariation. These may include maternal effects, a non-random common environment, or a common physiological condition of donor tissue used for cloning.

Cell-line: **13.** A culture or cultures that arise from an individual explant. See clone.

Chimera: **7.** An organism composed of genetically different tissues occurring adjacent to each other.

Chloramphenicol acetyl transferase: **15.** This enzyme inactivates the bacterial antibiotic chloramphenicol, and is thus used as an unselected DNA marker or reporter gene.

Chronogenes: **12.** Genes that are activated (read) in a specific sequence.

Cleavage polyembryony: **13.** Development of multiple (genetically identical) embryos by division and separation of embryonic cells from a single zygote. See simple polyembryony.

Clonal replication: **11.** A tool of genetic testing in which a given genotype is replicated across environments in time and space. See genetic replication.

Clone: **2, 8, 11.** (a) A group of vegetatively-propagated organisms consisting of an ortet and its ramets. (b) A cell line of single-cell origin. (c) A gene or piece of DNA replicated (usually) in a host bacterium. All imply genetically (nearly) identical material, and reproduction by mitotic division.

Coancestry: **4.** Relatedness, which would result in inbreeding if the genotypes concerned mated and produced offspring.

Codominant: **8.** The products of both genes at a locus are expressed, often nearly equally. Isozymes provide an example of codominant expression.

Coenocyte: **12, 13.** An organism, organ or cell with many nuclei within a common cytoplasm.

Combined testing: **11.** Clonal and family testing are combined in a single test, or may be carried out in parallel. See multiple-stage testing and sequential testing.

Competence: **3, 11.** A developmental term referring to a cell's (or tissue's) ability to produce certain structures, organs, or even entire organisms.
Commercialization: **11.** The transition from research to routine operational application (of a technology).
Conversion: **13.** See embryo conversion.
Cot-curve: **14.** A DNA hybridization (reassociation) curve, often used to show degree of relatedness of two organisms by the similarity of their DNA molecules.
Crown gall: **15.** A tumorous growth on plants, caused by *Agrobacterium tumefaciens.*
Cryopreservation: **10, 11, 12, 13.** Storage of cells, tissues, seeds etc. at low temperatures, usually in liquid nitrogen (− 196°C).
Culture medium: **13.** Nutritive medium for the growth of tissue cultures.
Curvilinear: **3.** Consisting of or enclosed by a curved line or lines. More complicated than a simple linear relationship.
Cyclophysis: **9.** The maturation of apical meristems may lead to differences in performance among vegetative propagules taken at different times from members of the same clone or even from the same donor plant. See periphysis, topophysis.
Cytokinins: **11, 13.** A class of plant growth regulators involved in cell division, and often effective in shoot organogenesis.
Cytophotometry: **14.** A technique of making measurements by means of photographs taken through a microscope.
Cytosine: **14.** One of the four bases that provide information in DNA, often methylated.
Dedifferentiation: **12.** Reversion of differentiated cells or nuclei to non-differentiated (often meristematic) cells or nuclei.
Densitometric scanning: **14.** A technique of measuring density across a region of transimittance and reflectance.
Deployment: **1, 4, 11.** The physical movement of clones or other genetic units from one site (usually a nursery) to another (usually plantations), including their locational configuration on the recipient site.
Developmental homeostasis: **6.** Canalization. The degree to which some "normal" phenotype is produced in spite of variation in the environment, often the result of control systems that cushion some elements of the internal environment from the impacts of fluctuations in the external environment.
Diallel: **5.** A mating design that uses each parent as both a male and a female, crossed to all other parents reciprocally and including selfs. A 5-parent (5 × 5) diallel requires 25 matings.
DNA Transformation: **15.** The uptake and expression of foreign DNA in a living cell. Originally defined as an inherited alteration of the phenotype of the transformed cell, see stable and transient.
Dominance: **5, 11.** An intralocus interaction in which the average value of the heterozygote is not exactly intermediate between the average values of the two homozygotes.
Dominance genetic variance: **5, 11.** The component of non-additive genetic variance that is due to within-locus dominance deviations. Often used as shorthand for the portion of non-additive genetic variance estimated by full-sib/half-sib mating designs.
Effective population size: **11.** An adjustment of the census population number based on the distribution of reproductive success, ranging from numbers much less than census number if there are large differences to double the census number if all parents contribute equally. Used to calculate rate of inbreeding.
Electrophoresis: **7.** A technique that allows separation of proteins and other large molecules based on their size, configuration and charge.
Electroporation: **15.** A process by which an electrical impulse is used to create transient pores in cellular membranes, thereby increasing the efficiency of uptake of DNA from solution.
Elite tree: **11, 12.** A tree verified as superior or desirable by appropriate testing.
Embling: **13.** See somatic seedling.
Embryo conversion: **13.** The transitionary process of embryo-to-plant development. May involve conversion from axenic to non-axenic environment. See hardening off.
Embryo maturation: **13.** A late phase of embryo development, leading to dormancy and germination competence.

Embryogenesis: **11, 12, 13, 14.** Process by which an embryo initiates and develops from a zygote or, asexually, from a somatic cell or group of cells.

Endonuclease: **14.** A class of enzymes that cut the DNA double helix at specific 4-to-8 nucleotide sequences. See Restriction fragment length polymorphisms.

Endopolyploidy: **6.** The chromosome complement of some cells is a multiple of the chromosome complement of the zygote, usually resulting from local or cell-specific endomitosis (chromosome duplication without division of the nucleus).

Endospermic: **7.** Originating from the endosperm. In angiosperms, this is generally but not always triploid tissue.

Epicormic: **12.** Initiating from the stem, as when dormant or suppressed buds elongate and form epicormic shoots following exposure or fire.

Epigenetic: **3, 7, 9, 12.** Interactions among developmental processes above the level of primary gene action. Epigenetic variation does not follow the rules of Mendelian inheritance, is often the result of changed gene expression and may be reversible. It may be somatically inherited, but it is not transmitted through meiosis.

Epistasis: **5, 9, 11.** An interlocus interaction in which the expression of combinations of specific genes or genotypes from different loci is not accurately predicted by a simple linear combination of their average effects.

Eukaryote: **14.** A cell with histone-containing chromosomes, and membrane-bound organelles and nucleus, or an organism composed of such cells. See prokaryote.

Expansion stock: **8.** Propagule donors that originate from foundation stock and that are used to produce large numbers of propagules for production outplantings.

Explant: **10, 11, 12, 13.** An organ or tissue excised from a donor plant, that is used to initiate an *in vitro* culture.

Ex situ: **10.** Off the site; away from the natural habitat.

F_1-, F_2-Progeny: **5.** Starting with a particular cross, the offspring of that cross are the first-filial or F_1 progeny, and the offspring of crosses among those F_1 progeny are the second-filial or F_2 progeny.

Factorial mating: **5.** A mating scheme in which m males are each crossed to the same n females is an $m \times n$ factorial. For example, if 2 trees designated male are crossed to each of 3 designated female, that is a 2×3 factorial of 6 total crosses.

Family forestry: **1.** Tested open-pollinated, polycross or full-sib families are deployed as single families or as prescribed family mixtures.

Fascicular needles: **3.** Secondary needles borne in clusters (fascicles). On seedlings, fascicular needles often develop in the axils of primary, acicular needles.

Fecundity: **5.** A measure of reproductive productivity (often restricted to female gametes, or eggs).

Foundation stock: **8.** The ultimate source of material for the clone. Often the ortet and double-labeled primary ramets. See basic material, expansion stock.

Free nuclear division: **13.** The first nuclear division of the zygote of most conifers occurs without cell division, resulting in a coenocyte.

Full-sib progeny: **5.** The offspring resulting from a cross of two unrelated parents.

Gametoclonal variation: **7.** Heritable variation arising in tissue-culture of gametic origin.

Gametophyte: **12.** In organisms such as trees that alternate haploid (n) and diploid ($2n$) phases or generations, a gametophyte is the haploid (n) gamete-producing phase or generation.

GCA: **11.** General combining ability. See additive genetic variation.

Gene amplification: **6.** The selective synthesis of DNA that results in multiple copies of a gene, for example the synthesis of many copies of ribosomal-RNA genes as a means of enhancing protein synthesis.

Genet: **6.** The set of all modules derived from a single zygote.

Genetic chimaera: **6.** A genetic mosaic; a tree composed of tissues of two or more genotypes. May arise as a result of somatic mutation or somatic crossing-over.

Genetic entries: **11.** Refers to the units of test or selection, ie, individuals, clones, families, provenances, or species.

Genetic gain: **5, 11.** The average (heritable) change from one generation to the next as a result of selection.

Genetic replication: **11.** (a) A copying or transmission of DNA. (b) A tool of genetic testing in which genetic entries are replicated across environments. Provides information about the average values of the genes contained in each genetic entry, with the confounding effects of environment being estimable.

Genome: **10, 14.** The genetic material of an organism. Its complete set of chromosomes (DNA), with their associated genes.

Germ line: **6.** Cells whose cell descendents eventually undergo meiosis and produce gametes.

Germplasm: **8, 10, 11.** Genetic material. See reproductive material.

GUS (gusA): **15.** Abbreviation for beta-glucuronidase.

G × E Interaction: **11.** Generally: the relative magnitudes of differences in performance of a set of clones (or of other genetic entries) are different in different places or years; Narrowly: the rankings of a set of such genetic entries are different in different places or years.

Hairy root: **15.** A disease caused by *Agrobacterium rhizogenes.* The bacterium can be disarmed for genetic engineering.

Half-sib progenies: **5, 11.** The offspring resulting from a cross of one parent to many other parents, each unrelated to each other or to the common parent. Usually produced in designs such as diallels or factorials. See full-sib progenies.

Hardening off: **11.** By exposing propagules to increasingly harsh environmental conditions, they are weaned from the abnormally favorable conditions of the nursery, greenhouse, or culture room, and become ready for field planting.

Hedge-orchard: **11.** An orchard of hedged plants that produces cuttings or other tissue for vegetative propagation. See hedging.

Hedging: **11.** The repeated clipping of an ortet and/or some of its ramets to produce a low hedged plant. The purposes of this procedure are to slow or halt maturation of the plant, and to control its size and shape for efficient operation.

Helper plasmid: **15.** A plasmid used to provide transfer functions that are not contained in the plasmid to be transferred.

Heritability: **5.** Strictly, the ratio of genetic variance to phenotypic variance. See also narrow-sense heritability and broad-sense heritability.

Heterosis: **5, 6.** The average of the progeny is better than the average of the parents. Sometimes called hybrid vigor.

Heterozygosity: **5.** See homozygosity.

Homozygosity: **5.** When both alleles at a locus are identical, *aa* or *AA*, the locus is homozygous. When there is one of each, *Aa*, it is heterozygous.

Hybridization: **5, 11.** Interspecific: a cross between species. Intraspecific: a cross between populations within a species, or even between individuals of contrasting genotype within a population. Introgressive: the moving of genes from one species or population to another by repeated backcrosses.

Hypervirulent: **15.** A term applied to specific strains of *Agrobacterium tumefaciens* that have a wide host range.

IAA: **15.** Indole-acetic acid, a naturally-occurring phytohormone.

Ideotype: **6.** An idealized multitrait characterization; for example, a crown that is long, narrow and dense with high branch angles.

Inverse PCR: **15.** A method for isolation of flanking sequences adjacent to a known DNA fragment, such as a transposon.

In vitro: **7, 10, 11, 13, 14.** "In glass"; in aseptic culture under laboratory conditions. See *in vivo.*

In vivo: **13, 14.** Grown in natural conditions (in the field, greenhouse, etc). see *in vitro.*

In situ: **10.** On site; within the natural habitat.

Inbreeding: **5.** A reduction in average heterozygosity resulting from a mating between relatives.

Inbreeding coefficient: **5.** A measure of inbreeding, F, from 0.0 (the base population) to 1.0 (fully homozygous). A negative F indicates outcrossing and greater heterozygosity than in the base population.

Indeterminate growth: **3.** Organs develop as soon as they are produced from the apical meristem, compared to determinate growth where organs are preformed in a bud and don't expand until the bud resumes growth.

Initiation: **13.** The first cellular proliferation, which results in the establishment of a culture from an explant.

Intragenet competition: **6.** Competition among meristems, organs or entire plants of the same genet.

Inversion: **14.** By a double break and rejoining, the base–order of a segment of a chromosome and the order of genes on that segment are reversed.

Isozymes: **7, 8.** Enzymes having similar structure, catalyzing the same reaction and (strictly) produced at different loci. Loosely, they may be produced by alleles (strictly, allozymes).

Kanamycin: **15.** An antibiotic frequently used to screen resistant cells from susceptible cells.

Kilobases: **14.** On a DNA or RNA strand, units of 1,000 bases.

Leaf-area index: **6.** A dimensionless parameter expressing the area of leaf surface over unit area of ground.

Locus: **5.** Plural loci. The gene, or the physical location of the gene within the DNA strand or chromosome. See alleles.

Luciferases: **15.** Enzymes that produce light as a byproduct of their reaction. Used as genetic markers, and as a possible genetic engineering project for Christmas trees that glow in the dark.

LUX: **15.** Abbreviation for bacterial luciferase genes.

Macropropagation: **10.** The *ex vitro* vegetative propagation of plants by rooted cuttings, rooted needles, grafting or root-suckers. See micropropagation.

Mannopine: **15.** An opine derived from glutamine and mannose.

Marker gene: **15.** A (selectable or unselectable) gene used to indicate the presence of foreign DNA, or to locate or identify a region of a chromosome, by its presence.

MAS: **15.** A gene for mannopine synthetase.

Maternal effects: **5.** A special case of "c-effects" due to a common maternal environment. Frequently associated with the nutrition that is provided by the mother to the embryos prior to their germination and independence.

Maturation: **1, 2, 3, 10, 11, 12.** A sometimes abrupt (heteroblastic) or, more common, unevenly gradual (homoblastic) process of orderly development from embryonic through juvenile and adolescent to mature.

Maturational gradient: **3.** In trees, maturation is often most advanced in the upper parts of a tree and least advanced near the base of the bole, with intermediate conditions in between.

Medium: **13.** See culture medium.

Megagametophyte: **13.** In a conifer, the haploid cells and tissues developing from the maternal gamete, with nutritive and other functions for the contained zygote.

Meiosis: **5.** A special cell division that results in a reduction of somatic-cell chromosome number ($2n$) to gamete chromosome number (n).

Mendelian rules of inheritance: **8.** A binary (for diploids) system of rules based on segregation of alleles at meiosis and recombination at fertilization.

Meristem: **10, 12.** An undifferentiated plant tissue, often with rapidly dividing cells, from which new tissues or organs arise.

Methylation: **3, 14.** Addition of a CH_3 group. For DNA, the methyl group is usually added to a cytosine residue.

Microcalorimetry: **10.** The (computerized) detection of small amounts of heat produced by live tissue.

Microinjectin: **15.** Direct physical injection of individual cells, particularly for DNA transfer.

Microprojectile bombardment: **15.** A method for direct physical transfer of DNA into a recipient cell at high velocity by means of a blank cartridge or other blast of gas or charged water.

Micropropagation: **9, 10, 12.** The *in-vitro* vegetative propagation of plants, often called plantlets, micropropagules, or somatic embryos.

Microsporogenesis: **12.** The development of pollen.

Mini-plasmid: **15.** A small plasmid DNA, usually the smaller of the two in a binary *Agrobacterium* DNA-transfer system.

Mobilization: **15.** The activation of a chromosomal segment to be transferred to a plant cell by a bacterial system of conjugal transfer.

Monoclonal blocks: **11.** A deployment option for clones (or, similarly for families in monofamily blocks) in which each clone (family) is established in a pure block. Diversity may be maintained by a mosaic of blocks of different genetic entries.

Morphogenesis: **7.** The process of differentiation of cells into different tissues or structures.

Multi-gene family: **3, 14.** Loci with identical or similar primary products and function, probably evolved by duplication and then, in some cases, some differentiation at the gene level.

Multiclonal mixture: **10.** A mixture of clones in (usually) equal proportions for use in forest plantations. See multiclonal variety.

Multilines: **2, 11.** Sublines of the general breeding population purposefully selected for different sets of traits or deployment destinations.

Multiple-population breeding: **11.** See multilines.

Mutation: **2, 10.** A heritable genetic change. Somatic: occurring in a somatic cell and thus not necessarily passed on to sexual offspring. Germinal: occurring in the germline, and thus may be passed on to sexual offspring.

Narrow-sense heritability: **5.** The ratio of additive genetic variance to phenotypic variance. Useful in predicting the response of a population to natural selection or to pick-the-winner selection. Related to breeding value.

Neomycin phosphotransferase: **15.** An enzyme that confers resistance to kanamycin and other aminoglycoside antibiotics. See npt II.

Neutral alleles: **11.** Alleles subject to no (or very weak) selection pressure, and thus useful as evolutionary clocks, for mating-system estimation, and similar selection-free chores.

Non-additive gene effects: **10.** See dominance, epistasis.

Non-additive genetic variation: **1, 5, 11.** The proportion of genetic variation that doesn't respond to simple mass selection, and that causes specific pairwise crosses to depart from performance values predicted by the breeding values of the parents.

Nopaline: **15.** An opine formed by the condensation of arginine and alpha-ketoglutaric acid, a reaction catalysed by nopaline synthetase.

Northern blot: **3, 14.** A technique by which RNA is transferred from an agarose gel to a nitrocellulose membrane on which it can be hybridized to radiolabeled complementary nucleic acid. Similar to southern and western blot techniques.

NPT II: **15.** The bacterial gene that encodes neomycin phosphotransferase II.

Nucleotide: **14.** A building block of DNA or RNA, consisting of a phosphate group, a sugar, and a purine or pyrimidine base.

Ontogenetic: **9.** Normal sequence of development; going through the usual stages of a life cycle.

Ontogenetic aging: **12.** Aging or maturation that occurs in a normal sequence.

Open-pollinated progeny: **1.** Progeny from a mating that is not controlled. A mixture of related and unrelated pollen is delivered by wind, or in some cases by insects etc. The o-p family is typically a mixture of full-sibs, half-sibs, and selfs, or, in a natural stand, of such relatives with various degrees of inbreeding.

Opines: **15.** An unusual class of compounds found most often in plant tissues infected by *Agrobacterium.*

Organalle: **14.** A body with specific structure and likely function in the cytoplasm of a cell.

Organogenesis: **7, 10, 11.** The process involving differentiation of organs from cells and tissues.

Organogenic capacity: **12.** Competence to form or give rise to an organ. See competence.

Ortet: **2, 4, 11.** The initial individual (usually from a zygotic embryo) that is vegetatively propagated to produce a clone. See ramet.

Orthotropic: **3, 12.** Growing vertically, usually with radial symmetry. Contrast to plagiotropic.

Overdominance: **6.** A special case of dominance in which the heterozygote is, on average, outside the range of the averages of the two homozygotes.

PCR: **15.** See polymerase chain reaction.

Pedigree: **8, 11.** A record of parentage, often including data on performance of the parents and of other relatives.

Periphysis: **9.** Effects of the environment in pre-conditioning tissue, which may lead to differences in performance of vegetative propagules taken from different members of the same clone or from different parts of the same donor plant. See cyclophysis, topophysis.

Phase change: **12.** The developmental change from one maturation state to the next.

Phasic development: **12.** The orderly development from one normal phase to the next. See phase change.

Phenotype: **5, 11.** The observed expression of a trait in an individual that is the result of a developmental interaction of the individual's genotype and its operational environment.

Phenotypic plasticity: **6.** The degree to which different phenotypes are produced by the same genotype (clone) in different environments.

Phloem: **3.** Conductive plant tissue that carries sugars and other organic nutrients and products, and is closely associated with xylem, which conducts water and inorganic nutrients.

Physiologically mature: **11.** Exhibiting "mature plant" traits; may or may not be associated with high chronological age.

Phytohormones: **13.** A broad class of naturally occurring chemicals, effective at very low concentrations for regulating plant growth and development. See auxin, cytokinin.

Plagiotropic: **3, 11.** A branchlike growth form of an independent plant, usually at an angle other than vertical and often bilaterally symmetrical.

Plagiotropism: **11.** A tendency for a propagule to grow at an angle and like a branch, in contrast to vertical, or orthotropic, growth.

Plantlet: **11, 13, 14.** Broadly, a plant produced *in-vitro*. Narrowly, the intensive-care stage of a plant produced in tissue-culture. See planting, vol 2. Glossary.

Plus-tree: **3, 4, 6, 9.** A tree or genotype selected on the basis of its outstanding single-copy performance, but not yet clonally tested or progeny tested.

Polygenic: **5.** Said of a trait controlled by several-to-many loci, each having relatively small effect on the expression of the trait.

Polylinker: **15.** A synthetic segment of DNA containing a large number of restriction sites, useful for cloning DNA fragments into or out of plasmid or viral vectors.

Polymerase chain reaction (PCR): **15.** Specific amplification of DNA sequences by preferential replication of the sequence between two sites defined by certain DNA polymerase primers.

Polymix crosses: **5.** Polycross. A mating scheme in which several (usually > 10) pollens (usually unrelated to each other) are mixed and crossed as that mix to a series of females that are (usually) unrelated to any of the males.

Polymorphism: **12.** Occurring in more than one form in a population, usually with the rarer morph at or above some minimum frequency (say 1%).

Precocious: **3.** In the reproductive sense, to predictably exhibit reproductive competence at an early age or small size.

Primer: **15.** An initiator for polymerase reactions on DNA, for example oligonucleotides paired to a single-stranded region and having an unblocked 3-prime hydroxyl group to which nucleotides can be added.

Primordium: **12.** A group of cells that gives rise to an organ, with some commitment and differentiation begun.

Probe: **3, 8, 14.** A small piece of radiolabeled nucleic acid used for detection of its complementary sequence by nucleic-acid hybridization.

Production population: **11.** Usually a highly selected subset of the breeding-population parents or clones that is propagated sexually or vegetatively for operational reforestation. See multilines, production clones, vol 2. Glossary.

Progeny test: **11.** Generally a common-garden test in which the breeding values of parents are evaluated and ranked on the basis of the performance of their offspring. See progeny trial, provenance test, vol 2. Glossary.

Prokaryote: **14.** Bacteria and other organisms lacking membrane-bound organelles and nucleus. See eukaryote.

Promoter: **3, 15.** The region of DNA at the upstream (5-prime) end of a gene or operon that serves as the initiation site for transcription. Promoters can be complex, containing several sequences that regulate level of expression as well as the temporal and spatial expression of the gene.

Protodermal: **12.** The origin of the skin, or epidermis.

Protoplast: **10, 13, 14, 15.** A plant cell without a cell wall; it may be produced by enzymatically removing the cell wall.

Provenance: **4, 11.** The geographic origin of a population. Mostly (but not always in European usage) the ultimate natural origin, implying where the population evolved prior to human intervention.

Quiescent centre: **12.** Within the rapidly dividing group of cells in the root meristem is a small group of cells that rarely divide, and these are carried passively within the rest of the meristem. It is currently controversial whether such exists in the shoot apical meristem.

Ramet: **4, 11.** All vegetative propagules of an ortet are ramets. A clone is composed of the ortet and its ramets.

Ramicorn: **9.** A large high-angled branch, often resulting from one member of a fork being partly suppressed by the more dominant member.

Random mating: **5.** In the ideal case, each individual in the population has the same probability of mating with every other individual. In practice, no selection influences the matings that occur.

Reciprocal recurrent selection: **5, 11.** A complicated breeding scheme in which selection within each of two independent lines is based on progeny tests of crosses between the lines. It is meant to select for both additive and non-additive genetic variation.

Recombinants: **7.** Individuals having new combinations of the genes (alleles) of the parent(s), the term often used when the genes are linked.

Recurrent breeding/improvement: **11.** A process aimed at achieveing long-term improvement in a set of value traits through selection, breeding and testing in repeated cycles (generations).

Regeneration: **13.** The reestablishment of a forest; in tissue culture, the production of organs, embryos or whole plants, usually as a morphogenic response to stimulus.

Reinvigoration: **12.** Reinvigoration occurs when plants or organs improve their external and internal environment, particularly their nutritional and water status, sometimes producing juvenile-appearing traits. See rejuvenation.

Rejuvenation: **10, 11, 12.** A change in a tissue or an organism from a more mature state to a more juvenile one.

Reporter gene: **15.** A marker gene used in transformation studies to indicate the presence of foreign DNA. See "gus" or "npt" for examples.

Restriction fragment length polymorphism (RFLP): **8, 10, 14.** Variation in the lengths of DNA fragments resulting from predictable cutting by restriction enzymes, which cleave the DNA double helix at specific nucleotide sequences.

Ribosome: **14.** The small organelle that is the site of protein synthesis.

SCA: **11.** Specific combining ability. Can refer to the degree to which the average performance of a specific (usually full-sib) family departs from the average of its parental breeding values; sometimes used in a similar sense to note the departure of an individual clone from the performance of its sibs; sometimes used to note the degree of non-additive genetic variation in a population.

Scale-up: **11.** The transition from research-scale to operational-scale application of a technology.

Seed orchard: **5, 11.** An orchard consisting of clones or seedlings from selected trees, isolated to prevent or reduce pollination from outside sources, and cultured for early and abundant production of seeds for reforestation.

Selectable marker: **15.** May be used in marker-assisted selection, or to select a transformed genotype from a mixture of transformed and nontransformed cells. RFLPs, antibiotic-resistant and herbicide-resistant sequences or genes are examples, the latter resulting in killing cells lacking the marker.

Selection intensity: **5, 11.** (This term has two conflicting meanings, and can cause confusion.) (a) The standardized selection differential between the average of the selected population and the average of the population selected from. (b) The percentage of individuals selected from a candidate population.

Senescence: **12.** An advanced state of aging, often associated with a sharply decreased ability to repair damage.

Sequential testing: **11.** A testing approach in which large numbers of genetic entries are tested in few replications in the early stages (say screening and then candidacy testing) and progressively fewer entries are then tested in more replications or in larger plots (say performance or compatibility testing). See combined testing, multiple-stage testing.

Serial propagation: **11.** The ortet donates cuttings or tissue that become primary ramets, some of which in turn later donate cuttings or tissue that become secondary ramets, etc.

Sex ratio: **3.** Generally, the ratio of males to females at some point in development. On a monoecious tree, the ratio of male to female strobili or flowers.

Simple polyembryony: **13.** The creation of (for example) 3/4-sibs by the fertilization of genetically identical eggs within a single gametophyte by genetically non-identical pollen from a single male. See cleavage polyembryony.

Somaclonal variation: **7, 9, 10, 15.** Variation arising in tissue-culture of somatic tissue, often at unusually high rates due to higher rates of gene mutation, transpositions, and various chromosome abnormalities. See gametoclonal variation.

Somatic crossing-over: **6.** An exchange of genetic material between homologous chromosomes in somatic cells, ie, that is not associated with their pairing at meiosis.

Somatic embryogenesis: **7, 9, 10, 11, 13.** A process by which somatic cells are differentiated into somatic embryos.

Somatic seedling: **10.** A propagule of somatic-embryo origin (sometimes called embling) that is past the intensive-care stage of early nursery, and can be or has been planted in normal field conditions.

Southern blot: **14, 15.** A method for analysis of DNA fragments separated by size on agarose gels, blotted onto a membrane, and hybridized to specific (usually radiolabeled) probes. See northern blot. This one is named after a person named Southern.

Split-plot: **5.** A field design in which not all entries or all treatments are in each plot. Check an experiment design text for detail.

Sporophyte: **12.** The diploid ($2n$) spore-producing phase or generation. See gametophyte.

SRIC: **6.** Short-rotation intensive culture.

Stability: **4, 11.** Stable clonal mixtures or ecosystems are less likely to respond dramatically to physical and biotic perturbations than unstable ones. Re clones: rankings of stable clones are fairly constant with modest changes in environment.

Stable transformation: **15.** As distinguished from transient expression of foreign DNA, a stable transformation is foreign DNA incorporated into the host genome (usually a chromosome) and it is inherited.

Steckling: **10.** A rooted cutting that is sufficiently developed so that it can be or has been planted in the field. See somatic seedling, plantling for parallel terminology.

Stocktype: **11.** The kind of propagule sent to the field, such as container-grown or bare-root; seedling, steckling or plantling; etc.

Storage protein: **13.** A class of proteins in seeds with no known enzymatic activity that accumulate in protein bodies and provide nitrogen during germination.

Subculture: **11, 12, 14.** To transfer from a prior medium to another medium and grow for a specific period of time.

Sublines: **5.** A more-or-less arbitrary separation of lines within a breeding population; breeding crosses are made within lines and production crosses between lines, resulting in no inbreeding in the production stock.

Suspension culture: **13.** The culture of plant cells in liquid medium, where cells or clumps of cells are aerated and maintained in a dispersed state by gyration of the vessel.

Suspensor cells: **11.** Cells that develop very early in ontogeny from the base of the embryo, which elongate to push the embryo into the megagametophyte or endosperm, and which are often embryogenic in culture.

T-DNA. **15.** That region of the Ti (tumor inducing) plasmid of *Agrobacterium tumefaciens* that is transferred to plant cells.

Technology transfer: **11.** The stage in development of a technology or process in which operational personnel learn about the technology, commit to it, and develop procedures for operational-scale application.

Terminator: **15.** A region of nucleotide sequence that defines the termination of transcription, important for efficient transcription.

Tissue culture: **11, 12.** A general term for aseptic cell, tissue, organ and protoplast culture.

Topophysis: **9.** Differences in maturation or developmental potential among apical meristems of different branch hierarchical order may result in differences in performance among vegetative propagules taken from diffferent parts of the same donor plant. See cyclophysis, periphysis.

Topworking: **3.** Grafting of scions into the tops of trees, usually in the hope of inducing flowering by the scions.

Totipotency: **10, 12, 13.** Said of a nucleus, cell or tissue that has the capacity to develop into a complete organism. See competence.

Transcription: **14.** The enzyme-mediated process of transcribing the information in a DNA strand into a complementary RNA strand.

Transgenic plant: **15.** A plant containing DNA inserted by some form of genetic engineering, usually by a transformation event in culture followed by regeneration and recovery of a plant.

Transient expression: **15.** Foreign DNA in a cell with expression that decays with time and is not inherited.

Translation: **14.** The process by which information in an mRNA strand creates the sequence of amino acids during polypeptide (protein) synthesis. See ribosome.

True-to-type: **7, 11, 12.** Individuals having essentially the same phenotype as the donor plant, cultivar or clone.

Truncation selection: **9.** All members of a candidate population equal to or greater than some minimum value in a trait or index are selected, and no members below that value are included in the selected set.

Unselectable marker: **15.** Markers such as "gus" or "lux" that are (usually histologically) detectable but can not be efficiently used to sort transformed from untransformed cells.

Value function: **11.** The goal of an organization or program expressed in terms of a set of value traits; the relationships among a set of traits expressed in terms of the defined values of those traits.

Value traits: **11.** The characteristics that are important contributors to the value of a genetic improvement program.

Variance: **5, 11.** A statistical measure of variation with the useful property of having additive components, for example genetic and environmental components of total phenotypic variation.

Variety: **11.** See cultivar.

Vector: **15.** A carrier of a disease, or, for genetic engineering, of foreign DNA.

Vegetative propagation: **10.** Asexual propagation. See clone, cutting, embryogenesis, grafting, tissue culture, macropropagation, micropropagation.

Vegetative multiplication: **1, 11.** When seeds of desired families or populations are in short supply, embryos or seedlings are clonally reproduced by rooting cuttings, tissue culture, grafting, or somatic embryogenesis. Clonal records are usually not kept.

VIR: **15.** The region of the *Agrobacterium* Ti plasmid that affects the host range and efficiency of infection.

Virulence: **15.** The relative infectiousness of a microorganism that causes disease or infection.

Xylem: **3.** See phloem.

Zygotic forestry: **1.** Whether seedlings or clonal propagules, individual genotypes are not well known or characterized. Performance knowledge is at the level of family, breed or population. See vegetative multiplication.

References

Curtis H, Barnes NS (1989) Glossary pp G1-G22. In: Biology, Fifth Edition, Part 2. Biology of organisms. Worth Publ. Inc. New York, 1192 pp.

Donnelly DJ, Vidaver WE (1988) Glossary of plant tissue culture. Belhaven Press, London 141 pp.

Futuyama DJ (1986) Evolutionary biology. 2nd Edition. Sinauer Assoc. Sunderland, Ma. 600 pp.

Grierson D, Covey SN (1988) Plant molecular biology. 2nd Edition. Blackie, London. 233 pp.

Guralnik DB (Ed.) (1976) Webster's New World dictionary of the American language. Second College Edition. Wm. Collins and World Publ. New York, 1692 pp.

Lackie JM, Dow JAT (Eds.) (1989) The dictionary of cell biology. Academic Press, London, 262 pp.

Pierik RLM (1987) Abbreviations and glossary. pp. 7–20. In: In vitro culture of higher plants. Martinus Nijhoff, Dordrecht, 344 pp.

Rieger R, Michaelis A, Green MM (1991) Glossary of genetics–classical and molecular. 5th Edition. Springer Verlag, Berlin, Heidelberg, New York, 553 pp.

Snyder EB (1959) Glossary of forest tree improvement workers. Soc. Am. Foresters by the S. Forest Expt. Sta., Forest Service, USDA. 22 pp.

Subject Index

Printing: Mercedesdruck, Berlin
Binding: Buchbinderei Lüderitz & Bauer, Berlin